AF595556

Site-Specific Nutrient Management

Site-Specific Nutrient Management

Editor

Witold Grzebisz

MDPI • Basel • Beijing • Wuhan • Barcelona • Belgrade • Manchester • Tokyo • Cluj • Tianjin

Editor
Witold Grzebisz
Department of Agricultural
Chemistry and Biogeochemistry
Poznan University of Life
Sciences
Poznań
Poland

Editorial Office
MDPI
St. Alban-Anlage 66
4052 Basel, Switzerland

This is a reprint of articles from the Special Issue published online in the open access journal *Agronomy* (ISSN 2073-4395) (available at: www.mdpi.com/journal/agronomy/special_issues/Specific-Nutrient).

For citation purposes, cite each article independently as indicated on the article page online and as indicated below:

LastName, A.A.; LastName, B.B.; LastName, C.C. Article Title. *Journal Name* **Year**, *Volume Number*, Page Range.

ISBN 978-3-0365-1344-7 (Hbk)
ISBN 978-3-0365-1343-0 (PDF)

Contents

About the Editor

Witold Grzebisz

Professor Dr. Witold Grzebisz has been working at the University of Life Sciences in Poznań, Poland, for 43 years as an academic teacher and scientist. Under his supervision, more than 100 students have achieved a Master of Science (Msc) degree and 28 students have achieved a PhD in the field of agriculture, with special emphasis on soil fertilization and plant nutrition.

Professor Grzebisz is the author of more than 200 original, peer-reviewed scientific papers, of which half have been indexed in world databases (Scopus). He is the author of numerous books and monographs concerning the basis of crop plant nutrition and fertilization addressed to students and advisory staff in agriculture.

Preface to "Site-Specific Nutrient Management"

The natural milieu of food production is a field that has to being treated as a base production unit. The production potential of the field depends not only on soil geological origin and the natural fertility level but also on its management history. Effective crop production is based on a farmer's extensive knowledge and skills to protect cropped plants against stresses. Most are of environmental origin, such as drought. The sustainable exploitation of soil productivity depends on those soil characteristics that are crucial for the effective management of water and nitrogen. The production effect of these two basic production factors, in fact, depends on the status of other, secondary factors. A farmer's knowledge is necessary, first, for their recognition, and second, for developing and implementing measures for their amelioration.

The evaluation of field production level requires thorough recognition of the reasons for its variability. An insight into the soil profile, as the first group of variabilities, cannot be limited to topsoil, as is the case in current agro-chemical diagnostic procedures. An effective management of soil resources requires data on soil characteristics in the entire rooted soil zone. The subsoil cannot be treated as a black box. Temporal variability is an inherent property of crop plants during the growing season. The proper development of yield depends on a synchronization of crop plant requirements for water, nitrogen, and their supply from soil. However, those nutrients that are responsive to water and nitrogen uptake and utilization by plants throughout the growing season cannot be neglected. It is also impossible to achieve the production effect, i.e., yield, without taking into account the spatial variability of yielding factors within a field during critical stages of yield component development. The fourth type of variability affecting field productivity is defined by cropping sequence. Effective management of soil fertility should be always oriented towards a crop within a particular crop rotation with the highest sensitivity to the current status of soil fertility.

A sound management approach to soil fertility is to ameliorate those production factors that are correctable to a level, maintaining maximum water and nitrogen efficiency. Any attempt by farmers that aims to fulfill this target results in an amelioration of the nitrogen gap. This is the only way to reach the twin objectives of the sustainable agriculture, i.e., yield increase, concomitant with the decreased pressure of reactive nitrogen compounds on the environment.

The effective recognition of variability in production factors and their impact on the efficiency of water and nitrogen requires the use of a broad range of diagnostic tools. Reliable diagnosis of the present soil fertility status cannot, however, be conducted successfully without classic soil sampling, and time-consuming soil and agrochemical analyses. Remote-sensing techniques provide a source of data on the dynamics of crop biomass and nitrogen content during the growing season. Spectral data form the basis of quick correction of a crop nutritional status within the growing season.

This book is addressed to four groups of potential readers, at least. The first are farmers who need practical knowledge of how to recognize and create a hierarchy of factors that limit the efficiency of their applied production measures. They also need good advisory support to develop the skills that are necessary to ameliorate the factors that limit soil productivity. Second, the multifactorial level of the evaluation of production factors, as presented in this book, can be helpful for researchers within the field of agronomy. The third group are those who need a broad knowledge of and who are now or will be involved in the agriculture sector. The fourth group are environmentalists. Any action oriented towards environmental protection and addressing correction in current agricultural practices cannot be successful without comprehensive knowledge of agro-ecosystem functioning.

The authors involved in the preparation of the papers included in this issue are specialists in agronomy, soil and plant diagnosis, crop plant fertilization, microbiology, and remote-sensing. As a guest editor, I would like to express my gratitude to the authors and co-authors of the selected papers for their efforts in transfering the achievements and knowledge in their work on producing food to those who need support in the conditions arising from the present global challenges.

Witold Grzebisz
Editor

Editorial

Site-Specific Nutrient Management

Witold Grzebisz 

Department of Agricultural Chemistry and Environmental Biogeochemistry, Poznan University of Life Sciences, Wojska Polskiego 28, 60-637 Poznan, Poland; witold.grzebisz@up.poznan.pl; Tel.: +48-618-487-788

Abstract: The editorial introduces to a Special Issue entitled "Site-Specific Nutrient Management. The concept of the nitrogen gap (NG) is as a core challenge for an effective realization of the so called "twin objectives" in sustainable agriculture. This special issue stresses on some hot spots in crop production, being responsible in the yield gap development, that farmers have to take control. The yield gap cannot be ameliorated without the synchronization of the in-season requirements of the currently grown crop for N with its three-dimensional variability in a supply on a field (temporal, spatial and vertical). A recognition of soil fertility status in the rooted zone, which includes availability of both mineral N and nutrients decisive for its uptake, is the first step in the NG amelioration. The sustainability in soil fertility, as a prerequisite of N fertilizer application, requires a wise strategy of organic matter management, based on farmyard manure, and/or cultivation of legumes. The soil fertility status, irrespectively of the World region determines ways of the N rate optimization. The division of a particular field into homogenous productive units is the primary step in the NG cover. It can be delineated, using both data on soil physico-chemical properties of the soil rooted zone, and then validated by using satellite spectral images of the crop biomass in a well-defined stage of its growth, decisive for yield. The proposed set of diagnostic tools is a basis for elaboration an effective agronomic decision support system.

Keywords: a field; crop production; sustainability; homogenous productivity units; soil fertility; nitrogen indicators: in-season; spatial; vertical variability of N demand and supply; spectral imagery; vegetation indices

Citation: Grzebisz, W. Site-Specific Nutrient Management. *Agronomy* **2021**, *11*, 752. https://doi.org/10.3390/agronomy11040752

Received: 17 March 2021
Accepted: 8 April 2021
Published: 13 April 2021

1. Introduction: On a Way to Reach Sustainability in Crop Production

The effective realization of the 'twin objectives' of the concept of sustainable agriculture assumes efficient food production and the simultaneous protection of both the local and global ecosystem. This concept is based on the optimization of the applied production inputs [1]. In crop production, the efficient use of inputs is defined by the size of the yield gap (YG), which results from inefficiency in the use of N fertilizer (N_f) under well-defined soil and weather conditions [2,3]. The full recognition of the production factors allows the climatic-yield potential (CYP) of the currently grown plant and its seasonal variability to be determined [4]. The size of the existing YG, considered as the deviation from the CYP, is the basis for the elaboration of an agronomic decision support system oriented towards its cover. In rain-fed agriculture, the first step towards YG cover is to increase the resistance of a grown plant to in-season weather variability. This action, as a rule, focuses on improvement in water management [5]. The key, and in fact the long-term strategy, of water management control should be oriented to soil fertility improvement (organic matter content, soil pH). This is required in order to decrease the yield variability. The efficient use of both indigenous fertilizer and N_f requires efforts that concentrate on the optimization of both the N_f rate and the factors responsible for its uptake and utilization by the currently-grown plant. Solving these challenges requires well-elaborated diagnostic methods which take into account both classical chemometric and remote-sensing tools [2].

2. Special Issue Overview: General Topics

2.1. In-Season Management of Nitrogen: A Challenge for the Present Generation

The first chapter of the Special Issue comprises two papers which focus the reader's attention on N management within a particular field, as a basic production unit [2,6]. Nitrogen, under conditions of ample water supply, is the main production factor. It affects growth and the exploitation of the yield potential of the currently-grown plant. This first temporal (in-season) variability is defined by a crop's requirement for N, and can be taken under control, provided there is a recognition of the crop's critical stages of yield formation. In the case of seed crops, the game for yield takes place during the linear period of a crop's biomass increase. For cereals, it covers the phase of stem elongation [7]. In spite of the in-season variability of the seed crop requirements for N, the yield for a particular field depends to a considerable degree on the factors which are responsible for its spatial variability. Hence, the main challenge for a farmer is to divide the whole field area into homogenous field productive units, HFPUs [8]. Spectral imagery is a useful tool to recognize the plants' nutritional status within a growing season. Its advantage over classical biometric methods is the quick determination and simultaneous discrimination of the difference in the rate of a crop biomass increase between HFPUs [9]. However, the accurate discrimination of a given HFPU boundary requires strict data on the soil's physical and chemometric characteristics which are decisive for water and nutrient content in the whole rooted soil zone [10].

The sound management of N, both indigenous to the soil and applied as N_f, relies on simultaneously maximizing yield and minimizing the negative impact of the N present in the soil/plant continuum of the environment. A basic set of operationally-required data comprises: (i) the N productivity in a particular HFPU, (ii) the size of the pool of mineral nitrogen (N_{min}) at the onset of the growing season or at a time of winter crop regrowth in spring, (iii) the total amount of applied N_f, and (iv) the in-season division of the whole N_f rate. The reliable indicators of the in-season N management are strongly correlated with N released from its organic pool during the growing season. This assumption, resulting from the study by Łukowiak at al. [6], can be fulfilled by using the N-balance (N_b) as the N management indicator. This only seemingly-simple index is based on the N input into the soil/crop system ($N_{in} = N_{min} + N_f$) and the total N content in a crop plant at harvest (TN). It allows the discrimination of HFPUs differing significantly in productivity, and consequently defines the requirement of a crop for N_f. Its practical advantage is to calibrate the variability in the N_f rate between neighboring HFPUs.

2.2. Soil Fertility Management/Improvement

The second chapter is devoted to the aspects of soil fertility management that can ensure a high efficiency of the applied production measures [10–12]. Sustainability in the crop production system is based on the stabilization of the nutrient supply to plants within a particular crop rotation. Sugar beet, in temperate climatic regions, is the most sensitive crop to the soil fertility level, which is decisive for the sugar yield [13]. The expected stabilization in the nutrient supply to this crop is mainly achieved through manure application, which is a source of both organic matter and nutrients. Sugar beet, as reported by Hlisnikovsky et al. [12], very efficiently exploits the applied N_f under conditions of simultaneous manure and NPK application. The high efficiency of N_f is mainly due to the increase in the content of the soil's available nutrients. In the studied case, it referred to P and K. A high sugar yield was obtained in treatments with a much lower input of mineral fertilizers (FYM + NPK2 vs. NPK4). A synergy between the effects of mineral fertilizers and manure is especially important in unfavorable weather conditions, for example drought.

A sustainable system of crop production should be based on legume crops, grown at least once in a crop rotation. The impact of these plants on soil fertility is well recognized, but is also frequently neglected in intensive farming systems [14]. The yield production functions of legumes, in the light of N_2 fixation, is out of discussion. However, farmers tend to overlook some other functions of legumes which are responsible for both the soil's fertility build-up and its long-term stabilization. The rhizodeposition of carbon by roots to the soil leads to an enormous increase in the activity of microorganisms. The in-season dynamics of C and N rhizodeposition are related to the stage of legume crop growth, reaching their maximum at the onset of flowering [15]. The increased activity of enzymes produced by both plants and microbes achieves its maximum just at the stage of the legume plant's flowering. The acid phosphatase (PAC) is an important factor, impacting the supply of phosphorus to plants [11]. The majority of the nutrients released from soil resources, especially N and P, are taken up by a legume plant during the phases of seed growth. The remaining portion increases the soil fertility level and become available to the succeeding crop [14].

The in-season dynamics of the nutrient uptake by a plant, especially of N_{min}, are a key factor affecting both (i) the rate of the crop plant growth, (ii) the formation of the yield components. As in the case of legumes, seed crops are also sensitive to the supply of nutrients during this period of yield component formation. In winter oilseed rape, the yield is significantly related to the nitrate nitrogen (NN) supply during the period extending from the rosette up to flowering [10]. The shortage of the N supply to plants at this time results in a significant yield decline. The supply of NN, as reported by Grzebisz et al. [10], depended not only on the amount of N_f applied to the crop, but also on the availability of other nutrients which are responsible for the NN uptake from the soil. In sandy soils, the most constraining nutrients for Winter Oilseed Rape (WOSR) growth and yield are potassium (K), magnesium (Mg), and calcium (Ca). The limiting effect of K on N uptake by plants, as the nutrient taken-up by WOSR in the highest amounts, persists through the entire growth of WOSR plants [10,16].

2.3. *Site-Specific Response of Crop Plants to Soil Fertility and Management*

The yield variability between cropped fields is affected by different factors. The main reasons for the observed variabilities not only result from differences in soil characteristics (soil type, agronomy class) but also from the field cropping history, or the current crop management. Such variability is discussed in this chapter.

The soil type is the key factor affecting yields of high-yielding crops, like winter wheat, as was clearly documented by Hlisnikovsky et al. [17]. The principle cause of huge differences in yields cannot only be attributed to the content of the main soil fertility indicators, like the total N content (N_t). It is also worth taking into consideration the content of the available P, K, and Mg, as well as Ca. The stabilization impact of FYM application has been revealed as an important factor, both in the Chernozem and in Cambisol, as documented by the authors. The advantage of naturally-fertile soils like Chernozem and Phaeozem is the much lower rate of N required to achieve the maximum yield of wheat. In the case of Cambisol, the same yield can be also obtained provided there is an application a significantly higher N_f rate. In contrast, the quality of the wheat grain requires an extra N_f dose, irrespective of the achieved yield. It can therefore be concluded that the N_f efficiency in wheat depends on the soil fertility level, but the grain quality is driven solely by the N_f rate.

The same problem, but related to maize, was discussed in the paper by Wang et al. [18]. The used index of N efficiency, i.e., the economic optimal N rate (EONR), in combination with the soil type, the course of the weather, and plant density, can be used as an advisory tool oriented towards increasing the efficiency of N_f. The year-to-year variability in yield trends, irrespective of the weather course in a particular growing season, was completely different for Black Earth and Aeolian sandy soil. As a rule, the yields of maize grown on Black Earth increased in response to an increased N_f rate in accordance with the Mitscherlich law. The trend of the maize yield grown on Aelioan sandy soil followed the linear-plateau model, showing a significant seasonal variability. The comparison of the EONR values clearly stresses the differences between soil types. The importance of other factors, leading to the increased efficiency of the applied Nf with respect to farmers' practices was significant, but at a much lower level. The Partial Factor Productivity of N fertilizer (PFP_{Nf}) was revealed as a good indicator, reflecting differences in the effects of maize production factors.

The yield of a seed crop is defined by two primary components, i.e., the seed density per unit soil area (SD) and the seed weight (1000 seed/grain weight, TSW/TGW). The effect of N application on WOSR depends on both (i) SD, which is established during inflorescence development and fixed during the phase of pod growth, and (ii) TSW, which is established during the stages of pod formation and seed maturation [19]. The study by Łukowiak and Grzebisz et al. [20] on WOSR clearly showed that yield was significantly affected by the N content of seeds (N_{se}). This seed characteristic was a decisive factor for seed survival during the post-anthesis WOSR growth, being responsible for SD. In addition, N_{se} was positively related to PFP_{Nf160}. It was found to be a criterion for the differing of the investigated fields with respect to yield in a particular year. An efficient N management strategy requires data on the N_{min} content at the onset of the growing season [21]. The interaction between a season and the field location allowed two strategies of N_f management to be distinguished, both of which were oriented towards SD maximization. The first strategy was based on current soil productivity. The second one was based on indigenous soil fertility. Both strategies resulted in a high SD as the prerequisite of a high seed/grain yield. This observation was also corroborated by the main conclusion, drawn from the study with wheat and maize discussed in chapter 2.

2.4. Field Spatial Variability of Soil/Plant Characteristics: Evaluation by Satellite Imagery

The efficient management of N in a field with a currently-grown plant requires a recognition of the real status of four variables: (i) the crop's in-season requirement for N, (ii) the spatial variability of the actual N status in a plant, (iii) the N_{min} resources at the onset of the growing season, and (iv) the amount of in-season N_{min} released from its soil resources in the rooted soil zone [10,22].

N management, limited to the current N plant nutritional status and N_f input, should be preceded by a field division into zones of homogenous productivity (HFPU) [8]. This is the first step in elaborating an effective soil fertility management system (SFMS). The basis of a particular fertilization system (FS) is data on the current status of the soil organic matter content, soil pH, and content of available nutrients. As reported by Cammarano et al. [23], remote sensing in connection with data on the content of inorganic N and organic carbon (OC) has been found to be an effective diagnostic tool for the delineation of HFPUs. The first criterion was the Green Normalized Index (GNDVI). This was determined on the basis of a set of 10 years of gathered Landsaft satellite images of wheat at its flowering. The second criterion was the Soil Brightness (SOB), which is based on RapidEye optical satellite images conducted on bare soil. A combination of both spectral indices allowed three or four HFPUs to be delineated. The three-zonal model explained the N and OC variability the best (45%). An indirect evaluation of the soil productivity was possible due to the fact that the wheat canopy at the onset of flowering fairly well reflects its potential productivity (grain density, photosynthetic area). The authors stressed the fact that a field division into three zones fulfills the requirement for the application of the commercial precision agricultural tools.

The in-season recognition of the N requirements of a currently-grown plant depends on the seed/grain density, but in practice on the number of spikes per unit area (SN, m^2) [24]. The most important is, however, the determination of the critical stage of a cereal plant growth with respect to the impact of SN on the grain yield (GY). As reported by Panek et al. [25], the observed relationship, irrespective of the field location, was the strongest during a period of 4–6 weeks before the onset of wheat or triticale milk maturity. The yield prediction was conducted using seven vegetation indices (VIs), based on satellite images from Sentinel-2. It is worth stressing that yield prediction based on NDVI was a strong yield predictor in the later stages of the growth of both cereals. The rule obtained corroborates the hypothesis that the rate of the cereal crop biomass increase during the shooting phase defines the grain yield [7].

The differences in a crop's nutritional status at the stages of its growth which are decisive for yield provides a basis to determine the applied N_f rate. The calculated N_f dose requires, however, correction, taking into account the spatial variability of a crop's requirement for N. A practical solution aimed at the effective management of N_f in accordance with a plant's need for N is to apply the concept of variable-rate nitrogen (VRN). A study, based on 1263 observations by Larson et al. [26] on cotton, showed that the net returns (NRs) increased in response to the applied VRN strategy under well-defined field areas, as determined by higher organic matter content, deeper profiles, or an erodible soil. These three different cases, representing field areas of contrasting productivity, corroborated the necessity for a field's delineation into homogenous productive zones.

3. Conclusions

A particular field is a basic unit for crop production. The key challenge for farmers and their advisors is to define and then elaborate a set of agronomic measures oriented towards covering the nitrogen gap. This is the most effective way to realize the 'twin objectives' of sustainable crop production management within a particular field. The spatial variability of yields can be ameliorated, at least, by the division of a field into homogenous productive units. The criteria of a field's division can be based on physical or chemical soil properties, including the whole zone rooted by grown plants. They can also be delineated using satellite spectral images of both bare soil and crop biomass in a well-defined stage, which is decisive for yield component development. The developed agronomic decision support system (ADSS) should be adjusted to the yield potential of the homogenous field unit in order to optimize the efficiency of the applied inputs, mainly focusing on nitrogen fertilizer.

Funding: This research received no external funding.

Conflicts of Interest: The authors declare no conflict of interest.

References

1. Smith, P. Delivering food security without increasing pressure on land. *Glob. Food Secur.* **2013**, *2*, 18–23. [CrossRef]
2. Grzebisz, W.; Łukowiak, R. Nitrogen gap amelioration is a core for sustainable intensification of agriculture–a concept. *Agronomy* **2021**, *11*, 419. [CrossRef]
3. Anderson, W.; Johansen, C.; Siddique, H.M. Addressing the yield gap in rainfed crops: A review. *Agron. Sustain. Dev.* **2016**, *26*, 18. [CrossRef]
4. Licker, R.; Johnston, M.; Foley, J.A.; Barford, C.; Kucharik, C.J.; Monfreda, C.; Ramankutty, N. Mind the gap: How do climate and agricultural management explain the "yield gap" of croplands around the world? *Glob. Ecol. Biogeogr.* **2010**, *19*, 769–782. [CrossRef]
5. Grafton, R.Q.; Williams, J.; Jiang, Q. food and water gaps to 2050: Preliminary results from the global food and water systems (GFWS) platform. *Food Secur.* **2015**, *7*, 209–220. [CrossRef]
6. Łukowiak, R.; Grzebisz, W.; Ceglarek, J.; Podolski, A.; Kaźmierowski, C.; Piekarczyk, J. Spatial variability of yield and nitrogen indicators–a crop rotation approach. *Agronomy* **2020**, *10*, 1959. [CrossRef]
7. Xie, Q.; Mayes, S.; Sparkes, D.L. Preanthesis biomass accumulation and plant organs defines yield components in wheat. *Eur. J. Agron.* **2016**, *81*, 15–26. [CrossRef]
8. Denton, O.A.; Aduramigba-Modupe, V.O.; Ojo, A.O.; Adeoyolanu, O.D.; Are, K.S.; Adelana, A.O.; Oyedele, A.O.; Adetayo, A.O.; Oke, A.O. Assessment of spatial variability and mapping of soil properties for sustainable agricultural production using geographic information system techniques (GIS). *Cogent Food Agric.* **2017**, *3*, 1279366. [CrossRef]
9. Gerstmann, H.; Möller, M.; Gläßer, C. Optimization of spectra indices and long-term separability analysis for classification of cereal crops using multi-spectral RapidEye. *Int. J. Appl. Earth Obs.* **2016**, *52*, 115–125. [CrossRef]
10. Grzebisz, W.; Łukowiak, R.; Kotnis, K. Evaluation of nitrogen fertilization systems based on the in-season variability in the nitrogenous growth factor and soil fertility factors–a case of winter oilsed rape (*Brassica napus* L). *Agronomy* **2020**, *10*, 1701. [CrossRef]
11. Sulewska, H.; Niewiadomska, A.; Ratajczak, K.; Budka, A.; Panasiewicz, K.; Faligowska, A.; Wolna-Maruwka, A.; Dryjański, L. Changes in *Pisum sativum* L. plants and in soil as a result of application of selected foliar fertilizers and biostimulators. *Agronomy* **2021**, *11*, 1558. [CrossRef]
12. Hlisnikovsky, L.; Menšik, L.; Křížová, K.; Kunzová, E. The effect of farmyard manure and mineral fertilizers on sugar beet beetroot and top yield and soil chemical parameters. *Agronomy* **2021**, *11*, 133. [CrossRef]
13. Barłóg, P.; Grzebisz, W.; Pepliński, K.; Szczepaniak, W. Sugar beet response to balanced nitrogen fertilization with phosphorus and potassium. Part I. Dynamics of beet yield development. *Bulg. J. Agric. Sci.* **2013**, *29*, 1311–1318.

14. Massawe, F.; Mayes, S.; Cheng, A. Crop diversity: An unexploited treasure trove for food security. *Trends Plant Sci.* **2016**, *21*, 365–368. [CrossRef] [PubMed]
15. Hupe, A.; Schulz, H.; Bruns, C.; Haase, T.; Heß, J.; Joergensen, R.G.; Wichern, F. Even flow? Changes of carbon and nitrogen release from pea roots over time. *Plant Soil* **2018**, *431*, 143–157. [CrossRef]
16. Szczepaniak, W.; Grzebisz, W.; Potarzycki, J.; Łukowiak, R.; Przygocka-Cyna, K. Nutritional status of winter oilseed rape in cardinal stages of growth as yield indicator. *Plant Soil Environ.* **2015**, *61*, 291–296. [CrossRef]
17. Hlisnikovsky, L.; Menšik, L.; Kunzová, E. The development of winter wheat yield and quality under different fertilizer regimes and soil-climatic conditions in the Czech Republic. *Agronomy* **2020**, *10*, 1160. [CrossRef]
18. Wang, X.; Miao, Y.; Dong, R.; Chen, Z.; Kusnierek, K.; Mu, G.; Mulla, D.J. Economic optimal nitrogen rate variability of maize in response to soil and weather conditions: Implications for site-specific nitrogen management. *Agronomy* **2020**, *10*, 1237. [CrossRef]
19. Grzebisz, W.; Szczepaniak, W.; Grześ, S. Sources of nutrients for high-yielding winter oilseed rape (*Brassica napus* L.) during post-flowering growth. *Agronomy* **2020**, *10*, 626. [CrossRef]
20. Łukowiak, R.; Grzebisz, W. Effect of site specific nitrogen management on seed nitrogen–a driving factor of winter oilseed rape (*Brassica napus* L.) yield. *Agronomy* **2020**, *10*, 1364. [CrossRef]
21. Olfs, H.-W.; Blankenau, K.; Brentrup, F.; Jasper, J.; Link, A.; Lammel, J. Soil- and plant-based nitrogen-fertilizer recommendations in arable farming. *J. Plant Nutr. Soil Sci.* **2005**, *168*, 414–431. [CrossRef]
22. Córdova, C.; Barrera, J.A.; Magna, C. Spatial variation in nitrogen mineralization as a guide for variable application of nitrogen fertilizer to cereal crops. *Nutr. Cycl. Agroecosystems* **2018**, *110*, 83–88. [CrossRef]
23. Cammarano, D.; Zha, H.; Wilson, L.; Li, Y.; Batchelor, W.D.; Miao, Y. A remote sensing–based approach to management zone delineation in small scale farming systems. *Agronomy* **2020**, *10*, 1767. [CrossRef]
24. Triboi, E.; Triboi-Blondel, A.-M. Productivity and grain or seed composition: A new approach to an old problem—Invited paper. *Eur. Agron. J.* **2002**, *16*, 163–186. [CrossRef]
25. Panek, E.; Gozdowski, D.; Stępień, M.; Samborski, S.; Ruciński, D.; Buszke, B. Within-field relationships between satellite-derived vegetation indices, grain yield and spike number of winter wheat and triticale. *Agronomy* **2020**, *10*, 1842. [CrossRef]
26. Larson, J.A.; Stefanini, M.; Yin, X.; Boyer, C.N.; Lambert, D.M.; Zhou, X.V.; Tubaña, B.S.; Scharf, P.; Varco, J.J.; Dunn, D.J.; et al. Effects of landscape, soils, and weather on yields, nitrogen use, and profitability with sensor-based variable rate nitrogen management in cotton. *Agronomy* **2020**, *10*, 1858. [CrossRef]

Review

Nitrogen Gap Amelioration Is a Core for Sustainable Intensification of Agriculture—A Concept

Witold Grzebisz * and **Remigiusz Łukowiak**

Department of Agricultural Chemistry and Environmental Biogeochemistry, Poznan University of Life Sciences, Wojska Polskiego 28, 60-637 Poznan, Poland; remigiusz.lukowiak@up.poznan.pl
* Correspondence: witold.grzebisz@up.poznan.pl; Tel.: +48-618-487-788

Abstract: The main reason for the development of the yield gap in crop production is the inefficient management of nitrogen (N). The nitrogen gap (NG) cannot be ameliorated without an indication and quantification of soil characteristics that limit N uptake by a crop plant. The insufficient supply of N to a plant during its cardinal stages of yield formation is a result of two major-variabilities. The first is spatial variability in the soil characteristics responsible for water supply to a plant, also serving as a nutrient carrier. The second is a vertical variability in soil factors, decisive for pools of available nutrients, and their in-season accessibility to the grown crop. The long-term strategy for NG cover should focus first on soil characteristics (humus stock, pH, nutrient content) responsible for water storage and its availability to the currently grown plant. Diagnostics of plant nutrient availability should deliver data on their contents both in the topsoil and subsoil. The combined use of both classical diagnostic tools and spectral imagery is a way to divide a single field into units, differing in productivity. Remote-sensing techniques offer a broad number of tools to define the in-season crop canopy requirement for fertilizer N in homogenous field units.

Keywords: climatic potential yield; yield gap; nitrogen use efficiency; soil constraints; subsoil; spatial variability; remote sensing-techniques; field

Citation: Grzebisz, W.; Łukowiak, R. Nitrogen Gap Amelioration Is a Core for Sustainable Intensification of Agriculture—A Concept. *Agronomy* **2021**, *11*, 419. https://doi.org/10.3390/agronomy11030419

Academic Editor: Claudio Ciavatta

Received: 21 January 2021
Accepted: 22 February 2021
Published: 25 February 2021

1. Food Gap and Sustainable Intensification of Agriculture

The global human population, depending on the scenario, will reach 9–10 billion in 2050. Some prognoses of food requirements in 2050, based on the level of food production in 2005, fluctuate from 50% to 110% [1,2]. The analysis made by Hunter et al. [3] showed a much lower level of food demand in 2050 compared to 2010, ranging from 25% to 70%. The effective management of global food demand in 2050 should be a result of the simultaneous implementation of three complementary strategies, termed "mega-wedges" [4]. They are as follows: (i) filling the food production gap, (ii) decreasing food losses in the entire food chain, (iii) reduction in the worldwide food demand. The relative contribution of each particular mega-wedge in food security control has been assessed as 46%, 33.6%, and 20.4%, respectively. The second strategy does not include only direct losses of energy and proteins in the food chain, but also reduced productivity of soil and water (10.4%). This means that above 55% of the future food demand directly depends on efforts oriented on covering the food production gap. With respect to food losses, it is necessary to stress that the net excess of the global protein supply is almost equal to its intake by humans (36% vs. 44%) [5]. It can be, therefore, concluded that the management of the food production chain requires significant changes in agriculture. The conceptual (diagnostic, management, techniques, technology) preparation of the agriculture sector for a considerable increase in food production in the coming 30 years is the key challenge for the present generation.

The current level of food production is a function of two main factors, i.e., (i) actual crop yield, (ii) crop yield improvement. The actual yield of a particular crop in a strictly defined locality (field) is a result of the efficiency of production inputs under the course of meteorological conditions during the growing season [6,7]. The realization of the second target depends on progress in (i) breeding of new cultivars, (ii) improvement in water efficiency (WUE), (iii) improvement in nitrogen use efficiency (NUE). It is necessary to remember that the success of the Green Revolution, resulting in the significant increase of yields of major crops (cereals, rice, maize) was a result of the induced synergism between new, high-yielding cultivars, high rates of applied fertilizer nitrogen (N_f), and a high level of crop protection, based on agrochemicals [8]. An insufficient level of the required synergism between these three main factors has resulted in stagnation in the world average yields of main crops during the last two decades [9]. The necessary annual rate of yield increase of four global crops. i.e., maize, rice, wheat, and soybean to cover the food gap in 2050 would have to reach 2.4%. The current yearly rate of yield increase for these four crops is far below the assumed threshold, being at the level of 1.6%, 1.0%, 0.9%, and 1.3%, respectively [10]. For example, the genetic progress in nitrogen use efficiency for wheat in the years 1985–2010 was only in the range of 0.30–0.37% y^{-1}. This was a result of the increased value of the Nitrogen Harvest Index (NHI), i.e., the relative amount of N accumulated in grain [11].

Future progress in food production, in fact, will depend on two key drivers. The first is high-tech intensification, based on highly productive cultivars of main crop plants, and the efficient use of mineral fertilizers and other agro-chemicals. This strategy has been responsible for about $\frac{3}{4}$ of the global food production increase in the last 85 years. The remaining $\frac{1}{4}$ was due to the increase in the area of arable land, mainly in less-developed countries. The productivity of the newly cultivated soils is to a great extent driven by their natural fertility. In the future, this option will be strongly limited due to the lack of high or even medium fertile soils. The primary resources of potential arable land are extensive pastures and tropical forests. The key disadvantage of the first is a shortage of water and of the second, low fertility, mainly due to high acidity and a shortage of essential nutrients. In addition, this option requires a considerable financial input to increase soil fertility, as a prerequisite of an economically and environmentally sound level of production [12,13].

The progressively increasing demand for food, resulting from the permanently growing human population, is in contradiction to the concept of sustainable development of agricultural production. The most advanced scenarios of a sustainable approach to food production assume a massive reduction or even the elimination of N fertilizers and pesticides. The main objective of this restrictive view of agriculture is to arrest the degradation of both local and global environments [14]. In the current and future reality, food demand is too high to completely abandon modern production measures [15,16]. A realistic view of sustainable agricultural development, termed as sustainable intensification of agriculture (SIA), defines this concept as a process or production system where yields are increased without an adverse impact on the environment and without the cultivation of low-quality land [17]. A less sophisticated, but at the same time a more practical definition of SIA, has been proposed by Smith [18]. The core of this definition is to produce more, and high-quality food per unit of used and applied production measures, taking into account both soil fertility and externally applied measures (fertilizers, pesticides, fuel) on the one hand, and protection of the existing ecosystems against damage on the other hand. These opinions are summarized in the concept of *twin objectives*, which relies on the assumption that agricultural development, including new tools, both implemented as new production technologies, production systems, and management, has to ensure the stability of the global ecosystem [19,20].

The area of arable land and the amount of available water during the growing season of a particular crop plant are key factors that limit food production for a country, region, or field. A rise in food production cannot be based solely on the increased efficiency of water usage. As reported recently by Grafton et al. [21], based on an analysis of numerous methods of crop plant irrigation, no direct substitution between water and nitrogen was observed. The key factor limiting yield, as results from this study, is nitrogen. The authors clearly stated that the greatest challenge to the progress of crop plant productivity depends on the improvement in N management. The required rate of food production increase to cover the food gap by 2050 can be achieved by an adequate supply of fertilizers, both nitrogenous, and those balancing N, i.e., containing P, K, Mg, S, and micronutrients. Scenarios of nitrogen fertilizers (N_f) consumption in agriculture to cover the food production gap vary 3-fold, i.e., from 85 Mt N y^{-1} to 260 Mt N y^{-1} [22]. In the model of N flows by Conijn et al. [23], the total consumption of N_f will increase to 181 Mt y^{-1}, i.e., by 76% in 2050 as compared to 2010 (103 Mt N y^{-1}). The increased N_f consumption will lead, however, to a simultaneous increase in losses of its active compounds into the environment. The N loss in 2050 is projected to lie in the range of 102–156% with respect to 2010 [24].

Any increase in the production efficiency of both key agronomic factors, i.e., water and nitrogen, depends on the soil status of all the other production factors decisive for their efficiency. The production of phosphorus fertilizers to cover crop plant requirements for this nutrient is set to increase in the period 2010–2050 by 32% (from 17.9 to 23.7 Mt P y^{-1}) [23]. An analysis made by Pradhan et al. [25] showed that to fulfill food production goals in 2050 the consumption of N, P, and K fertilizers will increase in the range of 45–73% for N, 22–46% for P_2O_5, and 2–3-fold for K_2O, respectively, compared to 2010. The sustainable intensification of agriculture cannot be realized without taking into account other nutritional factors that limit the productivity of water and nitrogen, such as magnesium, sulfur, and micronutrients [26,27].

The principal challenge to the concept of sustainable intensification in agriculture is to develop effective diagnostic and management tools oriented to the increased efficiency of the applied fertilizer nitrogen. Its realization is a prerequisite for decreasing the pressure of losses of its active forms to the local and, as a consequence to the global environment. These twin objectives can be successfully realized, provided there is a recognition and a simultaneous amelioration of factors constraining the productivity of nitrogen both in the critical stages of yield formation by the currently grown crop, and resulting from the spatial variability of its supply to plants.

2. Yield Potential and Yield Gap

2.1. Water Limited Yield—WLY

The yielding potential of a particular crop plant expresses its genetic potential for the exploitation of solar radiation and CO_2 fixation [28]. Yield potential (Y_p), as proposed by Evans and Fisher [29], defines the maximum attainable yield of a crop cultivar grown under conditions of the non-limiting supply of nutrients, and effective control of pests and diseases. These growth conditions can be achieved provided the implementation of irrigation to the currently grown crop [30]. Water and nitrogen are classified as factors limiting yield [28].

The importance of water supply to crop plants during the growing season is well-known to farmers. A temporary shortage of water is a natural feature of natural, i.e., non-irrigated agriculture [31]. Therefore, in practice a much more adequate term is water limited-yield (WLY, Y_w), i.e., Y_p defined under natural water supply to crop plants. The really attained yield, in fact, depends on the unit productivity of water (water-use-efficiency—WUE). This index expresses the amount of yield per total volume of evaporated and transpired water during a life-cycle of a currently grown crop [32]:

$$\mathrm{WUE} = \mathrm{Y_a}/\mathrm{ET_a} \tag{1}$$

where:

WUE—water-use-efficiency, kg yield mm^{-1} or m^{-3} of water;

Y_a—actual yield, kg, t ha^{-1};

ET_a—water use (transpired and evaporated water), mm, m^3.

Two methods can be used as simple and practical tools to calculate Y_w. The first is the FAS procedure (French and Schulz approach—FAS). This method is based on the assumption that the yield increase under the same environmental conditions is directly related to the increase in WUE [33]. The FAS method of water productivity calculation is composed of three components: (i) the maximum water productivity (TE), (ii) the quantities of water from current precipitation, (iii) the size of water resources in the soil volume rooted by the currently grown crop. The water-limited yield, (WLY) is calculated based on the equation:

$$WLY = TE\ (R\text{-}\Sigma E_s) + WR \quad (2)$$

where: TE refers to the transpiration efficiency TE (TE = k/VPD; k—biomass/transformation ratio; VPD—vapor pressure deficit; R—the sum of rainfall during the growth period; E_s—the seasonal soil evaporation, equal to 110 mm, WR -water reserves in the rooted soil volume.

The yield gap can be defined as the difference between the yield resulting from the effect of water available to plants during the growing season, i.e., the water-limited yield (WLY) and the actual yield (Y_a):

$$YG = WLY - Y_a \quad (3)$$

$$YG_f = 1 - (Y_a/CPY) \quad (4)$$

The key component in Equation (2) is the TE index. Originally, it was set for wheat grown in Australia at the level of 20 kg grain mm^{-1} of available water. Under favorable growth conditions, as stated by Passioura [34], this index can reach up to 30 kg grain mm^{-1}. As reported by Grzebisz et al. [26], TE is sensitive to the amount of supplied water and nutrients, ranging during the critical stages of spring triticale growth, from 14 to 39 kg grain mm^{-1}. The practical advantage of the Y_W calculation, based on the FAS approach, is to quantify the yield loss due to inefficient water management. The fraction of the yield gap (YG_f) is a relative measure (a value, extending from 0 to 1) of the yield gap (YG) due to unfavorable growth conditions with respect to those created by the potential water supply (WLY) to the currently grown plant. As shown in Figure 1, the highest WLY was recorded in the wet year (1997). Irrigation applied to winter wheat during the critical stages of yield component formation, i.e., stem elongation and heading did not result in a yield increase. The highest values of the YG_f, measured on the irrigated plots, indicate that water was not exploited by wheat due to the presence of other growth constraints. In contrast, the artificial reduction in the water supply (the treatment with imposed drought—D) to wheat during the critical period of yield component formation, resulted in a slightly lower grain yield. However, a much higher level of water exploitation was recorded, as indicated by much lower YG_f indices, especially on the K fertilized plot (0.11 vs. 0.29 on the irrigated plot, I). The other two years were characterized by a natural water shortage during the stem elongation period (1998), or during the grain filling period (1999). As a result, the YG_f, in general, approached zero, indicating that actual yields were close to the potential productivity of water. The negative values of YG_f stress the effect of other growth factors, which increased water productivity. This phenomenon was observed on treatments with the imposed drought, revealed most frequently on the K-treated plots. The presented data corroborates the main assumption of the FAS procedure of WLY calculation that WUE depends on the amount of water available to the currently grown crop. The YG_f index clearly reflects both the status of water management and soil fertility status.

2.2. Climatic Potential Yield—CPY

The maximum attainable yield of a particular crop, typical for the climatic region, is determined by the prevailing meteorological conditions. Based on this assumption Licker et al. [35], proposed to use the term, the climatic potential yield (CPY), instead of the less precisely defined, Yield Potential. The CPY can be defined as the maximum yield of a crop plant cultivar grown under a natural water supply, provided the optimization of other growth conditions. The yield gap (YG) is calculated based on the algorithm:

$$YG = CPY - Y_a \quad (5)$$

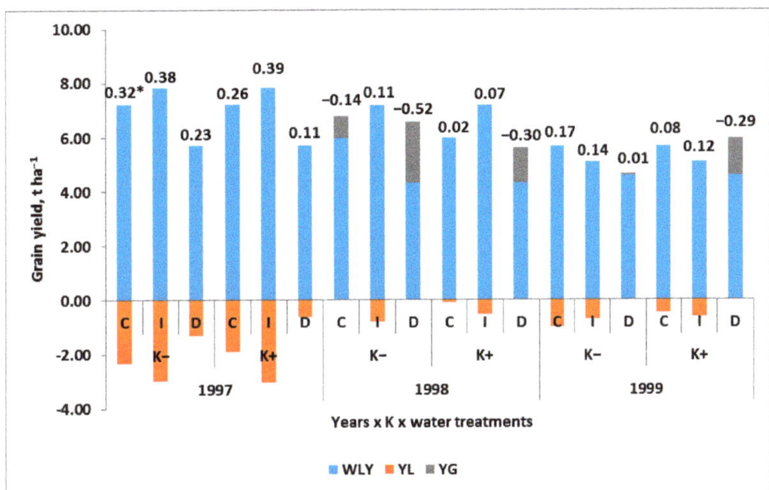

Figure 1. Graphical presentation of Water Limited Yield and loss/gain yield of winter wheat. Legend: WLY—water-limited yield; YL—yield loss; YG—yield gain; 0.32—the yield gap fraction; C—water control; I, D—plots irrigated or with imposed drought by water shortage during stem elongation and heading. (Author's own result; unpublished) * yield gain (−)/yield loss.

The fraction of the yield gap (YG_f) is a relative measure (a value, extending from 0 to 1) of the Y_a distance to the yield defined by the dominant weather conditions within a given climatic region (see Equation (4)). YG_f values approaching zero, indicate that meteorological conditions during the growing season were favorable, allowing the attainment of the climatic yielding potential by the grown crop. The best source of CPY data are experiments conducted by accredited Experimental Stations. An example of CPYs and their respective indices are shown in Table 1. The differences between CPY and Y_a for winter wheat were extremely pronounced. The calculated YG_f accounted on average for 58% for winter wheat as compared to 26% for sugar beets. The low year-to-year variability of this index for both crops indicates (i) stability of the CPY for these two crops, irrespective of weather variability during the growing season, (ii) the presence of other growth factors affecting the actual yield. The main reasons for the recorded difference between winter wheat and sugar beets are soil conditions. In Poland, winter wheat, despite high requirements for soil fertility, is cultivated on a broad range of soil agronomic classes [36]. In contrast, sugar beets are cultivated on very fertile soils [37]. The second growth factor, significantly impacting the CPY of wheat, is the level of crop protection and the level of applied N. The higher input of agronomic measures resulted in a CPY increase of 13%.

Table 1. The Climatic Yield Gap for basic seed crops in Poland, t ha^{-1}.

Years	Winter Wheat							Sugar Beet			
	CPY_{a1}	CPY_{a1}	Y_a	YG_{a1}	YG_{a2}	YG_{fa1}	YG_{fa1}	CPY	Y_a	YG	YG_f
2016	7.67	8.77	4.72	2.95	4.05	0.38	0.46	88.4	66.5	21.9	0.25
2017	8.21	9.76	5.11	3.10	4.65	0.38	0.48	88.9	67.9	21.0	0.24
2018	7.65	8.66	4.30	3.35	4.36	0.44	0.50	80.5	59.9	20.6	0.26
2019	8.36	9.17	4.64	3.72	4.53	0.44	0.49	80.0	57.5	22.5	0.28
2020	9.31	10.19	5.06	4,25	5.13	0.42	0.50	-	-	-	-
Mean	8.24	9.31	4.77	3.47	4.54	0.42	0.49	84.5	63.0	21.5	0.26
SD	0.68	0.65	0.33	0.52	0.40	0.04	0.02	4.86	5.00	0.9	0.02
CV, %	8.2	7.0	7.0	15.0	8.9	8.7	3.7	5.8	8.0	4.0	7.5

CPY—the climatic yield potential; Y_a—actual yield; YG—yield gap; YG_f—the yield gap fraction; a1, a2—medium and high input of production measures (crop protection + higher N rate); SD—standard deviation; CV—coefficient of variation.

2.3. Partial Factor of Productivity of Nitrogen—PFP_N

Nitrogen is, assuming the same meteorological conditions (precipitation) for a given locality (region), the key growth factor limiting yield [28,38]. Hence, the amount of N_f or the whole N input at the beginning of the growing season becomes the principal independent variable, affecting both the plant growth rate during the growing season and its yield [39,40]. The efficiency of N_f can be determined, using the concept of the partial factor productivity of fertilizer nitrogen, (PFP_N) [41]. This approach is frequently applied for making a country-to-country comparison [42]. The procedure to calculate the maximum attainable yield (Y_{attmax}), and in consequence, the nitrogen gap (NG) based on PFP_N, requires data on the unit productivity of the applied N_f under optimum growth conditions, i.e., the ample availability of nitrogen. The following set of equations can be used to calculate both indices:

$$\text{Partial factor productivity of } N_{in} \quad PFP_{Nf} = Y/N_f \text{ (kg seeds kg}^{-1} N_f) \tag{6}$$

$$\text{Maximum attainable yield} \quad Y_{attmax} = cPFP_{Nf} \cdot N_f \text{ (t, kg ha}^{-1}) \tag{7}$$

$$\text{Yield Gap} \quad YG = Y_{attmax} - Y_a \tag{8}$$

$$\text{Nitrogen Gap} \quad NG = YG/cPFP_{Nf} \tag{9}$$

To delineate the role of PFP_{Nf} on yield, the critical value of PFP_{Nf} has to be defined in the set of data obtained. The critical PFP_{Nf} ($cPFP_{Nf}$) is calculated as the average of the third quartile (Q3) of PFP_{Nf} values measured for each crop in a particular growing season. To determine the $cPFP_{Nf}$, the calculated PFP_{Nf} values are ranked in ascending order. The third quartile comprises values above the 75th percentile. The $cPFP_{Nf}$ is the average of the PFP_{Nf} values of the Q3. In the last step of the procedure, the nitrogen gap (NG) is calculated by transforming YG into the amount of the available N, but not used by the crop during the growing season [39]. The NG data can be used to prepare a graphical model of the efficiency of available N, i.e., mineral nitrogen present in the soil/crop plant system during the growing season of an actually grown crop. As shown in Figure 2, Y_{attmax} for 32 tested fields in 2020, amounted to 8.663 t ha^{-1}. Theoretically, at this yield level, the soil N_{min} content at wheat harvest is "zero". The excess of N_{min}, as indicated by negative values of NG, leads to a yield decline and vice versa. The key question is to indicate a reason for the appearance of the NG and its size. In most studied cases, it was the excess of applied N_f, concomitant with the low fertility of a soil agronomic class.

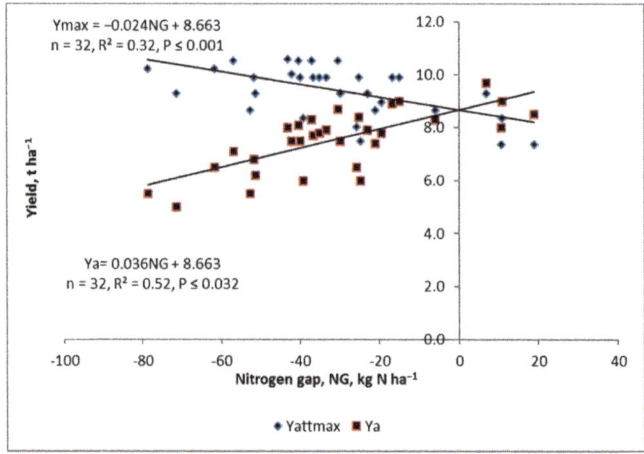

Figure 2. Trends in actual and maximum attainable yields in response to Nitrogen Gap in winter wheat. Legend: NG—nitrogen gap; Y_{attmax}—maximum attainable yield; Y_a—actual yield.

3. Soil Factors—Limiting Crop Plant Growth

3.1. Growth Factors Efficiency and Yield

A crop yield depends on its potential to take in water and nutrients in well-defined stages of its growth [43]. Three groups of factors are responsible for the exploitation of the yielding potential of a particular crop cultivar: (i) weather conditions during the growing season, (ii) soil fertility level, (iii) soil and crop management systems [35,43]. Weather conditions during the growing season in non-irrigated agriculture are the strongest environmental factor, significantly affecting year-to-year variability in yields [44]. Soil productivity has been indicated as one of the most important objectives listed in the sustainable development goals (SDGs) by the United Nations in the 2030 Agenda for Sustainable Development [45]. Mueller et al. [46] classified natural factors constraining plant productivity into three main groups, comprising: (i) soil moisture and its thermal regime—directly related to weather variability during the growing season and to in-season changes in water availability in the soil profile, (ii) root growth pattern and nutrient uptake patterns, (iii) field topography. Soil fertility can be defined as the inherent soil potential for supply of air, water, and nutrients to the currently grown crop plant in the required amounts and chemical forms, necessary for exploiting its yielding potential [47]. The key question with respect to the optimum ranges of growth conditions for a currently cultivated plant on a given field cannot be simply answered. The main reason is the huge number of factors that affect plant growth during the growing season, and its subsequent yield. Wallace and Wallace [43] suggest the existence of more than 60 factors, grouped into seven categories, including both biophysical soil properties (five groups), weather and soil, and finally crop management methods. Based on these evaluations, the production outcome, i.e., actual yield (Y_a) can be considered as a function of the climatic potential yield (CPY) and the efficiency of factors (E), affecting yield:

$$Y = CPY \cdot (E_1 \cdot E_2 \cdot E_3 \cdot \ldots E_{n-1} \cdot E_n) \tag{10}$$

The strength of each factor's impact on yield declines in accordance with its partial fraction, approaching 1.0. The value of 1.0 for a given factor indicates its optimum status with respect to the rate of plant growth, the formation of yield components, and the eventual yield.

3.2. Soil Fertility Constraints—Humus Content and Water Resources

During the first step in nitrogen gap (NG) amelioration, it is necessary to develop a set of effective diagnostic tools for recognizing the main soil characteristics—constraints that negatively affect plant growth. The second step should be oriented to working out a set of agronomic methods, allowing for the NG cover. Four groups of growth factors require a farmer's serious attention when evaluating the strength of soil factors that limit plant growth during the growing season. They are (i) total soil moisture amount and its availability to plants, (ii) the size of the in-season mineral nitrogen pool, (iii) pool of available nutrients responsible for N efficiency to plants during the growing season, (iv) plant accessibility to all sets of nutrients within the growing season.

The most important of these factors is water management before and during the growing season. World crop production systems to 60–80% depending on the amounts of water accumulated in the soil profile, strictly in the layer occupied by roots of the currently grown plant [48]. The quantity and availability of water stored in the soil profile to plants depend both on soil texture and its structure [49]). The capacity of the root zone (RZC) for water (the volumetric water capacity, VWC, at field capacity within a range of 10 and 1500 kPa suction) is measured for crop plants down to 100 cm [50]. In Europe, the RZC_{100} for sandy soils is in the range of 50 to 100 mm, and for loamy soils from 100 to 200 mm.

The amount of storage and available water depends to a great extent on the content of soil organic matter (SOM, humus). In regions, or even in fields with a high contribution of sandy soil, the best option to increase VWC is to raise the humus stock (HS) in the soil profile. This operation can be only successful if the depth of the organic layer is increased [51]. It has been well-documented that the amount and resistance of soil humus to degradation processes is a function of soil texture [52]. Loveland and Webb [53] in an extensive review showed that the minimum content of humus required to maintain soil productivity is 1.7% for sandy soil (4% of clay particles) and almost 6% for clay loam (38% clay particles). The humus content in a particular soil is, in fact, a constant value, defined by the content of total silt and total clay. Based on this assumption Piéri [54] developed a soil degradation index, termed as the humus stability index (S), whose formula is as follows:

$$S = (H/Si + C) \cdot 100 \tag{11}$$

where: S—index of soil humus stability, H—humus content (%), Si and C denote silt and clay content (%), respectively.

The S index is a simple index to determine the current status of soil degradation. Four classes of soil sensitivity to degradation, based on the S index can be applied:

(1) $S < 5$—structurally degraded soil;
(2) $5 < S < 7$—a great risk of soil structure degradation;
(3) $7 < S < 9$—a small risk of soil structure degradation;
(4) $S > 9$—no risk of soil structure degradation.

The S value above nine indicates that the humus content is at an optimum to protect the stability of the soil structure. In the next step in the procedure, evaluating the status of soil humus, the humus stock gap (HSG) can be calculated [55]:

$$SHS_{SD} = OC \cdot 1.7 \cdot 10 \tag{12}$$

$$HSG = HS_A - SHS_{SD} \tag{13}$$

where:

SHS_{SD}—Standardized Humus Stock, t ha^{-1}; HS_A—Actual Humus Stock, t ha^{-1};
HSG—Humus Stock Gap, t ha^{-1}; OC—mean value of organic carbon content, kg m^{-2};
1.7—constant used to recalculate the OC content into humus;
10—constant, recalculating kg m^{-2} into t ha^{-1}.

The yield gap/gain (YG/G), i.e., yield loss or gain, based on the humus content, can be calculated based on the formula:

$$YG/YG = HSG \cdot 15.6 \tag{14}$$

where: 15.6—constant, recalculating HSG into grain yield [56].

The OC data for the SHS_{SD} calculation with respect to European soils were based on data reported by Batjes [57]. The average content of humus in Luvisols in the soil layer of 0.0–0.3 m is 85 t ha^{-1}, increasing up to 154.7 t ha^{-1} in a soil profile of 1.0 m. In Cambisols, the respective values are 117.3 and 200.6 t ha^{-1}. Both figures are low as compared with Chernozems, for which the respective values are 153 and 374 t ha^{-1}. The simple calculation of HSG shows that a net increase in the humus stock of 1 g m^{-2}, which is equal to 17 t ha^{-1} of humus, results in a yield increase of 265.2 kg ha^{-1}. The potential yield increase can to a great extent be explained by the higher water-holding capacity of humus. As reported by Libohova [58] 1.0 g of humus holds up to 1.5 of water.

The main reason for the degradation of humus stock (HS) in arable soils is the rapid mineralization rate of the labile organic carbon pool, irrespective of soil management [59]. This process is accelerated by intensive NPK fertilization and soil plowing [60]. The regeneration of the HS in arable soil is a long-term process. The most effective amelioration strategies oriented towards an HS increase in arable soils, tested in long-term trials, rely mostly on intensive manure application. In European soils, as reported by Powlson et al. [61], expectations regarding the effect of manure are rather low. A yearly application of 10 t manure over 90 years can raise the humus content in a soil layer of 0.3 m only by 4.8%. As reported, however, by Szajdak et al. [62], a yearly application of 30 t ha^{-1} of manure to light soil over 38 years doubled the humus content. Despite the HS increase, no differences in rye yields were recorded compared to the effect of NPK application alone. The main practical disadvantage of this approach to HS increase, even in mixed farming, is its theoretical quality. This solution is not realized in farming practice due to a lack of manure. The option, applied in intensive agriculture, oriented only towards crop plant production, is the management of straw directly in the field. As frequently reported, this method of straw management can both increase the HS and yields of succeeding crops [63].

3.3. Nutrient Availability and Crop Plant Accessibility to the Nutrient Pools

The agronomic term nutrient availability refers to the amount of a particular nutrient taken up by the currently grown crop within a single growing season. Chemical tests of extractable nutrients are only an approximation of the amount of a given nutrient which potentially can be taken up by the crop plant [64]. In addition, the content of available nutrients is, as a rule, determined in the upper soil, limited usually to a layer of 0.2 m [65]. The term accessibility refers to the crop plant's access to soil pools of attainable nutrients within the growing season [64]. Plant access to a respective nutrient pool depends on the rate of root system growth, which is driven by the hormonal status of a plant, which to a great extent depends on plant access to nitrate nitrogen. A decreased supply of N-NO_3 to the aboveground plant parts affects the pattern of dry matter partitioning, leading subsequently to an increase of its investment into roots [66,67]. This crop plant strategy is oriented towards the capture of water, nitrogen, and nutrients supporting their use efficiency. It prevails in regions and soils sensitive to temporary water shortages [68,69].

The observed trends of crop plant response to irrigation are fully corroborated by the fact that ample water supply is a decisive yield factor, providing an optimum supply of nutrients, mainly N [46]. As shown in Figure 3, the yield of spring triticale decreased in accordance with the decreased amount of available water. However, the absolute and relative yield decrease, i.e., YG and YG_f were much lower under K fertilized treatment. The main reason for the observed trend variability was the impact of K application on WUE. As a rule, the WUE-Eta trend reflected the trend of Y_a, very well, but its steepness was lower on the K fertilized plots. The WUE-WLY indices showed different trends of dependence on K treatment. The index value, under conditions of K fertilizer application, increased in the opposite direction to the quantity of supplied water, i.e., from 23 kg grain mm^{-1} on the irrigated plot to almost 39 kg grain mm^{-1} on the plot with the artificial drought imposed during the stem elongation stage of triticale. It can be concluded that under natural precipitation, the yield gap can be ameliorated, at least partly, through the supply of nutrients like K, which exert a significant impact on water and nitrogen management by crop plants [26,70].

Fertilizer recommendations, except for mineral N (N_{min}), do not consider the content of available nutrients in the sub-soil. The total content of the majority of nutrients, taking K as a classic example, depends on the content of clay minerals [71,72]. The content of available nutrients is also sensitive to other soil characteristics, for example, soil pH (phosphorus, micronutrients); manure application, and the content of organic matter (nitrogen, phosphorus, micronutrients); fertilization, and cropping sequence (nitrogen, phosphorus) [65,71–75]. The sub-soil has numerous functions; as a source of water, nitrogen, available pools of other nutrients, and as a natural milieu for the plant root system [76,77]. As shown in Figure 4, crop plants can penetrate the soil for K down to a depth of 0.90 m of the soil profile. In the presented case, the strongest soil depletion with the $CaCl_2$-extractable K was recorded for winter oilseed rape. This phenomenon is corroborated by the fact that this crop has an extremely high requirement for K during the spring vegetation [78]. The same phenomenon was observed for phosphorus. As reported by Łukowiak et al. [79], crop plants can exploit the $CaCl_2$-extractable P down to 0.9 m of the soil profile. The authors of this study documented that the recovery of 60% of the available P pool was concomitant with the highest yield of winter oilseed rape (WOSR). This figure may seem shocking, but it refers to the P content in the soil solution. An open question remains as to the contribution of the sub-soil P pool in the total P taken up by the currently grown plant. In the case of seed crops, for example, WOSR, an important part of P is taken up by plants following the onset of flowering [80]. A permanent application of P fertilizers, as documented for long-term experiments, leads to the enrichment of the sub-soil with P. As a consequence of this operation, this pool becomes a considerable source of P for high-yielding crops [81]. It can be, therefore, concluded that chemical tests for the soil pools of nutrients in the subsoil cannot be limited only to the N_{min} content [82,83]. It has been recently documented that the simultaneous determination of N_{min} and some other key nutrients in the $CaCl_2$-solution is a source of knowledge of both their pools and the occurring relationships between them [82,83].

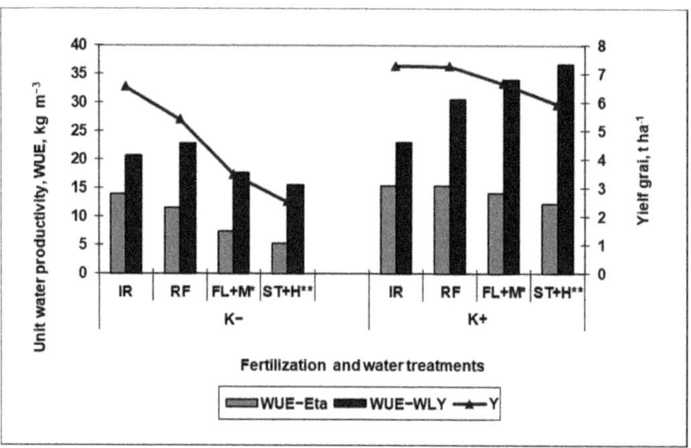

Figure 3. Effect of water treatments and potassium management on indices of water use efficiency (based on [26]. Legend: IR, RF—irrigated and rainfed water treatments; FK+M, ST+H—water shortage imposed at * stem elongation + heading; ** flowering + early milk. WUE—water Use Efficiency; Eta—calculated based on evapotranspiration; WLY—calculated based on FAS approach [33].

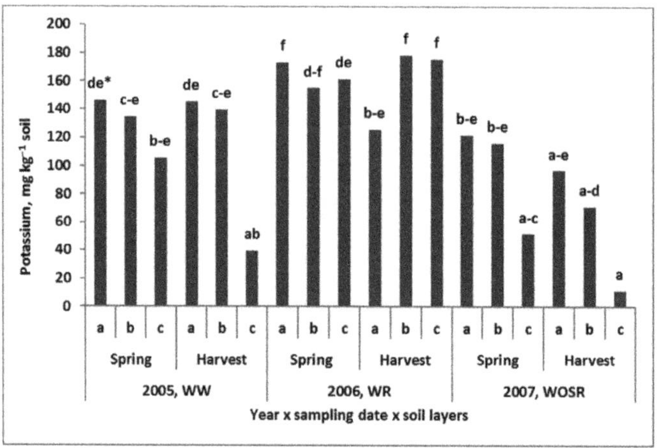

Figure 4. Status of $CaCl_2$ extractable K during the growing season in three consecutive years (author's own result; unpublished). Legend: a, b, c—soil layers of 0–0.30, 0.30–0.60, 0.60–0.90 m; WW—winter wheat; WR—winter rye; WOSR—winter oilseed rape; *the same set of letters indicates a lack of significant difference between the treatments.

The vertical trends in nutrient content variability need to be considered when working out fertilization recommendations. It has been well-documented in literature that the subsoil significantly impacts both the profile of crop rooting and in consequence the uptake of water and nutrients [76]. Rooting depth is not a constant pattern, even for the same crop. It changes due to the impact of numerous biophysical, environmental, and also soil and crop management factors [84]. The rate and habit of the root system, similarly to the shoot, undergo temporal changes during plant development and in response to its nutritional status [85]. The depletion of N-NO_3 at the onset of WOSR flowering in the soil layer of 0.9 m depth is a prerequisite of high yield [80]. The efficient uptake of NO_3^- ions depends on the respective concentration of K^+ in the soil solution [86]. For the low-yielding WOSR plantation P resources in the vegetative organs at the onset of flowering are sufficiently high to cover the requirements of the growing seeds. A high-yield of seeds can be achieved, but only provided there is efficient remobilization of P from vegetative tissues, and its simultaneous uptake from the subsoil [77,79,87]. It is necessary to take into account two other important facts. The first refers to the growth status of the root system. The onset of flowering results in the progressive root system dying, i.e., the rate of root mortality is higher than the appearance of new roots [64,88]. It is necessary to stress that the uptake of both, K and P depends to a much greater degree on the root elongation rate than on the concentration of both ions in the soil solution [89]. So far, the routine fertilizer recommendations have neglected the vertical variability of factors responsive to nutrient uptake by crop plants. This knowledge gap is one of the key reasons for the differences in the prognosis of crop production intensification in a sustainable way.

4. The In-Season Management of Nitrogen—Cardinal Growth Stages

The first task in the reorientation of the crop production system is to calculate the total requirement of a currently grown plant for nitrogen. This can be calculated for an average yield harvested on a particular field or based on the potential yield of a given cultivar in the same climatic region. Two additional components have to be taken into account to make a reliable estimation of the total N requirement by the currently grown cultivar. The first is nitrogen concentration in the main product, for example, seeds or grain. There is still ongoing scientific discussion with respect to the extent to which N concentration in seeds or grain is a conservative, i.e., genetically, or environmentally governed trait [90–92]. The second component refers to the partitioning of N taken up by the crop between the main yield, and its by-product, for example, between grain and straw. At harvest, this process refers to the value of the nitrogen harvest index (NHI), which is a conservative trait of seed crops [11,93].

Total N input (N_i) in the soil/crop system for an assumed yield of grain/seeds is calculated based on the algorithm:

$$N_i = (Y_{CPY} \cdot G_{Nc})/NHI \tag{15}$$

where:

N_i—nitrogen input, kg ha^{-1};

Y_{CPY}—climatic potential yield of the grown cultivar, kg ha^{-1};

G_{Nc}—grain/seed nitrogen concentration, kg t^{-1};

NHI—nitrogen harvest index, a value in the range of 0.6–0.8.

The key objective of nitrogen fertilizer (N_f) application is to synchronize its application time with the crop plant requirement. The dominant factor is the stage of plant growth and the required content of N, which progressively increases with the crop growth. The right determination of the N_f dose in the critical stage of yield formation is, therefore, the decisive factor in the exploitation of the yielding potential of the currently cultivated crop. The crop demand for N in a particular growth stage depends on the rate of plant biomass growth, primarily driven by temperature and water supply [94]. The sum of physiologically active temperatures (GDD), and water and nitrogen supply are major factors for the quantity of biomass produced by the crop during the respective phase [95]. The rate of seed crop growth throughout the vegetative season is not constant. Based on this criterion, three mega-phases can be distinguished, named as canopy foundation (CF), yield component construction (YCC), and yield realization (YR) [96]. The first two periods describe the vegetative part of the plant life cycle and the third one its reproductive phases (Figure 5). The shape of the dry matter accumulation curve can be described mathematically using very sophisticated models [97]. In cereals, the CF period, extending from sowing up to the end of tillering, is responsible for the number of tillers per plant. The course of dry matter accumulation is best described by the exponential regression model. The YCC period, extending from the beginning of stem elongation up to flowering, is responsible for the set of flowers. The rate of dry matter accumulation during this period is best described by the linear regression model. The dry matter yield of wheat at the end of the CK2 can be used to make a prognosis of grain yield [98].

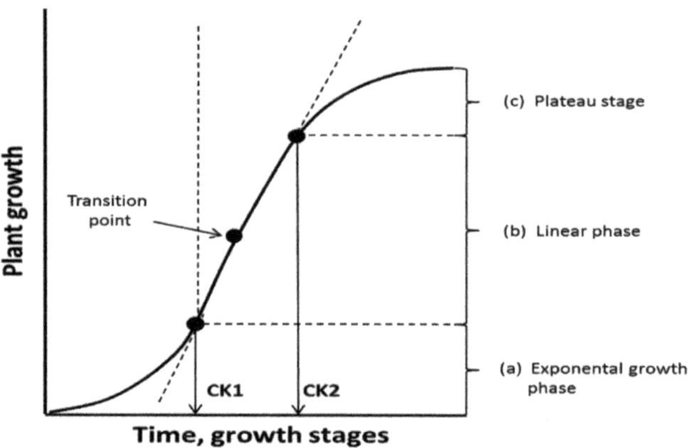

Figure 5. A conceptual pattern of dry matter accumulation by a typical seed/grain crop. Legend: CK1, CK2—cardinal stage 1 and 2, respectively (author's own concept).

The borderline stages of plant growth, which occur between the mega-phases of CF and CYC, and between the CYC and YR periods, can be named Cardinal Knots (CKs), i.e., CK1 and CK2, for the first and second borderline of consecutive mega-periods, respectively. These two CKs are decisive stages for yield component formation. For example, as shown in Figure 6, the content of soil nitrate-nitrogen (NN) during the YCC period of oilseed rape undergoes a strong depletion (=N uptake by WOSR during the YCC period). The recorded NN depletion significantly affected the seed density, which was the key yield component, determining seed yield (R^2 = 0.89, $p \leq 0.001$). What is most important, however, is the fact, that the dependence obtained clearly defines the rosette stage (BBCH 30) as the decisive stage for the yield prognosis (NN_{op} = 163.3 kg ha^{-1} for Y_{amax} = 77.6 · 1000 seeds m^{-2}). The result obtained indicates that in farming practice, the time of N_f application has to precede the BBCH 30. In maize, the second rate of N_f is applied, based on N_{min} determination, at the 5th leaf stage [99,100].

The yield realization (YR) period commences from the onset of flowering and persists up to the physiological maturity of the plant. It can, however, be divided into two parts. The first part extends from flowering up to the watery stage of the seed/grain plant growth, i.e., to BBCH 71 (15% of seed/grain DW). It has well been documented that nitrogen supply to a seed crop, like cereals and oilseed rape, significantly affects the degree of yield component expression. The N pool accumulated by the seed plant during its vegetative growth, i.e., before flowering, considerably impacts the number of seed/grain per field unit area (physiological sink capacity) [101–103]. The second part of the YR period, which begins from the early milk-stage (BBCH 72) and finishes at physiological plant maturity (BBCH 89/90), is termed as the grain/seed filling period (GFP, SFP). The course of dry matter accumulation can be described by different mathematical models, but the linear or quadratic dominate. During this particular period, grain/seed reaches its final individual weight [104].

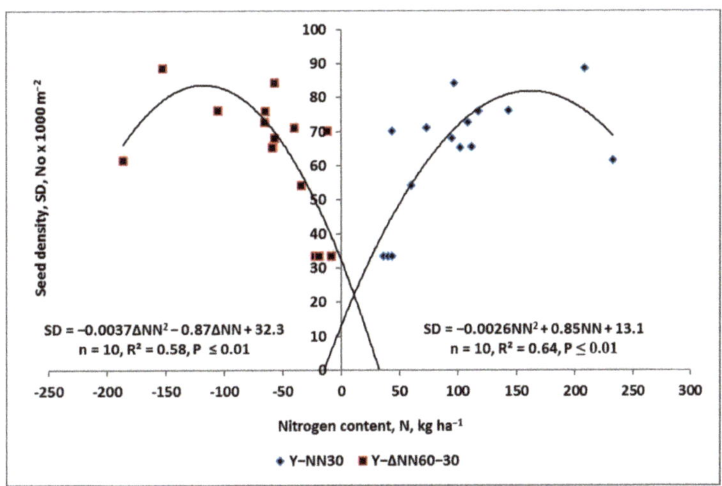

Figure 6. The response of seed density of winter oilseed rape to the content of nitrate nitrogen at the rosette stage and its uptake by the crop up to flowering. (based on [80]. Legend: NN- the content of nitrate nitrogen at BBCH 30; ΔNN—the change of the NN content during the period extending from the rosette up the flowering stage of WOSR growth.

Efficient in-season N management, including knowledge of N_{min} resources, should be oriented towards covering the plant N requirements during the period extending from the onset of flowering to the physiological plant maturity. For the seed crops, 75–85% of the N finally accumulated in seeds/grains originates from the pre-flowering resources, i.e., present in vegetative plant parts [90,105]. Post-flowering N management by the crop canopy can be described by three algorithms:

N remobilization quota: $$NRQ = N_{hv}\ N_{fl},\ kg\ ha^{-1} \quad (16)$$

N remobilization efficiency: $$NRE = NRQ/N_{fl} \cdot 100,\ \% \quad (17)$$

Post-flowering N uptake: $$PNUP = N_t - N_{fl},\ kg\ ha^{-1} \quad (18)$$

where:

N_{hv}—N amount in vegetative organs of crop canopy at harvest, kg ha^{-1};
N_{fl}—N amount in vegetative organs of crop canopy at the onset of flowering, kg ha^{-1};
N_t—total N amount in crop canopy at harvest, kg ha^{-1}.

The patterns of N accumulation by the plant between particular CKs during the growth of a crop are the basis for a build-up of an efficient strategy of the in-season N status correction. Any free choice of N_f timing and its dose, as frequently observed in classic or even modern fertilization programs, does not fit the crop requirements for N, being the main reason for its inefficiency. A sound strategy of N management in seed crops, despite an almost similar accumulation pattern throughout the life cycle, is crop-specific. For example, in bread wheat, the key period of yield component formation extends from CK1 to CK2. During this period, the requirement for N results from both (i) the number of grain per unit area, and (ii) the required protein content in grain [106]. Consequently, a nitrogen fertilization strategy based on the correction of the plant N status in both CK1 to CK2 should also take into account protein concentration in grain. As a rule, any increase in the number of grains per unit field area results in a protein concentration decrease [107]. Therefore, any fertilization strategy, oriented to the increase in the crude protein concentration, requires an extra N_f dose, which should be applied at the end of the CK2. In maize, its nutritional status at the 5th leaf stage, which slightly precedes CK1, is decisive for the degree of yield component expression [108,109]. Nitrogen fertilizer should be applied just at such a time preceding this cardinal stage of maize growth because it affects the potential number of leaves and cobs. Nitrogen status in maize at the CK2, before the beginning of flowering, is important for the yield development, but it has only a predictive value [110]. In farming practice, it makes no sense to apply N_f at this particular stage because the yield was already fixed in much earlier stages of maize growth.

Nitrogen efficiency depends on the supply of other nutrients needed for its uptake and utilization [70,111–113]. It is necessary to pay attention to the fact that the accumulation of K by major crop plants, like wheat, oilseed rape, sugar beets, or potato, is as a rule higher as compared to N [114,115]. The maximum K uptake by high-yielding WOSR reaches its maximum during the phase of the main stem elongation [77,78]. It can, therefore, be concluded that an efficient uptake of K from the subsoil by some crops, in comparison to N, is a necessary condition for the effective uptake of nitrogen. As recently reported by Grzebisz et al. [80], the efficient uptake of NO_3^- ions by WOSR depends on the adequate concentration of K and other nutrients, such as Ca and Mg in the sub-soil. All these nutrients are responsible for the development of yield components by crop plants.

One of the most important priorities in the breeding of crop plants is to increase the uptake efficiency of nutrients, especially of N and P from soil. Efficient acquisition of water and nutrients is required for the realization of both production and environmental goals. The efforts of key breeders have recently focused on the improvement of root system traits through [116–119]:

(1) Accelerating the early rate of root growth of short-season crops, for example, vegetables;
(2) Increasing root density in the topsoil to increase uptake of freshly applied fertilizers;
(3) Deeper rooting of crops grown in areas vulnerable to nitrate-nitrogen leaching.
(4) Stronger root branching to increase exploration of soil zones rich in mineral nitrogen;
(5) Increasing the density and length of root hairs for effective acquisition of phosphorus and potassium by a currently grown crop;
(6) Improvement in mycorrhiza association for effective acquisition of phosphorus and micronutrients.

5. Yield Gap Recognition and Diagnosis of Limiting Factors

An efficient N management strategy should be based on three major variables that affect the plant growth of a particular crop. Yield variability within a field is a result of (i) the stage of a crop plant growth, (ii) spatial variability in N uptake by plants, (iii) vertical variability in soil N pools, and a plant's access both to these pools and other nutrients responsible for N use efficiency. It is necessary to assume that the expression of yield components is a result of the growth pattern of a crop plant encoded in its early stages [80,120]. The principal difficulty in determining the right N_f dose is the number and strength of plant growth limiting factors with respect to their spatial variability on the field. Yield, in fact, its spatial variability within a field, is the ex-post result of (i) the degree of yield constraints recognized during the growing season, (ii) the efficiency of production measures applied to ameliorate constraints limiting plant growth and yielding. The spatial variability of yield is the main reason for the necessity for dividing the entire field area into zones that differ considerably in productivity [121]. The key challenge is to find a criterion for the particular field division into zones of the same level of productivity.

The target of modern agriculture is to work out and implement a set of highly reliable diagnostic tools which will be capable of defining the methods and approaches to the efficient use of N_f, that are in accordance with the concept of sustainable agriculture [21,122]. This concept, taking the field as the key production unit, is based on three main objectives:

(1) fulfilling the food production gap, in fact, ameliorating the nitrogen gap;

(2) achievement of a satisfactory level of applied production measures;
(3) a significant decrease in the negative impact of applied Nf on the environment.

5.1. Nitrogen Use Efficiency—Limiting Factors

The most important task for a farmer is to establish a hierarchy of factors affecting the yield gap in a particular field and recognizing their depth. Irrespective of the climatic zone, the plant growth rate and development of yield components depend on the supply of N and its use efficiency [38]. Nitrogen use efficiency (NUE) defines the amount of the main product, for example, seeds, grain, roots, tubers, per unit of supplied N [123]. This N index is composed of two sub-units, i.e., nitrogen uptake efficiency (NUtE) and nitrogen utilization efficiency (NUtE), presented as

$$NUE = NUpE \times NUtE \quad (19)$$

The simplified interpretation of the first part of this equation, frequently used by crop breeders, mainly focuses on the amount of soil available nitrogen, i.e., to the low- and high- N input growth environment [124]. In fact, the amount of available N is a strong factor discriminating plant crop genotypes [125]. Nitrogen uptake and its subsequent transformation into plant biomass, i.e., NUtE, depends both on the physiological potential of a plant to take up N from its soil resources, and on soil factors limiting the rate of its uptake and subsequent utilization by a plant [113]. The spatial variability of yield clearly indicates that the amount of N taken up by plants of the same cultivar, i.e., having an identical yielding potential, is also spatially variable. Plants suffering from an N shortage due to their lower supply during the cardinal stages of yield formation are not unable to capture the same amount of solar energy compared to those growing in conditions of ample water and N supply. As a result, these plants are not capable of achieving the rate of growth as determined by the supply of light energy [126]. The ex-post formulated questions are: (i) at what stage of crop plant growth does the yield formation become limited by a shortage of N? (ii) what is the reason for the insufficient N uptake? The first question has been thoroughly discussed in Part 4. of this review. The second question should focus the farmer's attention on factors pertaining to the plant potential at a particular stage of its growth to access water and nutrient resources.

5.2. Diagnostic Tools for the In-Season and Spatial Yield Gap Control

The recognition of the yield gap and its size requires an implementation of a set of analytical tools with a capacity to make a reliable diagnosis of factors that limit plant growth and yield. A significant improvement in NUE is possible provided there is a reliable quantification of soil characteristics limiting both N uptake and its utilization by the currently grown plant. Hence, a single field is considered a basic unit in any diagnostic procedure.

The key characteristic of the yield of a currently grown crop is its spatial variability (Figure 7). The hierarchy of factors decisive for NUE is well recognized, but not always taken into account by farmers and their advisors. The primary tasks, aimed at an improved strategy for spatial N management, are to polygonize the field based on:

(1) the climatic yield potential (CYP), and the water-limited yield (WLY);
(2) the total requirement of the crop for nitrogen;
(3) primary soil characteristics, assuming a homogenous regime of water and nitrogen supply to the
(4) currently grown crop during its critical stages of growth;
(5) soil pH and the content of active aluminum, as the key factor limiting the root system growth of crop plants;
(6) contents of available nutrients, assuming a homogenous regime of their supply to the currently grown crop.

It is necessary to create a broad set of zonation maps with respect to soil and crop characteristics to get an operational tool to establish homogenous field production units (HFPUs) [127,128]. Therefore, the soil water capacity and its availability to the currently grown crop determine the first criterion of the field zonation. The primary factor affecting the diffusion of ions toward the plant root in the soil is the content of available water [50,58]. This factor shows a strong spatial and vertical variability within the growing season, significantly affecting crop growth at cardinal stages of yield component formation [129]. As discussed in Part 3, the amount of available water is defined by the content of colloids and soil structure. Hence, the main reason for existing variabilities in the available water content is the content of mineral colloids. The primary data can be collected using the classic diagnostic procedures of soil analysis. The working out of zonation maps requires the implementation of geostatic methods [130,131]. The advantage of the classic methods of basic soil characteristic determination is the necessity to analyze water and nutrient resources present both in the topsoil and in the sub-soil which is rooted by a currently grown plant [80,120].

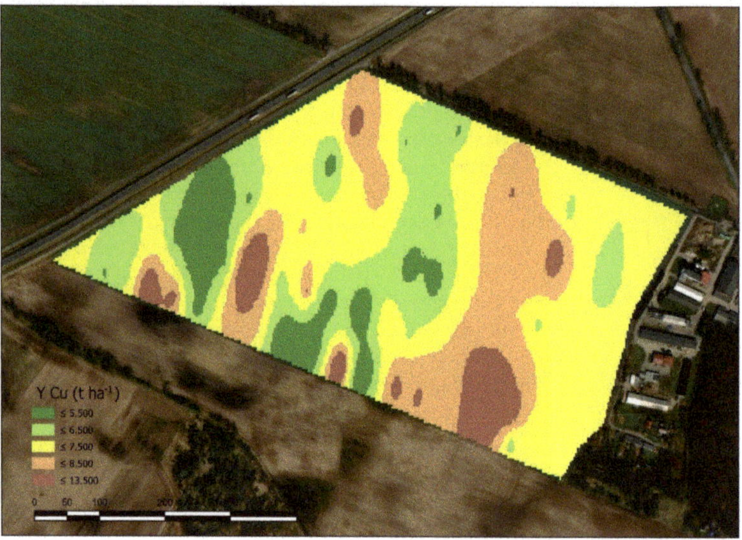

Figure 7. Spatial variability in winter oilseed rape yield expressed in cereal units (source: [128]).

The crop plant requirement for N_f is the key in-season nutritional factor, responsible for the development and status of yield components. The N_f dose can be calculated based on two characteristics of crop canopy:

(1) the rate biomass increase during a particular phase, in particular, within the period defined by cardinal knots (Figure 5);
(2) the trend in N concentration in plants during vegetative growth, with special attention to CKs.

A sequential determination of the N content along the growing season of a plant is a primary tool for its N status assessment. As a rule, during the vegetative growth of the seed crop, the proportion of active metabolic tissues (leaves) decreases stage-by-stage with respect to structural ones (stem). Consequently, the total N concentration in plant tissues decreases during vegetative growth, but its content in the plant progressively increases, reaching the highest value just before the onset of flowering (CK2). The temporal variability in relationships between crop dry matter biomass and N concentration is described by the following set of equations [132]:

(1) Nc—the critical N concentration, % or g N kg^{-1} DM

$$N_c = aW^{-b} \quad (20)$$

(2) NNI—Nitrogen nutrition index,

$$NNI = N_a/N_c \quad (21)$$

where: W—crop biomass, t ha^{-1}; a and b—estimated parameters.

The N_c curve is expressed as a power function, but its parameters are both crop and growth condition specific. As reported by Song et al. [133], an N_c based only on rice leaf dry matter was expressed as $N_c = 1.96L_{DM}^{-052}$. The in-season crop N nutritional status can be evaluated as proposed by Chen et al. [134] for winter wheat based on three classes:

(a) N deficient for NNI < 1.00,
(b) N optimum for $1.00 \leq NNI \leq 1.25$;
(c) N surplus: $NNI \geq 1.25$.

As discussed in Part 4, the yield is the outcome of the efficiency of crop production factors. A major characteristic of a crop plant is its time variability in requirements for water, nitrogen, and nutrients responsible for the efficiency of both key growth factors. In-season evaluation of plant nutritional status relies on plant destructive sampling and nutrient concentration determination in the lab. The key disadvantages of classic diagnostic methods are:

(1) ex-post data, delaying the diagnosis of plant nutritional status, and in consequence a recommendation of the N_f rate;
(2) lack of data about spatial heterogeneity in plant nutritional status, resulting in
    (a) uniform N fertilizer rate application, hardly related to the real, i.e., spatial variability in plant requirements for N,
    (b) low efficiency of applied nitrogen,

A significant increase in NUE requires the implementation of new diagnostic tools, capable of quantifying a crop plant requirement for N in real-time (a defined stage of plant development) and taking into account spatial differences in the productivity of field homogenous units. There has been considerable technological progress in developing different sorts of non-invasive instruments of potential use in plant nutritional status determination [135–137]. Remote sensing is a technique that offers a broad set of effective diagnostic tools to meet both production and environmental objectives. All spectral techniques rely on the plant's ability to absorb and simultaneously reflect solar radiation. Remote-sensing techniques can make a rapid assessment of plant biomass, leaf area index, nitrogen content, chlorophyll content, and finally yield [138]. The information capacity of hyper- and multi-spectral imagery is several times larger as compared to any classical diagnostic tool.

Based on the spectral imagery of a field, it is possible to create a zonal map of a temporary crop N status. The selection of the most reliable spectral indices depends on the sensitivity of spectral bands to both the total N course during plant growth (Nitrogen Dilution Curve—N_c) and canopy structure (biomass, density, N concentration). It has been confirmed that the intensity of solar radiation reflectance in visible light (waveband 400 to 720 nm) is negatively correlated with leaf N content, while NIR reflectance is positively correlated with leaf N content and/or crop biomass [139]. The Normalized Difference Vegetation Index (NDVI) is the most frequently used diagnostic tool for the N content and crop biomass determination [140–142]. In fact, NDVI allows a reliable diagnosis of N status in cereals only during the early stages of growth. This index reaches saturation status at dense canopies [143,144]. This weakness has been recently overcome by developing a new set of Vegetable Spectral Indices (VIs) worked out on bands lying in other spectral regions, including the red-edge region (700–740 nm). Scientific reports have published data on VIs which are capable of predicting an LAI extending from 0 to 6 [145].

6. Conclusions

The sustainable intensification of agriculture is a concept which relies on the assumption of realizing two, seemingly contradictory goals, i.e., increasing food production but without a negative impact on local or global ecosystems. This twin objective can be achieved if solutions strictly oriented towards amelioration of the yield gap are put into agricultural practice, which will involve a reliance on increasing efficient nitrogen use. The inefficiency of nitrogen in the soil/crop plant system, i.e., the Nitrogen Gap, requires a three-dimensional diagnosis. The first, as a matter of fact, the major variability refers to the in-season variability in the nitrogen requirements of the currently grown crop. This gap can be filled by a synchronization of the demand of a grown plant for nutrients, in fact, driven by nitrogen applied in the required N_f dose and at the right time. The recorded inefficiency of in-season applied nitrogen is due to the spatial variability of its supply. The main reason for the occurrence of the NG is both spatial and vertical variability in soil characteristics responsible for N uptake and its subsequent utilization by plants. The recognition of the size and strength of a plant's growth constraints, the disturbing formation of yield, is the prerequisite for dividing a field into zones of homogenous levels of productivity. Because the capacity of soil to soil water storage depends on the content of colloidal particles, the first goal of soil fertility amelioration is to increase the humus stock. This production objective is especially important for field zones naturally poor in mineral colloids. The subsequent steps in the action oriented to soil fertility amelioration depend on a farmer's ability to recognize the level of the subsoil fertility, including the content of available water and nutrients, and on the access of the currently grown crop to these resources during the growing season. A diagnosis of the soil fertility status is important for the development of application techniques of both nitrogen and nutrients, supporting its use efficiency. Classic and remote sensing techniques should be simultaneously applied to delineate the field into units of the homogenous level of productivity. Remote-sensing is the most efficient tool for the in-season diagnosis of the N plant status, provided it can discriminate the productivity of each field unit.

Author Contributions: Conceptualization, W.G. and R.Ł.; methodology, W.G.; software, R.Ł.; validation, R.Ł.; formal analysis, W.G.; investigation, R.Ł.; resources, W.G.; data curation, R.Ł.; writing—original draft preparation, R.Ł.; writing—review and editing, W.G.; visualization, R.Ł.; supervision, W.G.; project administration, R.Ł. All authors have read and agreed to the published version of the manuscript.

Funding: This research received no external funding.

Conflicts of Interest: The authors declare no conflict of interest.

References

1. Tilman, D.; Balzer, C.; Hill, J.; Befort, B.L. Global food demand and the sustainable intensification of agriculture. *PNAS* **2011**, *108*, 50. [CrossRef] [PubMed]
2. Le Mouël, C.; Forslund, A. How can we feed the world in 2050? A review of the responses from global scenario studies. *Eur. Rev. Agric. Econ.* **2017**, *44*, 541–591. [CrossRef]
3. Hunter, M.C.; Smith, R.G.; Schipanski, M.E.; Atwood, L.W.; Mortensen, D.A. Agriculture in 2050: Recalibrating targets for sustainable intensification. *BioScience* **2017**, *67*, 386–391. [CrossRef]
4. Keating, B.A.; Herrero, M.; Carberry, P.S.; Gardner, J.; Cole, N.B. Food wedges: Framing the global food demand and supply challenge towards 2050. *Glob. Food Sec.* **2014**, *3*, 125–132. [CrossRef]
5. Berners-Lee, M.; Kennelly, C.; Watson, R.; Hewitt, C.N. Current global food production is sufficient to meet human nutritional needs in 2050 provided there is radical societal adaptation. *Elem. Sci. Anth.* **2018**, *5*, 52. [CrossRef]
6. Brevik, E.C. The potential impact of climate change on soil properties and processes and corresponding influence on food security. *Agriculture* **2013**, *3*, 398–417. [CrossRef]
7. Iizumi, T.; Ramankutty, N. How do weather and climate influence cropping area and intensity? *Glob. Food Sec.* **2015**, *4*, 46–50. [CrossRef]
8. Taiz, L. Agriculture, plant physiology, and human population growth: Past, present, and future. *Theor. Exp. Plant Physiol.* **2013**, *25*, 167–181. [CrossRef]

9. FAOSTAT. Food and Agriculture Organization of the United Nations. Available online: http://faostat.fao.org/site/567/default.aspx#ancor (accessed on 25 October 2020).
10. Ray, D.K.; Mueller, N.D.; West, P.C.; Foley, J.A. Yield trends are insufficient to double global crop production by 2050. *PLoS ONE* **2013**, *8*, e66428. [CrossRef] [PubMed]
11. Cormier, F.; Foulkes, J.; Hirel, B.; Gouche, D.; Moënne-Loccoz, Y.; Le Gouis, J. Breeding for increased nitrogen-use efficiency: A review for wheat. *Plant Breed.* **2016**, *135*, 255–278. [CrossRef]
12. Smith, P.; Gregory, P.J.; van Vuuren, D.; Obersteiner, M.; Havlik, P.; Rounsevell, M.; Woods, J.; Stehfest, E.; Bellarby, J. Competition for land. *Phil. Trans. R. Soc. B* **2010**, *365*, 2941–2957. [CrossRef]
13. FAO. *The State of the World's Land and Water Resources for Food and Agriculture (SOLAW)—Managing Systems at Risk*; Food and Agriculture Organization of the United Nations and Earthscan: Rome, Italy, 2011.
14. Conway, G. *The Doubly Green Revolution: Food for All in the Twenty-First Century*; Penguin Books: London, UK, 1997.
15. Spiertz, J.H.J. Nitrogen, sustainable agriculture and food security. A review. *Agron. Sustain. Dev.* **2010**, *30*, 43–55. [CrossRef]
16. Pretty, J.; Bharucha, Z.P. Sustainable intensification in agricultural systems. *Ann. Bot.* **2014**, *114*, 1571–1596. [CrossRef]
17. The Royal Society. *Reaping the Benefits: Science and the Sustainable Intensification of Global Agriculture*; RS Policy document; The Royal Society: London, UK, 2009; p. 86.
18. Smith, P. Delivering food security without increasing pressure on land. *Global Food Sec.* **2013**, *2*, 18–23. [CrossRef]
19. Godfray, H.C.; Garnett, T. Food security and sustainable intensification. *Phil. Trans. R. Soc. B* **2014**, *369*, 20120273. [CrossRef] [PubMed]
20. Rockström, J.; Williams, J.; Daily, G.; Noble, A.; Matthews, N.; Gordon, L.; Wetterstrand, H.; DeClerck, F.; Shah, M.; Steduto, P.; et al. Sustainable intensification of agriculture for human prosperity and global sustainability. *Ambio* **2017**, *46*, 4–17.
21. Grafton, R.Q.; Williams, J.; Jiang, Q. Food and water gaps to 2050: Preliminary results from the global food and water systems (GFWS) platform. *Food Sec.* **2015**, *7*, 209–220. [CrossRef]
22. Mogollón, J.M.; Lasaletta, L.; Beusen, A.H.W.; van Grinsven, H.J.M.; Westhoek, H.; Bouwman, A.F. Assessing future reactive nitrogen inputs into global croplands based on the shared socioeconomic pathways. *Environ. Res. Lett.* **2018**, *13*, 044008.
23. Conijn, J.G.; Bindraban, P.S.; Schröder, J.J.; Jongschaap, R.E.E. Can our global food system meet food demand within planetary boundries? *Agric. Ecosys. Environ.* **2018**, *251*, 244–256. [CrossRef]
24. Bodirsky, B.L.; Popp, A.; Lotze-Campen, H.; Dietrich, J.P.; Rolinski, S.; Weindl, I.; Smitz, C.; Müller, C.; Bonsch, M.; Humpenöder, F.; et al. Reactive nitrogen requirements to feed the world in 2050 and potential to mitigate nitrogen pollution. *Nat. Commun.* **2014**, *5*, 3858. [CrossRef] [PubMed]
25. Pradhan, P.; Fischer, G.; van Velthuizen, H.; Reusser, D.E.; Kropp, J.P. Closing yield gaps: How sustainable can we be? *PLoS ONE* **2015**, *10*, e0129487. [CrossRef] [PubMed]
26. Grzebisz, W.; Gransee, A.; Szczepaniak, W.; Diatta, J. The effects of K fertilization on water-use efficiency in crop plants. *J. Plant Nutr. Soil Sci.* **2013**, *176*, 355–374. [CrossRef]
27. Wang, Z.; Hassan, M.U.; Nadeem, F.; Wu., L.; Zhang, F.; Li, X. Magnesium Fertilization improves crop yield in most production systems: A meta-analysis. *Front. Plant Sci.* **2019**, *10*, 1727. [CrossRef]
28. Rabbinge, R. The ecological background of food production. In *Crop Production and Sustainable Agriculture*; Rabbinge, R., Ed.; John Wiley and Sons: New York, NY, USA, 1993; pp. 2–29.
29. Evans, L.T.; Fischer, R.A. Yield potential: Its definition, measurement, and significance. *Crop Sci. Soc. Am.* **1999**, *39*, 1544–1551. [CrossRef]
30. Van Ittersum, M.K.; Cassman, K.G.; Grassini, P.; Wolf, J.; Tittonell, P.; Hochman, Z. Yield gap analysis with local to global relevance—a review. *Field Crops Res.* **2013**, *143*, 4–17. [CrossRef]
31. Van Wart, J.; Kersebaum, K.C.; Peng, S.; Milner, M.; Cassman, K.C. Estimating crop yield potential at regional to national scales. *Field Crops Res.* **2013**, *143*, 34–43. [CrossRef]
32. Liu, J.; Williams, J.R.; Zehnder, A.J.B.; Yang, H. GEPIC—modelling wheat yield and crop water productivity with high resolution on a global scale. *Agric. Syst.* **2007**, *94*, 478–493. [CrossRef]
33. Agnus, J.F.; van Herwaarden, A.F. Increasing water use and water use efficiency in dryland wheat. *Agron. J.* **2001**, *93*, 290–298.
34. Passioura, J. Increasing crop productivity when water is scares—from breeding to field management. *Agric. Water Manage.* **2006**, *80*, 176–196. [CrossRef]
35. Licker, R.; Johnston, M.; Foley, J.A.; Barford, C.; Kucharik, C.J.; Monfreda, C.; Ramankutty, N. Mind the gap: How do climate and agricultural management explain the „yield gap" of croplands around the world? *Glob. Ecol. Biogeogr.* **2010**, *19*, 769–782. [CrossRef]
36. Iwańska, M.; Paderewski, J.; Stępień, M.; Rodrigues, P.C. Adaptation of winter wheat cultivars to different environments: A case study in Poland. *Agronomy* **2020**, *10*, 632. [CrossRef]
37. Studnicki, M.; Lenartowicz, M.; Noras, K.; Wójcik-Gront, E.; Wyszyński, Z. Assessment of stability and adaptation patterns of white sugar yield from sugar beet cultivars in temperate climate environments. *Agronomy* **2019**, *9*, 405. [CrossRef]
38. Rubio, G.; Zhu, J.; Lynch, J. A critical test of the prevailing theories of plant response to nutrient availability. *Am. J. Botany* **2003**, *90-91*, 143–152. [CrossRef]
39. Grzebisz, W.; Łukowiak, R.; Sassenrath, G. Virtual nitrogen as a tool for assessment of nitrogen at the field scale. *Field Crops Res.* **2018**, *218*, 182–184. [CrossRef]

40. Łukowiak, R.; Grzebisz, W.; Ceglarek, J.; Podolski, A.; Kaźmierowski, C.; Piekarczyk, J. Spatial variability of yield and nitrogen indicators-a crop rotation approach. *Agronomy* **2020**, *10*, 1959. [CrossRef]
41. Cassman, G.; Dobermann, A.; Walters, D. Agro-ecosystems, nitrogen-use efficiency, and nitrogen Management. *Ambio* **2002**, *31*, 132–140. [CrossRef]
42. Grzebisz, W.; Diatta, J. Constrains and solutions to maintain soil productivity, a case study from Central Europe. In *Soil Fertility Improvement and Integrated Nutrient Management—A Global Perspective*; Whalen, J., Ed.; InTech: London, UK, 2012; pp. 159–183. ISBN 978-953-307-945-5.
43. Wallace, A.; Wallace, G.A. *Closing the Crop-Yield Gap through Better Soil and Better Management*; Wallace Laboratories: Los Angeles, CA, USA, 2003; p. 162.
44. Vogel, E.; Donat, M.G.; Alexander, L.V.; Meinshausen, M.; Ray, D.K.; Karoly, D.; Meinshausen, N.; Frieler, K. The effects of climate extremes on global agricultural yields. *Environ. Res. Lett.* **2019**, *14*, 054010. [CrossRef]
45. Tóth, G.; Hermann, T.; Da Silva, M.R.; Montanarella, L. Monitoring soil for sustainable development and land degradation neutrality. *Environ. Monit. Assess.* **2018**, *190*, 57. [CrossRef] [PubMed]
46. Mueller, L.; Schindler, U.; Mirschel, W.; Shepherd, T.G.; Ball, B.C.; Helming, K.; Rogasik, J.; Eulenstein, F.; Wiggering, H. Assessing the productivity of soil. A review. *Agron. Sustain. Dev.* **2010**, *30*, 601–614. [CrossRef]
47. Patzel, N.; Sticher, H.; Karlen, D.L. Soil fertility-phenomenon and concept. *J. Plant Nutr. Soil Sci.* **2000**, *163*, 129–140. [CrossRef]
48. Haberle, J.; Svoboda, P. Calculation of available water supply in crop root zone and the water balance of crops. *Contrib. Geophys. Geod.* **2015**, *45/4*, 285–298. [CrossRef]
49. Fatichi, S.; Or, D.; Walko, R.; Vereecken, H.; Young, M.H.; Ghezzehei, T.A.; Hengl, T.; Kollet, S.; Agam, N.; Avissar, R. Soil structure is an important omission in Earth System Models. *Nat. Commun.* **2020**, *11*, 522. [CrossRef] [PubMed]
50. Kristensen, J.A.; Balstrøm, T.; Jones, R.J.A.; Jones, A.; Montanarella, L.; Panagos, P.; Breuning-madsen, H. Development of a harmonized soil profile analytical database for Europe: A resource for supporting regional soil management. *Soil* **2019**, *5*, 289–301. [CrossRef]
51. Knebl, L.; Leithold, G.; Schulz, F.; Brock, C. The role of soil depth in the evaluation of management-induced effects on soil organic matter. *Eur. J. Soil Sci.* **2017**, *68*, 979–987. [CrossRef]
52. Baldock, J.A.; Skjemstad, J.O. Role of the soil matrix and minerals in protecting natural organic materials against biological attack. *Organic Chem.* **2000**, *31*, 697–710. [CrossRef]
53. Lovelland, P.; Webb, J. Is there a critical level of organic matter in the agricultural soils of temperate regions: A review. *Soil Tillage Res.* **2003**, *70*, 1–18. [CrossRef]
54. Piéri, C. *Fertilité des Terres de Savanes. Bilan de trente années de recherches et de développement gricole au sud du Sahara*; Ministère de la Coopération et du Développement, CIRAD: Paris, France, 1989; p. 444.
55. Spychalski, W.; Grzebisz, W.; Diatta, J.; Kostarev, D. Humus stock degradation and its impact on phosphorus forms in arable soils—a case of Ukrainian Forest-Steppe Zone. *Chem. Speciation Bioavail.* **2018**, *30*, 33–46. [CrossRef]
56. Bauer, A.; Black, A.L. Quantification of the effect of soil organic matter content on soil productivity. *Soil Sci. Soc. Am. J.* **1994**, *58*, 185–193. [CrossRef]
57. Batjes, N.H. carbon and nitrogen stocks in the soils of Central and Eastern Europe. *Soil Use Manag.* **2002**, *18*, 324–329. [CrossRef]
58. Libohova, Z.; Seybold, C.; Wysocki, D.; Wills, S.; Schoeneberger, P.; Williams, C.; Lindbo, D.; Stott, D.; Owens, P.R. Reevaluating the effects of soil organic matter and other properties on available water-holding capacity using the National Cooperative Survey Characterization Database. *J. Soil Water Conserv.* **2018**, *73*, 411–421. [CrossRef]
59. Ludwig, B.; Geisseler, D.; Michel, K.; Joergensen, R.G.; Schulz, E.; Merbach, I.; Raupp, J.; Rauner, R.; Hu, K.; Niu, L.; et al. Effects of fertilization and soil management on crop yields and carbon stabilization in soils. A review. *Agron. Sustain. Dev.* **2011**, *31*, 361–372. [CrossRef]
60. Wu, T.; Schoenau, J.J.; Li, F.; Qian, P.; Malhi, S.S.; Shi, Y.; Xu, F. Influence of cultivation and fertilization on total organic carbon and carbon fractions in soils from the Loess Plateau of China. *Soil Tillage Res.* **2004**, *77*, 59–68. [CrossRef]
61. Powlson, D.; Smith, P.; Coleman, K.; Smith, J.U.; Glendining, M.J.; Körschens, M.; Franko, U. A European network of long-term sites for studies on soil organic matter. *Soil Tillage Res.* **1998**, *47*, 263–274. [CrossRef]
62. Szajdak, L.; Życzyńska-Bałoniak, I.; Meysner, T.; Blecharczyk, A. Bound amino acids in humic acids from arable cropping systems. *J. Plant Nutr. Soil Sci.* **2004**, *167*, 562–567. [CrossRef]
63. Zhao, X.; Yuan, G.; Wang, H.; Lu, D.; Chen, X.; Zhou, J. Effects of full straw incorporation on soil fertility and crop yield in rice-wheat rotation for silty clay loamy cropland. *Agronomy* **2019**, *9*, 133. [CrossRef]
64. Noordwijk van, M.; Cadisch, G. Access and excess problems in soil nutrition. *Plant Soil* **2002**, *247*, 25–40. [CrossRef]
65. Ballabio, C.; Lugato, E.; Fernandez-Ugalde, O.; Orgiazzi, A.; Jones, A.; Borelli, P.; Montanarella, L.; Panagos, P. Mapping LUCAS topsoil chemical properties in European scale using Gaussian process regression. *Geoderma* **2019**, *355*, 113912. [CrossRef]
66. Bloom, A.J. Nitrogen dynamics in plant growth systems. *Life Support Biosph. Sci.* **1996**, *3*, 35–41. [PubMed]
67. Krouk, G.; Ruffel, S.; Gutiérrez, R.A.; Gojon, A.; Crawford, N.M.; Coruzzi, G.M.; Lacombe, B. A framework integrating plant growth with hormones and nutrients. *Trends Plant Sci.* **2011**, *16*, 178–182. [CrossRef]
68. Smucker, A.J.M.; Aiken, R.M. Dynamic root responses to water deficits. *Soil Sci.* **1992**, *154*, 281–289. [CrossRef]
69. Bodner, G.; Nakhforoosh, A.; Kaul, H.P. Management of crop water under drought: A review. *Agron. Sustain. Dev.* **2015**, *35*, 401–442. [CrossRef]

70. Zhang, F.; Niu, J.; Zhang, W.; Chen, X.; Li, C.; Yuan, L.; Xie, J. Potassium nutrition of crops under varied regimes of nitrogen supply. *Plant Soil* **2010**, *335*, 21–34. [CrossRef]
71. Blanchet, G.; Libohiva, Z.; Joost, S.; Rossier, N.; Schneider, A.; Jeangross, B.; Sinaj, A. Spatial variability of potassium in agricultural soils of the canton of Fribourg, Switzerland. *Geoderma* **2017**, 107 –121. [CrossRef]
72. Gao, X.-S.; Xiao, Y.; Deng, L.-J.; Wang, C.; Deng, O.; Zeng, M. Spatial variability of total nitrogen, phosphorus and potassium in Renshou County of Sichuan Basin, China. *J. Integr. Agric.* **2019**, *18*, 279–289. [CrossRef]
73. Boring, T.J.; Thelen, K.D.; Board, J.E.; Board, J.E.; De Bruis, J.L.; Lee, C.D.; Naeve, S.L.; Ross, W.J.; Kent, W.A.; Ries, L.L. Phosphorus and potassium fertilizer application strategies in corn-soybean rotations. *Agronomy* **2018**, *8*, 195. [CrossRef]
74. Penn, C.J.; Camberato, J.J. A critical review on soil chemical processes that control how soil pH affects phosphorus availability to plants. *Agriculture* **2019**, *9*, 120. [CrossRef]
75. Pasket, A.; Zhang, H.; Raun, W.; Deng, S. Recovery of phosphorus in soil amended with manure for 119 years. *Agronomy* **2020**, *10*, 1947. [CrossRef]
76. Kautz, T.; Amelung, W.; Ewert, F.; Gaiser, T.; Horn, R.; Jahn, R.; Javaux, M.; Kemna, A.; Kuzyakova, Y.; Munch, J.-C.; et al. Nutrient acquisition from arable subsoils in temperate climates: A review. *Soil Biol. Biochem.* **2013**, *57*, 1003–1022. [CrossRef]
77. White, C.; Sylvester-Bradley, R.; Berry, P.M. Root length densities of UK wheat and oilseed rape crops with implications for water capture and yield. *J. Exp. Bot.* **2015**, *66*, 2293–2303. [CrossRef]
78. Barraclough, P.B. Root growth, macro-nutrient uptake dynamics and soil fertility requirements of a high-yielding winter oilseed rape crop. *Plant Soil* **1989**, *119*, 59–70. [CrossRef]
79. Łukowiak, R.; Grzebisz, W.; Sassenrath, G. New insights into phosphorus management in agriculture—A crop rotation approach. *Sci. Total Environ.* **2016**, *542*, 1062–1077. [CrossRef] [PubMed]
80. Grzebisz, W.; Łukowiak, R.; Kotnis, K. Evaluation of nitrogen fertilization systems based on the in-season variability of the nitrogenous growth factors and soil fertility factors—a case of winter oilseed rape (*Brassica napus* L.). *Agronomy* **2020**, *10*, 1701. [CrossRef]
81. Bauke, S.L.; von Sperber, C.; Tamburini, F.; Gocke, M.I.; Honermeier, B.; Schweitzer, K.; Baumecker, M.; Don, A.; Sandhage-Hofmann, A.; Amelung, W. Subsoil phosphorus is affected by fertilization regime in long-term agricultural experimental trials. *Europ. J. Soil Sci.* **2018**, *69*, 103–112. [CrossRef]
82. Barłóg, P.; Łukowiak, R.; Grzebisz, W. Predicting the content of soil mineral nitrogen based on the content of calcium chloride-extractable nutrients. *J. Plant Nutr. Soil Sci.* **2017**, *180*, 624–635. [CrossRef]
83. Łukowiak, R.; Barłóg, P.; Grzebisz, W. Soil mineral nitrogen and the rating of $CaCl_2$ extractable nutrients. *Plant Soil Environ.* **2017**, *63*, 177–183.
84. Fan, J.; Mc Conkey, B.; Wang, H.; Janzen, H. Root distribution by depth for temperate agricultural crops. *Field Crops Res.* **2016**, *189*, 68–74. [CrossRef]
85. Forde, B.; Lorenzo, H. The nutritional control of root development. *Plant Soil* **2001**, *232*, 51–68. [CrossRef]
86. Marschner, P. (Ed.) *Marchnerr's Mineral Nutrition of Higher Plants*, 3rd ed.; Academic Press: London, UK, 2012; p. 672.
87. Grzebisz, W.; Szczepaniak, W.; Barłóg, P.; Przygocka-Cyna, K.; Potarzycki, J. Phosphorus sources for winter oilseed rape (*Brassica napus* L.) during reproductive growth—magnesium sulfate management impact on P use efficiency. *Arch. Agron. Soil Sci.* **2018**, *64*, 1646–1662. [CrossRef]
88. Grzebisz, W.; Kryszak, J. Effect of soil fertility on root morphology of winter rye. In *Root Ecology and Its Practical Application*; Kutschera, L., Hübl, E., Lichtenegger, E., Persson, H., Sobotik, M., Eds.; Verein fot Wurzelforschung: Klagenfurt, Austria, 1992; pp. 389–392.
89. Barber, S.A. *Soil Nutrient Bioavailability: A Mechanistic Approach*, 2nd ed.; Wiley and Sons: New York, NY, USA, 1995.
90. Gaju, O.; Allard, V.; Martre, P.; Le Gouis, J.; Moreau, D.; Bogard, M.; Hubbart, S.; Foulkes, M.J. Nitrogen partitioning and remobilization to leaf senescence, grain yield and grain nitrogen concentration in wheat cultivars. *Field Crops Res.* **2014**, *155*, 213–223. [CrossRef]
91. Hawkesford, M.J. Reducing the reliance on nitrogen fertilization for wheat production. *J. Cereal Sci.* **2014**, *59*, 276–283. [CrossRef] [PubMed]
92. Maeoka, R.E.; Sadras, V.O.; Ciampitti, I.A.; Diaz, D.R.; Fritz, A.K.; Lollato, R.P. Changes in the phenotype of winter wheat varieties released between 1920 and 2016 in response to in-furrow fertilizer: Biomass allocation, yield, and grain protein concentration. *Front. Plant Sci.* **2019**, *10*, 1786. [CrossRef]
93. Hawkesford, M.J.; Riche, A.B. Impacts of G x E x M on nitrogen use efficiency in wheat and future prospects. *Front. Plant Sci.* **2020**, *11*, 1157. [CrossRef]
94. Malhi, S.S.; Johnston, A.M.; Schoenaou, J.J.; Wang, Z.H.; Vera, C.L. Seasonal biomass accumulation and nutrient uptake of wheat, barley, and oak on a Black Chernozem soil in Saskatchewan. *Can. J. Plant Sci.* **2006**, *86*, 1005–1014. [CrossRef]
95. Zhou, G.; Wang, Q. A new nonlinear method for calculating growing degree days. *Sci. Rep.* **2018**, *8*, 10149. [CrossRef] [PubMed]
96. Yin, X.; Goudriaan, J.; Lantinga, E.A.; Vos, J.; Spiertz, H. A flexible sigmoid function of determinate growth. *Ann. Bot.* **2003**, *91*, 361–371. [CrossRef] [PubMed]
97. Zerche, S.; Hecht, R. Nitrogen uptake of winter wheat during shoot elongation phase in relation to canopy high and shoot density. *Agribiol. Res.* **1999**, *52*, 231–250.

98. Kahabka, J.E.; Van Es, H.M.; McClenahan, E.J.; Cox, W.J. Spatial analysis of maize response to nitrogen fertilizer in Central New York. *Precision Agric.* **2004**, *5*, 463–476. [CrossRef]
99. Luce, M.S.; Whalen, J.K.; Ziadi, N.; Zebarth, B.J. Nitrogen dynamics and indices to predict soil nitrogen supply in humid temperate soils. *Adv. Agron.* **2011**, *112*, 55–102.
100. Richards, R.A. Selectable traits to increase crop photosynthesis and yield of grain crops. *J. Exp. Bot.* **2000**, *51*, 337–458. [CrossRef]
101. Szczepaniak, W. A mineral profile of oilseed rape in critical stages of growth—nitrogen. *J. Elem.* **2014**, *19*, 759–778. [CrossRef]
102. Grzebisz, W.; Szczepaniak, W.; Grześ, S. Sources of nutrients for high-yielding winter oilseed rape (*Brassica napus* L.) during post-flowering growth. *Agronomy* **2020**, *10*, 626. [CrossRef]
103. Spiertz, J.; Vos, J. Grain Growth of Wheat and Its Limitation by Carbohydrate and Nitrogen Supply. In *Wheat Growth and Modelling*; Day, W., Atkin, R., Eds.; Plenum Press: New York, NY, USA, 1985.
104. Barraclough, P.B.; Lopez-Bellido, R.; Hawkesford, M. Genotypic variation in the uptake, partitioning and remobilization of nitrogen during grain-filling in wheat. *Field Crops Res.* **2014**, *156*, 242–248. [CrossRef]
105. Taulemesse, F.; Le Gouis, J.; Gouache, D.; Gibon, Y.; Allard, V. Bread wheat (*Triticum aestivum* L.) grain protein concentration is related to early post-flowering nitrate uptake under putative control of plant satiety level. *PLoS ONE* **2016**, *11*, e0149668. [CrossRef] [PubMed]
106. Bogard, M.; Allard, V.; Brancourt-Hulmel, M.; Heumez, E.; Machet, J.-M.; Jeufforoy, M.-H.; Gate, P.; Le Gouis, J. Deviation from the grain protein concentration—grain yield negative relationships is highly correlated to post-anthesis N uptake by winter wheat. *J. Exp. Bot.* **2010**, *15*, 4303–4312. [CrossRef]
107. Subedi, K.; Ma, B. Nitrogen uptake and partitioning in stay-green and leafy maize hybrids. *Crops Sci.* **2005**, *45*, 740–747. [CrossRef]
108. Grzebisz, W.; Wrońska, M.; Diatta, J.B.; Szczepaniak, W. Effect of zinc application at early stages of maize growth on the patterns of nutrients and dry matter accumulation by canopy. Part, I. Nitrogen uptake and dry matter accumulation patterns. *J. Elem.* **2008**, *13*, 17–28.
109. Kovacs, P.; Vyn, T.J. Relationships between earl-leaf nutrient concentration at silking and corn biomass and grain yield at maturity. *Agron. J.* **2017**, *109*, 2898–2906. [CrossRef]
110. Szczepaniak, W.; Barłóg, P.; Łukowiak, R.; Przygocka-Cyna, K. Effect of balanced nitrogen fertilization in four-year rotation on plant productivity. *J. Central Europ. Agroc.* **2013**, *14*, 64–77. [CrossRef]
111. Szczepaniak, W.; Grzebisz, W.; Potarzycki, J.; Łukowiak, R.; Przygocka-Cyna, K. Nutritional status of winter oilseed rape in cardinal stages of growth as yield indicator. *Plant Soil Environ.* **2015**, *61*, 291–296. [CrossRef]
112. Weih, M.; Hamnér, K.; Pourazari, F. Analyzing plant nutrient uptake and utilization efficiencies: Comparison between crops and approaches. *Plant Soil* **2018**, *430*, 7–21. [CrossRef]
113. White, P.J.; Brown, P.H. Plant nutrition for sustainable development and global health. *Ann. Bot.* **2010**, *105*, 1073–1080. [CrossRef]
114. Tan, Z.X.; Lal, R.; Wiebe, K.D. Global soil nutrient depletion and yield reduction. *J. Sustain. Develop.* **2015**, *26*, 123–146. [CrossRef]
115. Mi, G.-H.; Chen, F.-J.; Wu, Q.-P.; Lai, N.-W.; Yuan, L.-X. Ideotype root architecture for effcient nitrogen acqusition by maize in intensive cropping systems. *Sci. China Life Sci.* **2010**, *553*, 1369–1373. [CrossRef] [PubMed]
116. Ruiz Herrera, L.F.; Shane, M.W.; López-Bucio, J. Nutritional regulation of root development. *WIREs Dev. Biol.* **2015**, *4*, 431–443. [CrossRef]
117. Lynch, J.P. Root phenothypes for improved nutrient capture: An underexploited opportunity for global agriculture. *New Phytol.* **2019**, *223*, 548–564. [CrossRef]
118. Duque, L.O.; Villordon, A. Root branching and nutrient efficiency: Status and way forward in root and tuber crops. *Front. Plant Sci.* **2019**, *10*, 237. [CrossRef]
119. Fageria, N.K.; Moreira, A. The role of mineral nutrition on root growth of crop plants. *Adv. Agron.* **2011**, *110*, 3–83.
120. Metwally, M.S.; Shaddad, S.M.; Liu, M.; Yao, R.-J.; Abdo, A.; Li, P.; Jiao, J.; Chen, X. Soil properties spatial variability and delineation of site-specific management zones based on soil fertility using fuzzy clustering in a hilly field in Jianyang, Sichuam, China. *Sustainability* **2019**, *11*, 7084. [CrossRef]
121. Erisman, J.W.; Leach, A.; Bleeker, A.; Atwell, B.; Cattaneo, L.; Galloway, J. An integrated approach to a nitrogen use efficiency (NUE) indicator for the food production-consumption chain. *Sustainability* **2018**, *10*, 925. [CrossRef]
122. Moll, R.H.; Kamprath, E.J.; Jackson, W.A. Analysis and interpretation of factors which contribute to efficiency of nitrogen utilization. *Agron. J.* **1982**, *74*, 562–564. [CrossRef]
123. Swaney, D.P.; Howarth, R.W.; Hong, B. Nitrogen use efficiency and crop production: Patterns of regional variation in the United States, 1987–2012. *Sci. Tot. Environ.* **2018**, *635*, 498–511. [CrossRef]
124. Pan, W.L.; Kidwell, K.K.; McCracken, V.A.; Bolton, R.P.; Allen, M. Economically optimal wheat yield, protein and nitrogen use component responses to varying N supply and genotype. *Front. Plant Sci.* **2020**, *10*, 1790. [CrossRef] [PubMed]
125. Yin, X.; Struik, P.C. Constrains tot he potential Efficiency of converting solar radiation into phytoenergy in annual crops: From leaf biochemistry to canopy physiology an crop ecology. *J. Exp. Bot.* **2015**, *66*, 6535–6549. [CrossRef] [PubMed]
126. Papadopoulos, A.; Kalivas, D.; Hatzichrostos, T. GIS modeling for site-specific nitrogen fertilization towards soil sustainability. *Sustainability* **2015**, *7*, 6684–6705. [CrossRef]
127. Gulser, C.; Ekberli, I.; Candemier, F.; Demir, Z. Spatial variability of soil physical properties in a cultivated field. *Eurasian J. Soil Sci.* **2016**, *5*, 192–200. [CrossRef]

128. Córdova, C.; Barrera, J.A.; Magna, C. Spatial variation in nitrogen mineralization as a guide for variable application of nitrogen fertilizer to cereal crops. *Nutr. Cycl. Agroecosyst.* **2018**, *110*, 83–88. [CrossRef]
129. Denton, O.A.; Aduramigba-Modupe, V.O.; Ojo, A.O.; Adeoyolanu, O.D.; Are, K.S.; Adelana, A.; Oyedele, A.O.; Adetayo, A.O.; Olubukola, O.A. Assessment of spatial variability and mapping of soil properties for sustainable agricultural production using geographic information system techniques (GIS). *Cogent Food Agric.* **2017**, *3*, 1279366. [CrossRef]
130. Vasu, D.; Singh, S.K.; Sahu, N.; Tiwary, P.; Chandran, P.; Duraisami, V.P.; Ramamurthy, V.; Lalitha, M.; Kaliselvi, B. Assessment of spatial variability of soil properties using geostatical techniques for farm level nutrient management. *Soil Tillage Res.* **2017**, *169*, 25–34. [CrossRef]
131. Lemaire, G.; Jeuffroy, M.-H.; fstal, F. Diagnosis toll for plant an crop N status in vegetative stage: Theory and practices for crop N management. *Eur. J. Agron.* **2008**, *28*, 181–190. [CrossRef]
132. Song, L.; Wang, S.; Ye, W. Establishment and application of critical nitrogen dilution curves for rice based on leaf dry matter. *Agronomy* **2020**, *10*, 367. [CrossRef]
133. Chen, Z.; Miao, Y.; Lu, J.; Zhou, L.; Li, Y.; Zhang, H.; Lou, W.; Zhang, Z.; Kusnierek, K.; Liu, C. In-season diagnosis of winter wheat nitrogen status in smallholder farmer fields across a village using unmanned aerial vehicle-based remote sensing. *Agronomy* **2019**, *9*, 619. [CrossRef]
134. Stamatiadis, S.; Schepers, J.S.; Evangelou, E.; Tsadilas, C.; Glampedakis, M.; Dercas, N.; Spyropoulos, N.; Dalezios, N.R.; Eskridge, K. Variable-rate nitrogen fertilization of winter wheat under high spatial resolution. *Precision Agric.* **2018**, *19*, 570–587. [CrossRef]
135. Sharma, L.K.; Bali, S.K. A review of methods to improve nitrogen use efficiency in agriculture. *Sustainability* **2018**, *10*, 51. [CrossRef]
136. Li, H.; Zhang, Y.; lei, Y.; Antoniuk, V.; Hu, C. Evaluating different non-destructive estimation methods for winter wheat (*Triticum aestivum* L.) nitrogen status based on canopy spectrum. *Remote Sens.* **2020**, *12*, 95. [CrossRef]
137. Nigon, T.J.; Yang, C.; Paiao, G.D.; Mulla, D.J.; Knight, J.F.; Fernández, F.G. Prediction of early season nitrogen uptake by maize using high-resolution aerial hyperspectral imagery. *Remote Sens.* **2020**, *12*, 1234. [CrossRef]
138. Rodrigues, F.A.; Blasch, G.; Defourny, P.; Ortiz-Monasterio, J.I.; Schulthess, U.; Zarco-Tejada, P.J.; Taylor, J.A.; Gérard, B. Multi-temporal and spectra analysis of high-resolution hyperspectral airborne imagery for precision agriculture: Assessment of wheat grain yield and grain protein content. *Remote Sens.* **2018**, *10*, 930. [CrossRef] [PubMed]
139. Cabrera-Bosquet, L.; Molero, G.; Stellacci, A.M.; Bort, J.; Nogués, S.; Araus, J.L. NDVI as a potential tool for predicting biomass, plant nitrogen content and growth in wheat genotypes subjected to different water and nitrogen conditions. *Cereal Res. Comm.* **2011**, *39*, 147–159. [CrossRef]
140. Vizzari, M.; Santaga, F.; Benincasa, P. Sentinel 2-based nitrogen VRT fertilization in wheat: Comparison between traditional and simple precision practices. *Agronomy* **2019**, *9*, 278. [CrossRef]
141. Argento, F.; Anken, T.; Abt, F.; Vogelsanger, E.; Walter, A.; Liebisch, F. Site-specific nitrogen management in winter wheat supported by low-altitude remote sensing and soil data. *Precision Agric.* **2020**. [CrossRef]
142. Feng, W.; Wu, Y.; He, L.; ren, X.; Wang, Y.; Hou, G.; Wang, Y.; Liu, W.; Guo, T. An optimized non-linear vegetation index for estimating leaf area index in winter wheat. *Precision Agric.* **2019**, *29*, 1157–1176. [CrossRef]
143. Naser, M.A.; Khosla, R.; Longchamps, L.; Dahal, S. Using NDVI to differentiate wheat genotypes productivity under dryland and irrigated conditions. *Remote Sens.* **2020**, *12*, 824. [CrossRef]
144. Prey, L.; Schmidhalter, U. Sensitivity of vegetation indices for estimating vegetative N status in winter wheat. *Sensors* **2019**, *19*, 3712. [CrossRef] [PubMed]
145. Gerstmann, H.; Möller, M.; Gläßer, C. Optimization of spectra indices and long-term separability analysis for classification of cereal crops using multi-spectral RapidEye. *Int. J. Appl. Earth Obs.* **2016**, *52*, 115–125. [CrossRef]

Article

Changes in *Pisum sativum* L. Plants and in Soil as a Result of Application of Selected Foliar Fertilizers and Biostimulators

Hanna Sulewska [1], Alicja Niewiadomska [2], Karolina Ratajczak [1,*], Anna Budka [3], Katarzyna Panasiewicz [1], Agnieszka Faligowska [1], Agnieszka Wolna-Maruwka [2] and Leszek Dryjański [4]

1. Department of Agronomy, Poznań University of Life Sciences, 11 Dojazd Str, 60-632 Poznań, Poland; hanna.sulewska@up.poznan.pl (H.S.); katarzyna.panasiewicz@up.poznan.pl (K.P.); agnieszka.faligowska@up.poznan.pl (A.F.)
2. Department of General and Environmental Microbiology, Poznan University of Life Sciences, 50 Szydłowska Str., 60-656 Poznań, Poland; alicja.niewiadomska@up.poznan.pl (A.N.); agnieszka.wolna-maruwka@up.poznan.pl (A.W.-M.)
3. Department of Mathematical and Statistical Methods, Poznan University of Life Sciences, 28 Wojska Polskiego Str., 60-637 Poznań, Poland; budka@up.poznan.pl
4. Agrii Poland Ltd., 233 Obornicka Str., 60-650 Poznań, Poland; info@nu-agrar.pl

* Correspondence: karolina.ratajczak@up.poznan.pl; Tel.: +48-0618487403

Received: 14 September 2020; Accepted: 8 October 2020; Published: 13 October 2020

Abstract: The aim of this study was to assess the effect of selected biostimulators and foliar fertilizers on plant development, plant yield, soil fertility and soil biochemical activity (dehydrogenases, phosphatases, catalases) during the cultivation of pea (*Pisum sativum* L.). A field experiment was conducted between 2016 and 2018 at the Gorzyń Experimental and Educational Station, Poznań University of Life Sciences in Poland. The following treatments were tested: (1) control; (2) Titanit; (3) Optysil; (4) Metalosate potassium; (5) Rooter; (6) Bolero Mo; (7) Adob Zn IDHA; (8) Adob B and (9) Adob 2.0 Mo. Adob Zn IDHA stimulated yields, especially under average moisture conditions and less so in drought conditions, and the differences compared to control amounted 8.36 and 4.3%, respectively. The results showed a close relationship between the effects of the biostimulators and foliar fertilizers and weather conditions during the study. It was not possible to determine whether any of the biostimulators or foliar fertilizers had a positive effect on pea seed yield in any year. Similarly, it was difficult to clearly determine the effect of the biostimulators and fertilizers on biochemical activity in the soil, although soil enzyme activity was influenced most by application of the Bolero Mo fertilizer. In all study years, biological nitrogen fixation was always greater after the application of a biostimulator/fertilizer treatment.

Keywords: maximum photochemical efficiency of photosystem II; chlorophyll content index; soil enzymatic activity; biological index fertility; nitrogenase activity; microelements fertilization (Ti; Si; B; Mo; Zn)

---

1. Introduction

The value of pea (*Pisum sativum* L.) as a crop can be assessed in two ways. Firstly, as the seeds contain 20–24% protein, they are a valuable food and feed source [1]. Secondly, the crop residues that remain in the field after cultivation favourably affect the physical, chemical and biochemical properties of the soil [2]. Currently, European Union (EU) rules for integrated plant cultivation and the so-called greening [3] are perfectly tailored for this species in respect to the above requirements. According to the Agriculture Restructuring and Modernization Agency (ARIMR), the area of pea cultivation in Poland was 56,164 ha in 2019.

For many years, the agricultural practice in EU countries, adapting to introduced directives, has been to use environmentally friendly technologies to reduce the use of pesticides and to eliminate their active substances in the environment [4]. Thus, it is increasingly difficult for producers to limit biotic plant stress, such as pests, disease and weed infestation.

In addition, climate change, especially the periods of drought that increasingly occur during the growing season, has created many problems for growers [5], although some solutions are available, e.g., improvement of water retention by increasing the proportion of organic matter in the soil, limiting (unproductive) evaporation from the soil with agrotechnical methods, as well as the use of biostimulators [6,7]. While many definitions exist, a biostimulator is generally defined as "any substance or microorganism applied to plants with the aim to enhance nutrition efficiency, abiotic stress tolerance and/or crop quality traits, regardless of its nutrient content" [8]. However, it should be noted that biostimulators are not fertilizers in the sense that they do not contain nutrients intended to be delivered to the plant. Nevertheless, they may facilitate nutrient acquisition, e.g., by mobilizing elements in the rhizosphere or by developing new routes of nutrient acquisition, such as fixation of atmospheric nitrogen through the recruitment of bacterial endosymbionts [8].

The use of biostimulators in the cultivation of various plant species, including *Fabaceae*, has been shown to have contradictory effects, with some studies reporting a beneficial effect [9,10], while others have shown no effect [8,11]. As such, it can be assumed that the variation in effects come not only from the composition of the individual products but also from the timing of the application and the time between application and the occurrence of biotic or abiotic stress [12,13].

The sustainability of the soil ecosystem can be evaluated with biologically-based indicators, and soil enzymes have been effectively utilized as indicators of soil quality across a range of farming systems [14]. Improved knowledge of how soil enzymes function, and the factors that influence activity is vital to enhance soil management and quality, and food production. Soil enzymes catalyze and expedite organic matter decomposition and regulate nutrient cycling, and can, therefore, be used as a biological index for soil quality. In practice, soil enzymes can be simply integrated, are easily quantified, and are much more sensitive to soil management changes (than other soil quality indicators). Their activities are influenced by a range of factors, e.g., soil depth, type, temperature, moisture content, pH level, quality and quantity of available substrates, and management regimes. However, the activity of an individual enzyme is not reflective of soil quality as single enzyme activities are not representative of the rate of all metabolic processes (except if they catalyze a single specific reaction). Therefore, to accurately determine the level of soil quality, a number of enzyme activities should be evaluated. Catalases and dehydrogenases are found in the soil as essential parts of complete living microbial cells. They can be used as a measure of general microbial activity in the soil and, therefore, can be employed to derive a biological indicator of fertility (BIF). As members of the oxidoreductases class, these enzymes fulfil the most important functions in the environment [15]. Hydrolases are another important group of soil enzymes and include phosphatases, which participate in the phosphorus cycle.

The vast availability of fertilizers and biostimulants leads to an independent assessment of their value in terms of plant-soil interaction. This is the reason of the aim of the study to determine the effect of selected biostimulators and foliar fertilizers on the development and yield of pea, and to evaluate the fertility and biochemical activity (dehydrogenases, phosphatases, catalases) of the soil that they are grown on.

2. Material and Methods

2.1. Experimental Design

Between 2016–2018, experiment was conducted each year at the Gorzyń Experimental and Educational Station, Poznań University of Life Sciences (N—52.56692, E—015.90933, 69 m AMSL) to assess plant reactions and soil microbiological changes after the application of selected biostimulators and foliar fertilizers during the cultivation of pea ('Tarchalska' variety). The experiment was a randomised block design with four replications and 36 plots (plot size 14 m × 1.5 m (21 m^2)).

The research factor was the use of biostimulators or foliar fertilizers with 9 levels:

(1) Control—plants were not treated with biostimulators or foliar fertilizers.
(2) Titanit.
(3) Optysil.
(4) Metalosate potassium.
(5) Rooter.
(6) Bolero Mo.
(7) Adob Zn IDHA.
(8) Adob B.
(9) Adob 2.0 Mo.

Each biostimulator and fertilizer was applied in a timely manner, according to the manufacturer's recommendations (Table 1).

Table 1. Timing of application and dosage of biostimulators and fertilizers applied in the study.

	Biostimulator and Foliar Fertilizer	Term and Dose of Biostimulators	Biostimulator and Fertilizer Characteristics
Biostimulators	Tytanit	I: BBCH 13–14 II: BBCH 31–32 0.3 0.3 dm^3 ha^{-1}	Liquid, mineral stimulant containing titanium (Ti). It increases the yield, volume and development of plants, improves yield quality parameters and increases plants' natural resistance to stress. Composition: 8.5 g Ti $(dm^3)^{-1}$
	Rooter	BBCH 13–14 1 dm^3 ha^{-1}	Biostimulator—it stimulates the growth of the root system, accelerates regeneration and improves the uptake of soil minerals. Composition: P_2O_5 13.0%; K_2O 5.0%
Foliar fertilizers	Optysil	I: BBCH 16–18 II: BBCH 52–55 III: BBCH 71–73 0.5 dm^3 ha^{-1}	Liquid, silicon antistressor stimulating the growth and development of plants, activating their natural immune system, and increasing tolerance to unfavourable cultivation conditions. Composition: 200 g SiO_2 $(dm^3)^{-1}$
	Metalosate potassium	I: BBCH 11–13 II: BBCH 18–20 III: BBCH 31–32 3 dm^3 ha^{-1}	Liquid foliar fertilizer containing an easily absorbable form of potassium, which supplements potassium deficiency in plants with amino acids. Composition: K_2O 24%
	Bolero Mo	BBCH 58 1.5 dm^3 ha^{-1}	Liquid foliar fertilizer containing boron and molybdenum to supplement deficiency in plants. Composition: B 8.2%; Mo 0.8%
	Adob Zn IDHA	BBCH 58 1 dm^3 ha^{-1}	Foliar fertilizer containing zinc (Zn) fully chelated by biodegradable chelating agent IDHA. Composition: Zn 100 g kg^{-1} (weight percentage content 10, chelated by IDHA)
	Adob B	BBCH 55–58 2 dm^3 ha^{-1}	Liquid, highly concentrated foliar fertilizer containing boron that regulates auxin activity and participates in cell division. Composition: N 78 g kg^{-1}; B 150 g kg^{-1}
	Adob 2.0 Mo	BBCH 11–13 0.15 dm^3 ha^{-1}	Liquid, single-component fertilizer which increases the rate and efficiency of use of nitrogen by plants and improves interaction with iron. Composition: Mo 20%

BBCH—A uniform decimal code for growth stages of crops and weeds, IDHA—chelating agent.

A chlorophyll fluorescence meter (OS5p, Opti-Sciences, Inc., Hudson, NY, USA) with a photosynthetic active radiation (PAR) clip was used to measure the following parameters: F0—minimum fluorescence, F_m—maximum fluorescence, F_v—variable fluorescence and Y—quantum yield of photosynthetic energy, which are necessary to calculate the maximum photochemical efficiency of photosystem II (F_v/F_m), according to the formula ($F_v/F_m = F_m - F_0/F_m$). Settings for the fluorometer protocols were selected according to Sulewska et al. [16] as follows: modulation source: red; modulation intensity: 25; detector gain: 08; saturation flash intensity: 30; flash count: 001; flash rate: 255 (s). A chlorophyll meter CCM-200 was used to determine the Chlorophyll Content Index (CCI). Leaf Area Index (LAI) was determined with a SunScan Canopy Analysis System type SSI (Delta-T Devices, Cambridge, UK). The agrotechnical and cultivation treatments were carried out in accordance with the principles of good agricultural and experimental practice for this species [17]. Agrotechnical treatments and the dates of their implementation in the individual years of the study are presented in Table 2.

Table 2. Type and date of agrotechnical treatments carried out in the study.

Treatment	2016	2017	2018
Tilling set	04.04	03.04	05.04
Sowing date	04.04	04.04	07.04
Herbicide spraying	05.04 Afalon Dyspersyjny 1.1 L/ha	04.04 Stomp Aqua 455 CS 2.6 L/ha	09.04 Stom Aqua 455 CS 2.6 L/ha
Herbicide spraying	13.05 Basagran 480 SL 2.6 L/ha	26.05 Panthera 040 EC 1.75 L/ha	16.05—Fusilade forte 150 EC 1.7 L/ha
Insecticide spraying	-	17.05 and 29.05 Dursban 480 EC 1.25 L/ha	15.05 and 13.06 Superkill 500 EC 0.06 L/ha
Fungicide spraying	30.05 Gwarant 500 SL 2.0 L/ha	05.06. and 24.06 Azoksystrobina 250 SC 1.0 L/ha	-
Fungicide spraying	30.06 Korazzo 250 SC 1.0 L/ha	12.07 Signum 33 WG 0.8 kg/ha + Piorun 200 SC 0.2 L/ha	22.05 Korazzo 050 SC 1.2 L/ha
Harvest date	19.07	31.07	29.06

CS—capsule suspension, SL—soluble concentrate, EC—emulsifiable concentrate, SC—concentrate in the form of a concentrated suspension.

We used the white-flowered pea variety 'Tarchalska' from Danko (Poland). According to the FAO/WRB classification, the soil at the study site was classified as a typical luvisol soil formed from light loamy sands, deposited in a shallow layer on light loam (*Haplic Luvisols*) [18]. Potassium, manganese, copper and iron contents were average; phosphorus, magnesium, boron and zinc contents were high; and molybdenum content was very low (Table 3). Soil pH was 6.5, which indicates that the soil was slightly acidic, and humus content was also low.

2.2. Weather Conditions

Weather conditions during the growing seasons (2016–2018) are presented using the hydrothermal index, according to Sielianinov [19] (Figure 1). Variability in weather conditions during the study was reflected in the index values. Low index values (multi-year average) were recorded in March 2016 and 2017, April 2016, 2017 and 2018, May 2016 and 2018, June and July 2018, August 2018 and September 2016 and 2018. More favourable moisture conditions during the growing season were observed in 2016 and 2017, compared to the much drier 2018 growing season (May: K = 0.43, June: K = 0.41).

Table 3. Soil characterization at the study site.

Mineral Component	mg·kg^{-1}	Soil Abundance
Manganese	164.1	average
Zinc	15.4	high
Copper	2.6	average
Iron	728.0	average
Boron	10.2	high
Molybdenum	below testing limits	very low
Phosphorus	8.64 mg P/100 g soil	high
Potassium	12.28 mg K/100 g soil	average
Magnesium	5.3 mg mg/100 g soil	high
Humus content	0.8%	poor
C-org %	0.48	
pH in 1 M KCl	6.5	slightly acid

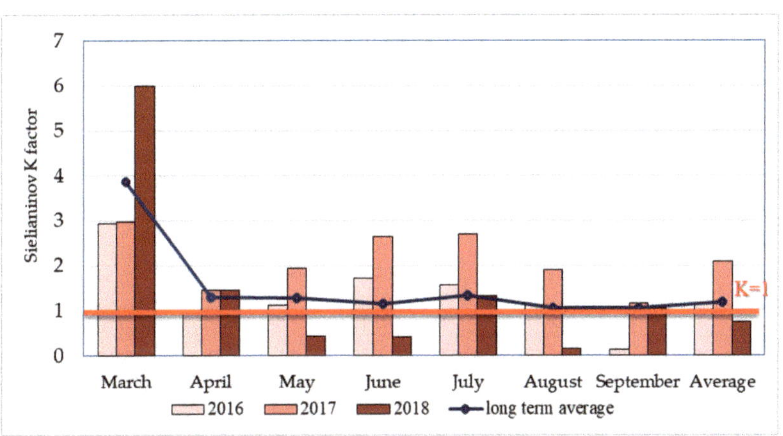

Figure 1. Sielianinov index (K) during the growing seasons 2016–2018 (recorded at the Agrometeorological Observatory in Gorzyń, Poland). Sielianinov K Index: <0.5—drought, 0.5–1.0—semi-drought, 1.0–1.5—zone of optimal moisture, >1.5—excessive moisture.2.3. Influence of Fertilizers on Nitrogenase Activity (Diazotrophy).

At the commencement of the flowering phase, the level of diazotrophy (as expressed by nitrogenase activity) was measured with the acetylene-ethylene reduction (ARA) method [20]. Five plants were randomly selected from the experimental treatment plots and were placed into tightly sealed 2000 mL vials. To achieve a 10% (*v*/*v*) acetylene concentration in the gas phase (air), purified acetylene (C_2H_4) was then injected into each vial. After one hour, 1 mL of gas was withdrawn and stored in small glass vials (each vial sealed with a rubber septum and aluminum seal). Ethylene concentration was quantified using a gas chromatograph CHROM 5 (Laboratorni Přistroje, Praha, Czech Republic). Dinitrogenase activity was quantified as the amount of acetylene reduced to ethylene (expressed as nMC_2H_4 plant^{-1} h^{-1}) and is presented here as the average value of five replications (from each measurement).

2.3. Soil Sampling for Biochemical Analysis

The soil samples that were collected from the upper 0–20 cm layer were used for biochemical analyzes. During the study period, the samples were collected at four terms during each growing season: 1st term—plant emergence (BBCH 5–10), 2nd term—full plant growth (BBCH 35–40), 3rd term—plant inflorescence (BBCH 51–59), 4th term—after harvest.

Soil samples were collected from 5 locations in each experimental plot (in four replications for each of the nine treatments). In total, 36 soil samples (weight per sample: 1 kg) were collected.

2.4. Soil Enzymatic Activity

Soil enzyme activity in the treatments was measured as follows:

- Dehydrogenases (EC 1.1.1.)—with 1% TTC (triphenyltetrazolium chloride) as a substrate after 24-h incubation at 30 °C, at a wavelength of 485 nm, expressed as μmol triphenylformazan (TPF) 24 h^{-1} g^{-1}dm of soil; colorimetry measurements [21];
- Acid and alkaline phosphatase (EC 3.1.3.2)—with sodium p-nitrophenyl phosphate as a substrate after 1-h incubation at 37 °C, at a wavelength of 400 nm; expressed as μmol para-nitrophenol (PNP) h^{-1} g^{-1}dm of soil (Novospac spectrophotometer); spectrophotometry measurements [22];
- Catalase (EC 1.11.1.6)—with 0.3% hydrogen peroxide as a substrate after 20 min incubation at room temperature (about 20 °C); titrated with 0.02 M potassium permanganate until its colour was light pink; expressed as μmol H_2O_2 min^{-1} g^{-1}dm of soil; manometrically measurements [23].

2.5. Biological Index of Fertility (BIF)

BIF was determined by employing dehydrogenase activity (DHA) and catalase activity (CAT) [24] using the formula: (DHA + kCAT)/2, where k is the proportionality factor (= 0.01).

2.6. Statistical Analyses

R and Statistica 12.0 (StatSoft Inc., Poland, Krakow) software packages were used for all statistical analyses. The effects of the experimental factor (biostimulant/fertilizer), and the development phase term (based on the BBCH scale) on enzymatic activity in the soil were tested with three-way ANOVA (Tables S1–S5). Nitrogenase activity and agronomic parameters were tested using two-way ANOVA (Tables S6–S16). Homogeneous subsets of mean were identified by means of Duncan's test, at a significance level of $\alpha = 0.05$.

As *year* was a variable, we used soil biochemical activity parameters (model 1) and agronomic parameters, together with biological nitrogen fixation (BNF) (model 2) ANOVA mixed models. The impact of two or three explanatory variables on the response variable, respectively, was assessed. In both cases, the following models were used:

$$y_{ijkl} = \mu + \alpha_i + \beta_j + \gamma_k + (\beta\gamma)_{jk} + (\alpha\beta)_{ij} + (\alpha\gamma)_{ik} + (\alpha\beta\gamma)_{ijk} + e_{ijkl} \quad \text{(model 1)} \tag{1}$$

$$y_{ikl} = \mu + \alpha_i + \gamma_k + (\alpha\gamma)_{ik} + e_{ikl} \quad \text{(model 2)} \tag{2}$$

where: μ—is the overall average value, α_i—is the effect of the operation of the random factor of *year* at level *i* ($i = 1, 2, 3$), β_j—is the effect of the action of the fixed factor *term* at level *j* ($= 1, 2 \ldots, 4$), γ_k—is the effect of the fixed fertilization factor at level *j* ($= 1, 2 \ldots, 9$ j ($= 1, 2 \ldots, 9$), with appropriate interactions of these factors, and e_{ijkl}—is the residual error.

In cases where the interaction of *year* with the other factors was significant, an analysis was carried out for each year separately. To estimate the cause-and-effect relationship between the studied soil biochemical activity parameters and agronomic parameters, principal component analysis (PCA) was used for each year separately, as well as for combined years. PCA was performed with the use of an appropriately scaled correlation matrix. PCA analysis was used to demonstrate the similarities between independent variables and determines the components that are a linear combination of the variables considered. Accurate analysis of the principal components allows the identification of the initial variables that are the reference system for the remaining variables.

A heat map (using the heatmaply function in R), was proposed as a graphical presentation of appropriately transformed data of soil biochemical activity parameters, agronomic parameters, and biological nitrogen fixation (BNF). Data transformation using 'normalise' was used to compare and group different data.

Fertilization data were represented by colours. Cluster analysis allowed for the grouping of both soil biochemical activity and agronomic parameters after the application of biostimulators, and the effect of the fertilizers/biostimulators so that the degree of connection between the applied fertilization treatments *within* one group was the largest, while the degree of connection *between* groups was the smallest.

Grouping of tree diagrams was obtained by using the Ward Hierarchical Clustering method and the Euclidean distance measurement.

3. Results

3.1. Yield, Biometric and Physiological Traits of the Pea Plants

The effects of the biostimulators and foliar fertilizers on the pea plants were variable between years and depended on weather conditions (Table 4). In 2016, which was characterized by average precipitation levels during the growing season, the greatest increase in yield (compared to the control) was observed after application of Bolero Mo or Adob Zn IDHA and amounted to 0.36 and 0.28 t ha^{-1}, respectively. In 2017, which was characterized by good water availability throughout the entire growing season, a significant increase in seed yield (compared to the control) was observed after the application of Rooter, Adob 2.0 Mo, and Optysil, and amounted to 0.39, 0.62 and 0.80 t ha^{-1} respectively. In 2018, which was characterized by very poor water supply during the critical period for pea development, seed yields were very low. In the event of drought stress, none of the tested biostimulators and fertilizers contributed to a significant increase in yield (Table 4).

Table 4. The influence of biostimulator and fertilizer treatments on crop yield (t ha^{-1}).

Experimental Combination	Year of Analysis		
	2016	2017	2018
Control—no biostimulator or foliar fertilizers applied to the plants	3.07 c	5.21 cd	1.33 ab
Tytanit	3.34 ab	5.31 cd	1.26 ab
Optysil	3.00 cd	6.01 a	1.21 b
Metalosate potassium	3.12 c	5.15 d	1.30 ab
Rooter	3.13 bc	5.60 b	1.34 ab
Bolero Mo	3.43 a	5.42 bc	1.34 ab
Adob Zn IDHA	3.35 a	5.40 bc	1.39 a
Adob B	3.33 ab	5.33 cd	1.29 ab
Adob 2.0 Mo	2.79 d	5.83 a	1.30 ab

Different letters denote significant differences at level $\alpha = 0.05$.

Features such as seed moisture content at harvest (Table 5), pod weight per plant and seed weight in one pod were affected more by weather conditions during the study period than by the application of biostimulators and fertilizers. A significant reduction or increase (compared to the control) in the values of these features was not found across years as a result of the use of a biostimulator or foliar fertilizer, therefore, tables with these results are not included in the paper.

Table 5. The influence of biostimulator and fertilizer treatments on seed moisture content (%).

Experimental Combination	Year of Analysis		
	2016	2017	2018
Control—no biostimulators or foliar fertilizers applied to the plants	12.6 ab	14.0 ab	15.0 a
Tytanit	11.6 b	15.2 a	14.3 a
Optysil	12.5 ab	14.5 ab	14.7 a
Metalosate potassium	12.4 ab	14.4 ab	14.3 a
Rooter	12.6 ab	13.2 b	14.3 a
Bolero Mo	13.0 a	15.0 a	14.6 a
Adob Zn IDHA	12.7 ab	14.5 ab	14.6 a
Adob B	12.9 a	15.3 a	15.0 a
Adob 2.0 Mo	12.8 a	14.4 ab	14.6 a

Different letters denote significant differences at level α = 0.05.

Similarly, the 1000 seed weight changed more strongly under the influence of weather conditions during the study period than from the treatments (Table 6). In 2016, a significantly higher 1000 seed weight value was observed after the pea crop was fertilized with Adob Zn IDHA, with an increase in yield of 11.5 g (compared to the control). In 2017, the increase in yield was 18.1 g, after application of Bolero Mo, while no differences were observed between treatments in the drier 2018.

Table 6. The influence of biostimulator and fertilizer treatments on the 1000 seed weight (g).

Experimental Combination	Year of Analysis		
	2016	2017	2018
Control—no biostimulators or foliar fertilizers applied to the plants	258.9 bcd	284.7 b	189.6 a
Tytanit	255.3 cd	285.2 b	195.3 a
Optysil	264.7 abc	290.5 ab	197.1 a
Metalosate potassium	255.6 cd	296.9 ab	190.7 a
Rooter	265.4 ab	290.0 ab	185.6 a
Bolero Mo	259.7 bcd	302.8 a	185.3 a
Adob Zn IDHA	270.4 a	299.8 ab	186.6 a
Adob B	254.4 d	298.1 ab	189.9 a
Adob 2.0 Mo	261.7 abcd	289.0 ab	201.4 a

Different letters denote significant differences at level α = 0.05.

In 2016, seed weight per plant was significantly greater than in the control treatment after the application of Metalosate Potassium, Optysil and Rooter. In the drier 2018, seed weight per plant was significantly greater after the application of Bolero Mo and Optysil, while in 2017, when the plants were supplied with sufficient water, seed weight per plant increased after application of the treatments, although none were statistically significant (Table 7).

The share of seed in whole plant weight changed significantly under the influence of the biostimulators and fertilizers in two (2017 and 2018) of the three years (Table 8). In the wetter 2017, no differences were found after the application of any of the treatments (compared to the control), while in the drier 2018, a significant increase of 20.2 percentage points occurred only after the application of Bolero Mo. Plant height was another feature that changed more between years than from the influence of the treatments (Table 9). A significant increase in plant height was only found in 2016, which followed the application of Adob Zn IDHA, Adob B, and Metalosate potassium, and was 10.3, 10.8 and 13.5 cm respectively. Plant dry mass was significantly greater in 2016 after Adob B application, in 2017 after Adob Zn IDHA application, Adob B, and Metalosate potassium applications, and in the drier 2018 after spraying with Optysil (Table 10).

Table 7. The influence of biostimulator and fertilizer treatments on seed weight per plant (g).

Experimental Combination	Year of Analysis		
	2016	2017	2018
Control—no biostimulators or foliar fertilizers applied to the plants	1.56 b	4.33 a	1.55 c
Tytanit	3.08 ab	4.40 a	0.98 c
Optysil	3.78 a	5.01 a	3.16 a
Metalosate potassium	3.89 a	5.39 a	1.51 c
Rooter	3.54 a	5.42 a	1.44 c
Bolero Mo	2.90 ab	5.71 a	2.31 b
Adob Zn IDHA	3.56 ab	5.79 a	1.20 c
Adob B	3.51 ab	5.96 a	1.24 c
Adob 2.0 Mo	3.20 ab	6.49 a	1.32 c

Different letters denote significant differences at level $\alpha = 0.05$.

Table 8. The influence of biostimulator and fertilizer treatments on the share of seed in whole plant weight (%).

Experimental Combination	Year of Analysis		
	2016	2017	2018
Control—no biostimulators or foliar fertilizers applied to the plants	37.9 a	48.6 ab	41.4 b
Tytanit	53.8 a	51.4 ab	30.4 b
Optysil	55.2 a	68.9 a	49.2 ab
Metalosate potassium	62.3 a	48.0 ab	44.9 ab
Rooter	66.0 a	60.5 ab	33.8 b
Bolero Mo	50.9 a	48.4 ab	61.6 a
Adob Zn IDHA	53.1 a	38.0 b	33.6 b
Adob B	45.0 a	44.8 ab	34.2 b
Adob 2.0 Mo	51.5 a	46.6 ab	38.8 b

Different letters denote significant differences at level $\alpha = 0.05$.

Table 9. The influence of biostimulator and fertilizer treatments on plant height (cm).

Experimental Combination	Year of Analysis		
	2016	2017	2018
Control—no biostimulators or foliar fertilizers applied to the plants	49.8 b	92.3 ab	54.0 abc
Tytanit	56.9 ab	89.1 ab	55.7 ab
Optysil	59.3 ab	85.6 ab	43.8 d
Metalosate potassium	63.3 a	86.5 ab	47.7 cd
Rooter	58.9 ab	87.3 ab	51.5 bcd
Bolero Mo	59.4 ab	83.0 b	59.3 a
Adob Zn IDHA	60.1 a	88.6 ab	59.1 ab
Adob B	60.6 a	100.0 a	53.4 abc
Adob 2.0 Mo	55.2 ab	96.3 ab	56.1 ab

Different letters denote significant differences at level $\alpha = 0.05$.

Application of biostimulators and foliar fertilizers significantly changed the CCI (Table 11). In 2016 and 2017, the Rooter, Optysil and Bolero Mo treatments significantly stimulated CCI compared to the control treatment, while in the drier conditions of 2018, the application of all treatments, except Rooter and Adob B, significantly increased the CCI value. It should also be noted that the CCI was slightly modified by the weather conditions between years.

Table 10. The influence of biostimulator and fertilizer treatments on plant dry mass (g).

Experimental Combination	Year of Analysis		
	2016	2017	2018
Control—no biostimulators or foliar fertilizers applied to the plants	4.57 b	8.93 bc	3.78 b
Tytanit	5.71 ab	8.47 c	3.19 b
Optysil	6.85 ab	8.45 c	6.43 a
Metalosate potassium	6.24 ab	13.89 a	3.43 b
Rooter	5.36 ab	9.29 bc	4.25 b
Bolero Mo	5.70 ab	12.25 abc	3.90 b
Adob Zn IDHA	6.70 ab	12.94 ab	3.59 b
Adob B	7.78 a	12.98 ab	3.59 b
Adob 2.0 Mo	6.18 ab	11.66 abc	3.39 b

Different letters denote significant differences at level $\alpha = 0.05$.

Table 11. The influence of biostimulator and fertilizer treatments on the Chlorophyll Content Index (CCI).

Experimental Combination	Year of Analysis		
	2016	2017	2018
Control—no biostimulators or foliar fertilizers applied to the plants	16.6 cd	12.8 cd	13.5 c
Tytanit	16.6 cd	15.1 cd	20.3 b
Optysil	22.4 a	23.8 ab	20.9 ab
Metalosate potassium	18.8 bc	16.3 c	21.4 ab
Rooter	20.9 ab	27.4 a	15.2 c
Bolero Mo	22.8 a	22.5 b	23.1 a
Adob Zn IDHA	14.3 d	17.1 c	19.2 b
Adob B	16.0 cd	16.9 c	14.5c
Adob 2.0 Mo	16.2 cd	11.0 d	21.4 ab

Different letters denote significant differences at level $\alpha = 0.05$.

Chlorophyll fluorescence (F_v/F_m parameter) describes the physiological state of a plant and provides a measure of its stress level. In the measurements carried out at the beginning of the maturation phase (BBCH 78) in 2016, plants fertilized with Adob Zn IDHA or Adob B were in a significantly better condition than the control plants. In 2017, plants fertilised with Bolero Mo or Metalosate potassium or were also less stressed, as were the plants that received an application of Optysil, Rooter or Bolero Mo in 2018 (Table 12).

Table 12. The influence of the biostimulator and fertilizer treatments on chlorophyll fluorescence (F_v/F_m).

Experimental Combination	Year of Analysis		
	2016	2017	2018
Control—no biostimulators or foliar fertilizers applied to the plants	0.743 cd	0.738 c	0.723 c
Tytanit	0.763 a-d	0.763 abc	0.760 abc
Optysil	0.783 abc	0.773 abc	0.813 a
Metalosate potassium	0.718 d	0.795 ab	0.748 bc
Rooter	0.755 bcd	0.740 bc	0.780 ab
Bolero Mo	0.763 a-d	0.805 a	0.798 ab
Adob Zn IDHA	0.808 a	0.750 abc	0.768 abc
Adob B	0.793 ab	0.770 abc	0.760 abc
Adob 2.0 Mo	0.755 bcd	0.745 bc	0.758 abc

Different letters denote significant differences at level $\alpha = 0.05$.

The LAI value was also significantly modified following the application of the biostimulators and fertilizers, and also differed between years, with the highest values observed in 2017 (Table 13), which was characterized by good water availability throughout the entire growing season. The use of Optysil and Adob 2.0 Mo in each of the years of the study significantly increased the LAI value compared to the control. In 2016, the use of all treatments, with the exception of Adob Zn, IDHA and Adob B, resulted in significantly increased LAI values compared to the control, and the largest increase in LAI in the study was observed after application of Optysil. In 2017, Adob 2.0 Mo performed even better than Optysil.

Table 13. The influence of the biostimulator and fertilizer treatments on leaf area index (LAI).

Experimental Combination	Year of Analysis		
	2016	2017	2018
Control—no biostimulators or foliar fertilizers applied to the plants	1.78 e	3.44 cd	2.61 e
Tytanit	2.40 ab	3.33 d	2.86 cd
Optysil	2.60 a	3.78 b	3.19 b
Metalosate potassium	2.05 cd	3.52 bcd	2.79 cde
Rooter	2.28 bc	3.33 d	2.80 cde
Bolero Mo	2.30 b	3.58 bcd	3.39 a
Adob Zn IDHA	1.88 de	3.39 d	2.63 e
Adob B	1.75 e	3.71 bc	2.73 de
Adob 2.0 Mo	2.28 bc	4.48 a	2.93 c

Different letters denote significant differences at level $\alpha = 0.05$.

The two-way analysis of variance showed that the application of foliar fertilization/bio-stimulants had a significant influence on enzymatic activity and on BIF values. Only the term of the study (development phase, based on BBCH scale) had a highly significant influence on enzymatic activity and on BIF. Two-way analysis of variance showed that the application of foliar fertilization/bio-stimulants had a significant influence on nitrogenase activity.

3.2. Biological Nitrogen Fixation

Field analyses of BNF showed that the fertilizer and biostimulant applications significantly enhanced nitrogenase activity during pea cultivation (Table 14). In all the treatments, nitrogenase exhibited higher activity than in the control. During the study period, greatest BNF activity was recorded in 2017, while the greatest nitrogenase activity was noted after the application of Tytanit, when the activity of the enzyme was five times higher than in 2016 and 2018, and six times higher in 2017 than in the control plot. Apart from the control treatment, the lowest BNF value was noted after the application of Adob 2.0 Mo.

Table 14. The influence of biostimulator and fertilizer applications on biological nitrogen fixation (BNF; nMC_2H_4 plant^{-1} h^{-1}).

Experimental Combination	Year of Analysis		
	2016	2017	2018
Control—no biostimulators or foliar fertilizers applied to the plants	46.5 d	62.3 f	34.0 d
Tytanit	244.0 a	351.0 a	169.0 a
Optysil	183.5 b	268.5 c	147.5 b
Metalosate potassium	161.5 b	263.8 c	143.0 b
Rooter	95.8 c	192.3 d	74.5 c
Bolero Mo	168.8 b	328.0 b	136.3 b
Adob Zn IDHA	61.0 cd	86.8 e	51.5 d
Adob B	164.3 b	353.4 a	141.0 b
Adob 2.0 Mo	57.3 cd	82.4 ef	43.9 d

Different letters denote significant differences at level $\alpha = 0.05$.

3.3. Soil Enzymatic Activity

Soil enzymatic activity analysis, which was closely related to the plant development phase (BBCH) significantly influenced the level of dehydrogenase activity in the soil. In all study years, the lowest value was recorded during the emergence of the plants, while the greatest value was observed at the beginning of the flowering phase (Table 15).

Table 15. The influence of biostimulator and fertilizer applications on dehydrogenase activity (μmol triphenyl formazane (TPF) 24 h^{-1} g^{-1}dm of soil.

Experimental Combination	Term of Analysis/BBCH			
	I/BBCH 5–10	II/BBCH 35–40	III/BBCH 51–59	IV/After Harvest
		2016		
1	0.0021 no	0.0046 mn	0.0195 c	0.0071 i–m
2	0.0021 no	0.0073 h–m	0.0195 c	0.0060 j–m
3	0.0018 no	0.0056 lm	0.0153 de	0.0088 h–k
4	0.0020 no	0.0066 i–m	0.0183 cd	0.0059 k–m
5	0.0015 o	0.0089 h–k	0.0200 bc	0.0088 h–k
6	0.0018 no	0.0095 f–i	0.0295 a	0.0091 h–j
7	0.0015 no	0.0091 g–i	0.0189 c	0.0124 ef
8	0.0016 no	0.0078 h–l	0.0267 a	0.0103 f–h
9	0.0018 no	0.0056 lm	0.0230 b	0.0122 fg
		2017		
1	0.0054 a–c	0.0005 c	0.0068 a–c	0.0022 bc
2	0.0062 a–c	0.0037 a–c	0.0095 a–c	0.0003 c
3	0.0031 a–c	0.0010 c	0.0109 ab	0.0014 bc
4	0.0062 a–c	0.0014 bc	0.0099 a–c	0.0003 c
5	0.0021 bc	0.0018 bc	0.0024 bc	0.0007 c
6	0.0035 a–c	0.0019 bc	0.0111 ab	0.0123 a
7	0.0023 bc	0.0057 a–c	0.0054 a–c	0.0081 a–c
8	0.0017 bc	0.0024 bc	0.0014 bc	0.0054 a–c
9	0.0076 a–c	0.0086 a–c	0.0008 c	0.0080 a–c
		2018		
1	0.0041 no	0.0090 mn	0.0379 d	0.0141 i–m
2	0.0042 no	0.0144 h–m	0.0381 d	0.0119 j–m
3	0.0036 no	0.0109 lm	0.0288 e	0.0173 h–k
4	0.0039 no	0.0130 i–m	0.0352 d	0.0116 k–m
5	0.0029 o	0.0175 h–k	0.0393 cd	0.0173 h–k
6	0.0034 no	0.0188 f–i	0.0578 a	0.0179 g–j
7	0.0030 no	0.0181 g–i	0.0358 d	0.0247 ef
8	0.0032 no	0.0154 h–l	0.0506 b	0.0204 f–h
9	0.0035 no	0.0109 lm	0.0446 bc	0.0240 e–g

Different letters denote significant differences at level α = 0.05, (1) control—no biostimulators or foliar fertilizers applied to the plants; (2) Tytanit; (3) Optysil; (4) Metalosate potassium; (5) Rooter; (6) Bolero Mo; (7) Adob Zn IDHA; (8) Adob B; (9) Adob 2.0 Mo.

In addition, the greatest dehydrogenase activity was observed in all years after Bolero Mo application (compared to the control). In contrast, the lowest value was noted after the application of Optysil foliar fertilizer in 2016 and 2018, as well as in 2017 after the application of the Rooter biostimulator. It was also observed across all study years that, dehydrogenase activity was always greater after the application of the treatments (compared to the control) in the second term of the analysis, when the plants were in full vegetation phase. Similar relationships were also observed during the flowering period for most treatments, however, reduced activity was noted after the application of Optysil in 2016, Adob B in 2017, and Optysil, Metalosate potassium or Adob Zn IDHA in 2018.

Analyses of acid phosphatase activity showed that the function of this enzyme was also closely related to the developmental phase of the plant, and its greatest activity was observed during the third term of analysis; at the beginning of the flowering phase in 2016 and 2018 (Table 16). At this time, a greater level of enzyme activity was only observed (compared to the control) after the application of Adob B foliar fertilizer. In 2017, the greatest values were recorded after the application of Metalosate potassium during the second term of analyses (the development phase BBCH 35–40). It should be noted that the lower level of activity of phosphatase (compared to the control) at the beginning of the flowering phase (i.e., when demand for phosphorus is greatest) was observed after the application of most of the treatments, Specifically, the lowest activity level was observed after the application of the Bolero Mo biostimulator in 2016 and 2017, and the foliar fertilizer Adob Zn IDHA in 2018.

Table 16. The influence of biostimulator and fertilizer applications on acid phosphatase activity μmol (*p*-nitrophenol) PNP h^{-1} g^{-1}dm of soil.

Experimental Combination	Term of Analysis/BBCH			
	I/BBCH 5–10	II/BBCH 35–40	III/BBCH 51–59	IV/After Harvest
		2016		
1	0.424 k	1.059 g–j	3.476 b	0.954 ij
2	0.353 k	1.198 f–j	2.654 e	0.968 h–j
3	0.384 k	1.049 g–j	3.170 b–d	0.974 h–j
4	0.372 k	1.217 f–j	2.793 de	0.868 j
5	0.355 k	1.398 fg	3.198 bc	1.122 f–j
6	0.476 k	1.340 f–h	2.541 e	1.268 f–l
7	0.379 k	1.463 e	2.901 c–e	1.182 f–j
8	0.329 k	1.233 f–j	3.929 a	1.378 fg
9	0.290 k	0.940 ij	2.850 c–e	1.399 fg
		2017		
1	0.195 h–k	0.398 ab	0.353 a–d	0.216 g–k
2	0.191 i–k	0.384 a–c	0.283 c–j	0.281 c–j
3	0.151 k	0.310 a–g	0.279 c–j	0.233 e–k
4	0.166 k	0.403 a	0.279 c–j	0.277 d–j
5	0.203 g–k	0.396 ab	0.246 d–k	0.236 e–k
6	0.202 g–k	0.289 b–j	0.234 e–k	0.199 h–k
7	0.166 k	0.333 a–e	0.331 a–f	0.184 jk
8	0.221 f–k	0.303 a–h	0.298 a–i	0.216 g–k
9	0.311 a–g	0.303 a–h	0.338 a–e	0.223 f–k
		2018		
1	0.181 l	0.904 kl	2.454 b	1.119 i–k
2	0.181 l	1.440 e–k	1.905 b–g	1.407 f–k
3	0.244 l	1.092 jk	2.125 b–f	1.464 e–k
4	0.196 l	1.298 g–k	1.965 b–g	1.132 h–k
5	0.190 l	1.753 b–j	2.233 b–d	1.581 c–k
6	0.197 l	1.882 b–h	2.337 bc	1.892 b–h
7	0.186 l	1.806 b–j	1.877 b–i	1.817 b–j
8	0.230 l	1.537 d–k	3.983 a	2.015 b–g
9	0.224 l	1.087 jk	2.183 b–e	2.236 b–d

Different letters denote significant differences at level $\alpha = 0.05$, (1) control—no biostimulators or foliar fertilizers applied to the plants; (2) Tytanit; (3) Optysil; (4) Metalosate potassium; (5) Rooter; (6) Bolero Mo; (7) Adob Zn IDHA; (8) Adob B; (9) Adob 2.0 Mo.

The greatest alkaline phosphatase activity level was observed in 2016 (Table 17). Similar relationships were observed as with acid phosphatase. In 2016 and 2018, the greatest activity was also noted at the beginning of the flowering phase (BBCH 59), while the greatest activity in 2017 was observed during the BBCH 35–40 phase.

Table 17. The influence of the biostimulators and fertilizers on alkaline phosphatase activity μmol (*p*-nitrophenol) PNP h^{-1} g^{-1}dm of soil.

Experimental Combination	Term of Analysis			
	I/BBCH5–10	II/BBCH35–40	III/BBCH51–59	IV/After Harvest
		2016		
1	0.316 m	0.760 kl	2.711 b	0.834 i–l
2	0.283 m	0.964 g–k	2.302 d	0.786 j–l
3	0.282 m	0.803 i–l	2.352 d	0.925 h–l
4	0.286 m	0.939 g–l	2.311 d	0.729 l
5	0.254 m	1.144 e–g	2.601 bc	1.002 f–j
6	0.326 m	1.145 e–g	2.743 b	1.088 f–h
7	0.267 m	1.189 ef	2.393 cd	1.213 ef
8	0.244 m	1.007 f–i	3.299 a	1.205 ef
9	0.234 m	0.749 kl	2.574 bc	1.308 e
		2017		
1	0.173 i–k	0.319 b–d	0.250 c–i	0.238 d–i
2	0.169 i–k	0.345 b	0.262 b–h	0.205 f–k
3	0.178 h–k	0.201 g–k	0.234 d–j	0.237 d–i
4	0.222 e–j	0.485 a	0.294 b–g	0.269 b–h
5	0.124 k	0.310 b–e	0.238 d–i	0.174 i–k
6	0.227 d–j	0.343 bc	0.202 g–k	0.210 f–k
7	0.142 jk	0.246 d–i	0.208 f–k	0.274 b–g
8	0.141 jk	0.296 b–f	0.244 d–i	0.276 b–g
9	0.168 i–k	0.297 b–f	0.244 d–i	0.280 b–g
		2018		
1	0.013 h	0.017 h	0.163 b–d	0.043 f–h
2	0.013 h	0.007 h	0.124 cd	0.023 h
3	0.011 h	0.022 h	0.119 b–e	0.025 h
4	0.016 h	0.017 h	0.151 cd	0.035 gh
5	0.017 h	0.027 h	0.119 b–e	0.048 e–h
6	0.005 h	0.028 gh	0.190 bc	0.099 d–g
7	0.005 h	0.023 h	0.113 d–f	0.109 d–f
8	0.003 h	0.021 h	0.225 ab	0.133 cd
9	0.010 h	0.018 h	0.139 cd	0.270 a

Different letters denote significant differences at level $\alpha = 0.05$, (1) control—no biostimulators or foliar fertilizers applied to the plants; (2) Tytanit; (3) Optysil; (4) Metalosate potassium; (5) Rooter; (6) Bolero Mo; (7) Adob Zn IDHA; (8) Adob B; (9) Adob 2.0 Mo.

The lowest level of activity occurred during the emergence of the plants (BBCH 5–10). In both 2016 and 2018, the highest phosphatase activity level was observed after the application of Adob B fertilizer; 3.299 and 0.225 μmol PNP h^{-1} g^{-1}dm of soil, respectively, while in 2017 the greatest value was observed after the application of Metalosate potassium; 0.485 μmol PNP h^{-1} g^{-1}dm of soil (Table 17).

The greatest catalase enzyme activity value was also recorded in 2016 and 2018 at the beginning of flowering phase. In 2017, the highest metabolic activity occurred in the period after plant harvest (Table 18). The biostimulators and foliar fertilizers used in most of the experimental treatments stimulated the level of catalase activity in relation to the control treatment. The highest level of catalase enzyme activity was recorded after the application of the Tytanit biostimulator in 2016, and after the application of the Adob B foliar fertilizer in 2018. In 2017, the greatest activity was observed after the applications of Tytanit and Bolero Mo (Table 18).

Table 18. The influence of the biostimulators and fertilizers on catalase activity (μmol H_2O_2 min^{-1} g^{-1} dm of soil).

Experimental Combination	Term of Analysis/BBCH			
	I/BBCH5–10	II/BBCH35–40	III/BBCH51–59	IV/After Harvest
		2016		
1	22.692 gh	28.276 gh	44.846 ef	19.767 gh
2	20.580 gh	42.620 f	72.124 a	20.974 gh
3	25.691 gh	23.572 gh	55.345 c–f	33.000 gh
4	21.507 gh	35.703 gh	72.429 a	23.249 gh
5	27.440 gh	39.026 gh	67.390 a–c	33.525 gh
6	22.283 gh	26.695 gh	57.982 b–d	30.047 gh
7	27.466 gh	34.392 gh	68.860 ab	32.881 gh
8	23.013 gh	34.076 gh	59.076 b–d	41.569 g
9	14.933 h	21.821 gh	54.870 de	25.644 gh
		2017		
1	22.726 z	21.723 z	25.257 y	33.962 k
2	22.666 z	32.811 l	28.155 p	44.407 a
3	14.366 z	17.502 z	26.240 w	36.803 f
4	26.387 u	16.629 z	29.289 o	38.730 e
5	14.196 z	27.569 r	39.211 c	36.601 h
6	17.732 z	16.554 z	27.184 s	39.308 b
7	10.839 z	9.009 z	39.033 d	36.630 g
8	18.267 z	26.443 t	32.289 m	35.145 i
9	10.784 z	13.189 z	31.404 n	34.014 j
		2018		
1	13.477 t	21.286 p	31.711 k	10.259 w
2	14.370 t	37.776 hi	38.910 g	19.011 r
3	23.299 o	26.861 m	37.363 i	22.729 o
4	8.146 y	38.546 gh	52.811 d	16.344 s
5	16.411 s	36.202 j	47.716 e	24.500 n
6	14.388 t	40.710 f	61.607 b	27.577 m
7	14.346 t	41.326 f	52.351 d	24.559 n
8	12.352 u	30.921 kl	64.465 a	27.635 m
9	4.7124 z	30.413 l	54.256 c	22.517 o

Different letters denote significant differences at level $\alpha = 0.05$, (1) control—no biostimulators or foliar fertilizers applied to the plants; (2) Tytanit; (3) Optysil; (4) Metalosate potassium; (5) Rooter; (6) Bolero Mo; (7) Adob Zn IDHA; (8) Adob B; (9) Adob 2.0 Mo.

The BIF value was determined based on dehydrogenase and catalase activity. A high value was observed during the flowering phase of the plants compared to the emergence phase, in all experimental treatments. However, the highest level was recorded in 2016 and 2017 after the harvest of the plants; 33.907 and 33.306, respectively (Table 19). In 2018, the highest BIF value was 55.813 and was observed at the beginning of the flowering phase.

Principal Component Analysis (PCA) in Figure 2a–c shows that the applied biostimulators/fertilizers differed in their influence on the agronomic and microbiological parameters in individual years. Figure 2d illustrates the influence of all the factors under analysis (treatment) on the indicators in a given year. In 2016 all the treatments applied in the experiment had strong positive influence on the activity of catalase (CAT) and alkaline phosphatase (PAL), as compared with the other parameters. In 2017 the biostimulants/fertilizers significantly influenced the BNF, H, PDN, Y and BIF, whereas in 2018 they influenced only the dehydrogenase activity (DHA). PCA explained a significant part of the variability in each study year, and also over the three-year study period. In 2016, approximately 60% of the total variability was explained by the first two principal components (Axis 1: 30.4%, Axis 2: 28.5%) (Figure 2a). It was observed that, to a greater or lesser extent, each of the biostimulators/fertilizer treatments affected the studied parameters (Figure 2a). In addition, there was a strong correlation between the basic soil biochemical parameters studied here [dehydrogenase activity (DHA), alkaline phosphatase activity (PAL), biological index fertilizer BIF, acid phosphatase activity (PAC) and soil moisture (M), and the strong effect of the applied Adob B (t8) on the parameters indicated above. In turn, BNF was closely correlated with catalase activity (CAT), LAI and share of seed in whole plant weight (SSPW). In 2016, the foliar fertilizers Optysil (t3) and Metalosate Potassium (t4), and the Rooter biostimulator (t5) had the greatest impact on the above-mentioned parameters.

Table 19. The influence of the biostimulators and fertilizers on the biological indicator of fertility (BIF).

Experimental Combination	Term of Analysis			
	I/BBCH5–10	**II/BBCH35–40**	**III/BBCH51–59**	**IV/After Harvest**
		2016		
1	2.288 hi	4.080 g–i	14.337 c–e	7.223 f–i
2	1.940 i	4.112 g–i	15.117 cd	8.370 f–i
3	1.343 i	4.810 g–i	17.358 bc	5.401 g–i
4	2.688 hi	4.310 g–i	16.451 cd	2.561 hi
5	3.338 g–i	5.935 f–i	11.409 de	9.616 f–i
6	2.518 i	5.029 g–i	17.223 bc	13.776 c–e
7	1.038 i	5.299 g–i	17.445 bc	14.522 c–e
8	0.758 i	4.982 g–i	28.795 b	18.148 bc
9	1.773 i	5.730 g–i	15.690 cd	33.907 a
		2017		
1	17.047 z	16.292 z	18.946 y	25.472 k
2	17.002 z	24.610 l	21.121 p	33.306 a
3	10.776 z	13.127 z	19.686 w	27.603 f
4	19.794 u	12.472 z	21.972 o	29.047 e
5	10.648 z	20.678 r	29.409 c	27.451 h
6	13.300 z	12.416 z	20.393 s	29.487 b
7	8.130 z	6.759 z	29.277 d	27.477 g
8	13.701 z	19.833 t	24.217 m	26.361 i
9	8.092 z	9.896 z	23.553 n	25.514 j
		2018		
1	2.899 h	3.881 h	37.256 bc	9.892 f–h
2	2.877 h	1.609 h	28.433 cd	8.802 gh
3	2.451 h	4.944 h	27.258 d	5.918 h
4	3.708 h	3.903 h	34.924 cd	10.260 e–h
5	3.785 h	6.069 h	27.597 d	11.215 e–h
6	1.223 h	6.479 h	43.747 ab	22.826 d–g
7	1.115 h	5.272 h	26.276 de	24.911 d–f
8	0.799 h	4.694 h	51.813 a	30.487 cd
9	2.290 h	4.205 h	32.277 cd	31.161 cd

Different letters denote significant differences at level $\alpha = 0.05$, (1) control—no biostimulators or foliar fertilizers applied to the plants; (2) Tytanit; (3) Optysil; (4) Metalosate potassium; (5) Rooter; (6) Bolero Mo; (7) Adob Zn IDHA; (8) Adob B; (9) Adob 2.0 Mo.

In 2017, PCA explained approximately 50% of the variation (Axis 1: 25%, Axis 2: 24.4%) (Figure 2b). A strong relationship between BIF, catalase activity and share of seed in whole plant weight (SSPW) was noted. The level of activity of these parameters was influenced by the Tytanit and Rooter biostimulators. In 2017, a strong relationship was also observed between biochemical soil activity parameters and agronomic parameters, such as plant dry mass (PDM), seed weight per plant (SW), seed moisture (M) and the physiological parameter F_v/F_m.

In 2018, as in 2016, all tested parameters were more or less affected by the applied foliar fertilizer and biostimulator treatments. In that year, PCA explained almost 60% of the total variability (Axis 1: 31.4%, Axis 2: 26%) (Figure 2c). The studied biochemical soil parameters (DHA, PAL, PAC, BIF) were closely correlated with each other. The most important influence on soil metabolism was the

application of Bolero B (t6), Adob B and Adob 2.0 Mo fertilizers. In turn, Adob Zn IDHA and Adob 2.0 Mo. foliar fertilizers influenced agronomic parameters, such as yield (Y) and seed moisture (M).

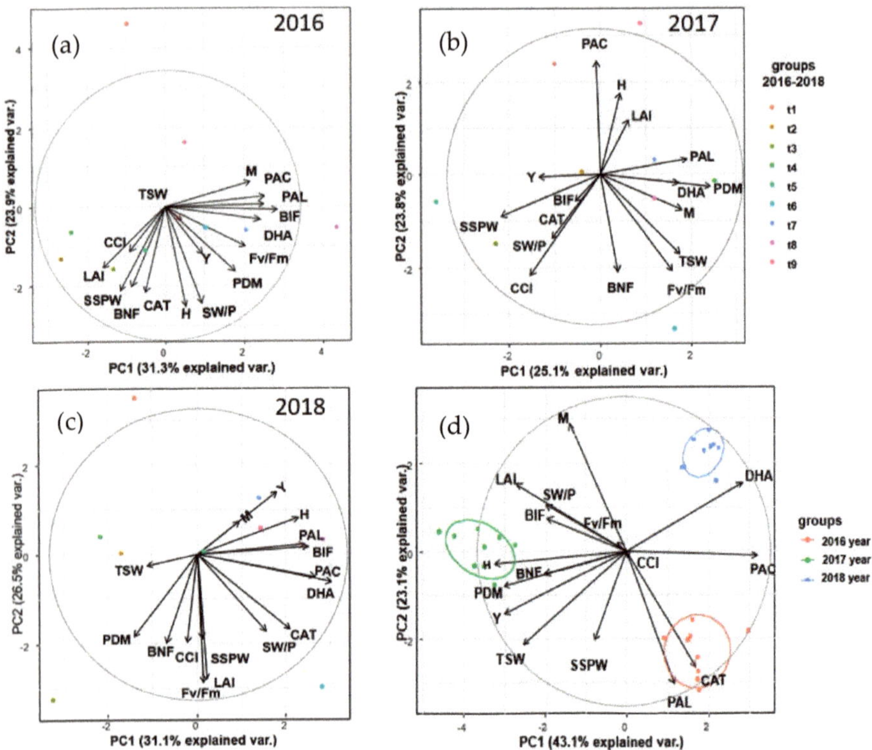

Figure 2. Principal components analysis for pea fertilization treatments in: (**a**) 2016, (**b**) 2017, (**c**) 2018, (**d**) 2016–2018 years. Abbreviations: Y—Yield, M—Seed moisture %, TSW—1000 seed weight, SW—Seed weight per plant, SW/P—seed weight in 1 pod PW—Pod weight per plant, SSPW—Share of seed in whole plant weight, H—plant height, CCI—chlorophyll content index, F_v/F_m—maximum photosynthetic efficiency of PSII, PDM—plant dry mass, LAI—leaf area index, BIF—biological index of fertility, BNF—biological nitrogen fixation, DHA—dehydrogenase activity, CAT-catalase activity, PAL—alkaline phosphatase level, PAC—acid phosphatase level. Treatment: t1—control—no biostimulators, t2—Tytanit; t3—Optysil; t4—Metalosate potassium; t5—Rooter; t6—Bolero Mo; t7—Adob Zn IDHA; t8—Adob B; t9—Adob 2.0 Mo; PC1—first principal component; PC2—second principal component.

PCA for each pea fertilization treatment in all the study years is shown in Figure 2d, and highlights the differentiated effect of the applied fertilizers/biostimulator treatments on the tested soil biochemical and agronomic parameters during the study period. In 2018, regardless of the fertilization treatment, a high level of dehydrogenase activity was observed, while in 2016, a high level of phosphatase and catalase activity was observed.

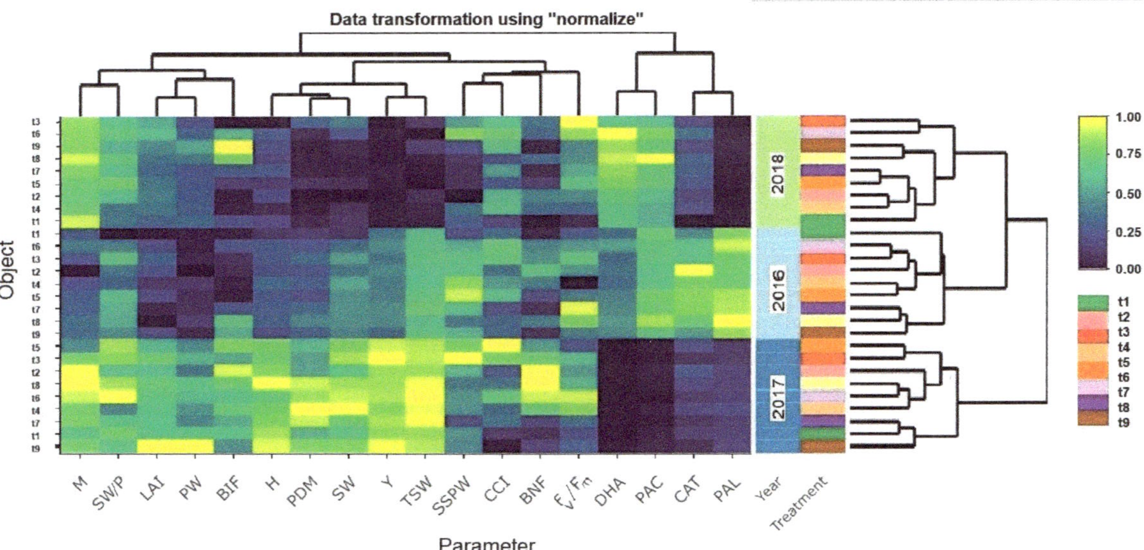

Figure 3. Relationship between the biometric, physiological characteristics of the pea plants and enzymatic activity in the soil, biological index of fertility (BIF), and nitrogenase activity. Abbreviations: M—Seed moisture %, SW/P—seed weight in 1 pod, LAI—leaf area index, PW—Pod weight per plant, BIF—biological index of fertility, H—plant height, PDM—plant dry mass, SW—Seed weight per plant, Y—Yield, TSW—1000 seed weight, SSPW—Share of seed in whole plant weight, CCI—chlorophyll content index, BNF—biological nitrogen fixation F_v/F_m—maximum photosynthetic efficiency of PSII, DHA—dehydrogenase activity, PAC—acid phosphatase level, CAT—catalase activity, PAL—alkaline phosphatase level. Treatment: t1—control—no biostimulators, t2—Tytanit; t3—Optysil; t4—Metalosate potassium; t5—Rooter; t6—Bolero Mo; t7—Adob Zn IDHA; t8—Adob B; t9—Adob 2.0 Mo.

The heat map and dendrogram presented in Figure 3 illustrates the large variation in activity level of the studied parameters and clearly demonstrates the correlation between the biochemical soil activity parameters and the examined agronomic parameters. Taking into account all the parameters tested in the study, the effect of the fertilizers/biostimulators was similar in 2016 and 2018, but differed in 2017. The research conducted in 2017 indicated that the fertilizer/biostimulator treatments had a stimulating effect on plant yield at low soil enzymatic activity (catalase, phosphatase, dehydrogenase). Only BIF, based on the level of metabolic activity of dehydrogenase and catalase, and BNF were highest in 2017. Different relationships were observed in 2016 and 2018, where a correspondingly higher level of biochemical activity in the soil was observed, and agronomic parameters exhibited lower values.

Regardless of the study year, the heat map indicates similar relationships between the soil biochemical activity parameters and the agronomic parameters. Soil fertility index (BIF) was correlated with pod weight per plant (PW) and LAI, as well as with seed weight in one pod (SW/P). In turn, yield (Y) was correlated with 1000 seed weight (TSW) and plant height (H), and BNF was correlated with the share of seed in whole plant weight (SSPW), CCI, and F_v/F_m. Lastly, phosphatase activity was correlated with catalase, dehydrogenase and acid phosphatase activity.

4. Discussion

4.1. Yield, Biometric and Physiological Traits

In this study, the yields of the 'Tarchalska' variety were higher in 2017, and lower in 2016 and 2018 than yields in the official Research Centre for Testing Plant Varieties and Registration (COBORU) trials [25], which in 2016, 2017 and 2018 amounted to 3.69, 4.93 and 3.00 t ha^{-1}, respectively.

The effectiveness of biostimulators and foliar fertilizers was closely related to weather conditions, because the different treatments significantly stimulated pea yield each year. Only the Adob Zn IDHA fertilizer increased pea seed yield in two out of the three study years; by 9.1% in 2016, and by 4.5% in 2018.

In 2016, the application of chelated zinc in the Adob Zn IDHA fertilizer increased 1000 seed weight and plant height, and reduced stress in the plants (as expressed by the F_v/F_m parameter). With an ample water supply in 2017, the effect of its action was to increase plant dry mass, but in the drier 2018 it only affected CCI. Zinc is taken up by a plant in small amounts and according to Stevenson and Cole [26] there are 27 to 150 mg of Zn kg^{-1} biomass in healthy plants. It participates in all major functions in the plant and is the microelement that most limits crop yield [27,28].

Indeed, Niewiadomska et al. [29] noted an increase in white lupin yield by 13.0% after the application of Adob Zn IDHA. Our results are partly consistent with the findings of Raj and Raj [30] regarding the beneficial effects of Zn on yield, physiological parameters, plant height in legumes, and with other research that showed a slight increase in yield after application of this fertilizer [31,32]. In addition, foliar spraying of Zn on *Vigna sinensis* [33] and *Celosia* [34] plants was shown to cause a significant increase in chlorophyll content. Also, El-Sallami and Gad [35] found that foliar spraying of zinc on plants of the *Asteraceae* family increased plant height, the number of leaves, and fresh and dry matter, which was partially confirmed by our own research. Mostafi et al. [36] also reported that the greatest 1000 seed weight value for soybean seeds was recorded after spraying with a zinc-iron mixture. Gomaa [37] also observed the stimulating effects of boron or zinc on plant growth.

In our study, pea seed yield in 2016 (a year with average weather conditions for the region) increased by 11.7%, after using the fertilizer containing boron and molybdenum (Bolero Mo), by 8.8% after application of Titanit and by 8.5% after application of Adob B (compared to the control). In turn, the molybdenum contained in Adob Mo stimulated yield in the wetter 2017. In earlier studies on white lupin, the application of boron with molybdenum (Bolero Mo), and also molybdenum alone (Adob 2.0 Mo) did not affect the seed yield [29], although, as in our study, it clearly stimulated LAI.

In the above-mentioned white lupin studies, the addition of molybdenum alone in a foliar fertilizer significantly increased the F_v/F_m parameter value, indicating a reduction in plant stress (compared to the control), and the CCI index. In our study, the F_v/F_m parameter value was stimulated by boron and molybdenum fertilization (Bolero Mo) in two out of three years, and the CCI increased in value each year after using Bolero Mo, and after application of Adob Mo in the drier 2018. Molybdenum is a cofactor for the nitrate reductase enzyme involved in nitrogen assimilation [38]. In bean plants living with *Rhizobium* bacteria, the demand for molybdenum is greater than in other plants, and its deficiency limits the number and dry mass of root papillae [39]. Our results are contrary to Omer et al. [40], where the various molybdenum applications did not modify any of the studied characteristics of lentil (*Lens culinaris* Medik.) plants, except for the height of the plant.

Fertilization of peas with boron alone (Adob B) gave weaker effects than in combination with molybdenum (Bolero Mo), both in terms of yield and in influencing physiological parameters, which in turn stimulated plant height in 2016 and dry plant weight in two of the years. These results are partly consistent with the work by Madna [41], where boron application increased the height of field bean plants, the leaf area of a single plant, total dry plant weight, number of pods per plant, number of seeds in a pod, seed weight of one plant and 1000 seed weight. Also, Mahmoud et al. [42] found that the application of boron significantly increased field bean plant height, total dry matter, number of pods per plant, number of seeds in the pod and seed yield.

In a previous study conducted by Sulewska et al. [43], a nitrogen and boron fertilizer (Adob B) had the strongest stimulatory effect on the growth of pea plants, and also derived the highest CCI values. Boron application was also shown to lead to an increase in the height of seed pea plants, flowering and the number of pods [44]. Moghazy et al. [45] demonstrated the stimulating effect of boron on the vegetative growth of green peas, i.e., plant length, number of leaves, number of stems, fresh plant mass, as did Niewiadomska et al. [29] for height, CCI and LAI values in white lupin plants. Fageria et al. [46] showed that boron application can significantly increase common bean yield. Boron is an essential microelement in the cultivation of bean plants, because it plays an important role in the development of flowers, reduces the fall of pod buds [47], and also increases the establishment of nodules [48]. The uptake of boron by crops is small, but its deficiency has been reported in many soils globally [49,50].

While Titanit was shown to increase seed yield in 2016, it did not change biometric features, although it increased LAI values in two of the study years, and also increased the CCI values in 2018. These results are consistent with research by Malinowska and Kalembasa [51], in which titanium was shown to activate metabolic processes, as well as accelerate the process of photosynthesis and nutrient uptake. Our results are partly in agreement with Grenda [52], where titanium application led to an increase in chlorophyll content and photosynthesis efficiency in rape plants, yield and thousand seed weight in wheat plants, and sugar content in sugar beet roots. The use of titanium (Titanit) with white lupin [29] produced better results than with peas: increased the height of white lupin plants, the number of pods with seeds per 1 m^2, reduced stress (F_v/F_m), and similarly increased the CCI and LAI values.

In 2017, when the plants were well supplied with water the best results were obtained with Optysil, Adob 2.0 Mo and Rooter, and the increase in yield was 15.4%, 11.9% and 7.5% respectively. Silicon can affect the metabolism and physiological functions of plants, especially under stress [53]. However, the active silicon contained in Optysil strongly stimulated pea yield only in the year (2016) with the average weather conditions for the region. A similar yield increase (15.1%) was shown in earlier studies on white lupin [29]. Other authors have also reported the beneficial effect of silicon on the yield of other plant species: monocotyledons that include rice [53], and sugar beet roots [54]. Silicon can be taken from the soil solution, constituting up to 10% dry matter (DM). However, rejective uptake of this micronutrient has been found in some species, especially bean species [55]. These plants are not able to accumulate silicon, and they absorb it more slowly than water, hence they contain less than when passively taking the element from the soil, and so cannot benefit from its positive effect. Therefore, it is possible that foliar application of a treatment containing active silicon enabled the uptake of this microelement by the peas plants in the wet 2017, and by white lupin in previous studies [29], which was externalized by a higher seed yield. In our study, the use of Optysil, aside from crop yield, significantly increased plant height by 16.0% (compared to the control) but did not improve any of the other assessed traits.

The drought that prevailed during the growing season in 2018 greatly limited the development and yield of the pea plants. Its influence was so pronounced that it limited the effectiveness of the biostimulators and fertilizers. The positive effect of the Rooter biostimulator in terms of root growth, biomass accumulation has been demonstrated by Kowalczyk and Zielony [56] and the product has also been shown to strongly affect the yield of tomatoes [57] and nappa cabbage [58]. Our research partly confirms these findings, because Rooter stimulated LAI and CCI values, increased the F_v/F_m parameter values and seed weight per plant, and also lowered the seed moisture content before harvest, but did not modify the other parameters tested.

4.2. Biological Nitrogen Fixation

Biostimulators are a source of substances and/or microorganisms that have an impact on the metabolic processes that occur in the plants [11]. Foliar fertilizers and biostimulators, which contain micro and macroelements caused an increase in BNF in the pea plants in this study. Various microelements, including boron, cobalt, molybdenum and iron, influence the level of molecular nitrogen binding by bacteria entering symbiosis with plants.

Molybdenum and iron are two metalloproteins that form a bacteroid enzyme complex – nitrogenase, which allows plants to bind molecular nitrogen [59]. During the process of nitrogen binding, molybdenum has two yield-forming functions: stimulation of (a) the number of root nodules and (b) the number of flowers/pods on the plant. Molybdenum is most readily available for plants in alkaline soils, while uptake by plants is prevented by drought and by an excessive concentration of sulphate anions in the soil [60]. In our study, a significantly increased level of BNF was observed after application of foliar fertilizers containing molybdenum.

Boron is an important microelement in BNF. In our study, a significantly increased BNF was observed after application of the foliar fertilizer that contained boron (Adob B), although, it was the lowest BNF value recorded, in comparison to the other fertilizers/biostimulators used.

The Tytanit biostimulator highly stimulated the BNF process and exhibited the best effect in all study years. The effect of titanium on plants is to modify the activity of enzymes, such as catalase, peroxidase, lipoxygenase or nitrate reductase [61]. It increases chlorophyll content in the leaves, which translates into increased cereal and vegetable crop yields. Kováčik et al. [62] noted that the use of Mg-Titanite (MgTi) in the form of titanium ascorbate in the growth phases BBCH 29 and BBCH 32 stimulated the formation of terrestrial winter wheat phytomass and increased the content of total chlorophyll in the leaves. It is known that the process of BNF is closely correlated with photosynthetic intensity, which in turn is dependent on chlorophyll content. The biological fixation of atmospheric nitrogen depends on many environmental conditions, such as water content and temperature [63].

4.3. Biochemical Activity

Enzymatic activity is one of the best indicators for assessing biological activity in the soil, and the following enzymes are widely used: dehydrogenase, phosphatase, urease and protease [64]. In the literature, numerous authors [65–67] report that dehydrogenases do not accumulate in the soil, but only in the cytoplasm and characteristic structures formed from the cytoplasmic membranes of living microorganisms. Being intracellular enzymes, they can be indicative of the presence of physiologically active soil microorganisms, which provides information on the respiratory activity of the entire soil microbiota, especially bacteria and Actinomycetes [68]. Numerous studies on dehydrogenases and their association with soil factors indicate that they are useful and sensitive indicators of soil changes [69–71].

In our study, a significant stimulating effect from the use of biostimulators and foliar fertilizers on dehydrogenase activity was observed. High dehydrogenase activity was observed during the full growing season of the pea crop (onset of flowering), and may be associated with increased secretion by the root system during this period, which subsequently leads to increased numbers of microorganisms. Similar results were obtained by Siwik-Ziomek and Szczepanek [72], who analyzed the impact of inorganic fertilization (NPK and S) and the Kelpak biostimulator on dehydrogenase activity in the soil during the cultivation of winter rapeseed. In their study, they observed an increase in dehydrogenase and catalase activity in the period from flowering to ripening. In addition, they showed that the use of the Kelpak biostimulator and sulphur application resulted in a significant stimulating modification of catalase and dehydrogenase activity. Moreover, Brzezińska et al. [68] noted that dehydrogenase activity is associated with the activity of other soil enzymes, for example catalase and β-glucosidase, as well as with the presence of nitrogen. Prashantha [73] indicated that, among other elements, boron increases the activity of dehydrogenases. In contrast, Niewiadomska et al. [29] reported the positive effect of molybdenum on the production of these enzymes by white lupin plant root nodules, and discussed the possibility of a positive correlation between titanium concentration and soil biochemistry.

Phosphatases are also important enzymes for the soil environment. Many authors report that phosphatases are good indicators of the potential for organic phosphorus mineralization and soil biological activity [74,75], and these enzymes are characterized by high susceptibility to changing soil conditions [76]. The main sources of phosphatase in the soil environment are mainly soil microorganisms, plant roots and soil fauna.

Aon and Colaneri [77] observed correlations between organic matter content and acid/alkaline phosphatase activity. Margalef et al. [78] noted that phosphatase activity is higher in soils with a low phosphorus content. A lack of phosphorus in our fertilization treatments increased acid phosphatase activity. A study conducted by Niewiadomska et al. [79], which assessed the effect of fertilizer application (PRP SOL) containing phosphorus, potassium, zinc, boron and molybdenum in the cultivation of yellow lupin, showed that the decrease in acid phosphatase activity was caused by the activation of compounds inaccessible to the plants. Fukuda et al. [80] reported that plants typically produce acid phosphatases when the amount of available phosphorus in the soil is low.

In our study, the largest significant decrease in the level of acid phosphatase activity was observed after application of the Bolero Mo biostimulator in 2016 and 2017, and by the Adob Zn IDHA foliar fertilizer in 2017. The results observed for most of the other experimental treatments at the beginning of the flowering phase were also promising in terms of phosphorus availability due to the action of fertilizers/biostimulators used, as evidenced by the reduced level of the enzyme activity in the soil compared to the control treatment.

In our study, different levels of alkaline phosphatase activity were observed after the application of foliar fertilizers and biostimulators. The increase in the activity of this enzyme could have been the result of the increased activity of soil microorganisms on organic phosphorus compounds (e.g., phytins) secreted into the soil by the pea plants. Phosphatase activity in the soil depends on available phosphorus content, which suggests that alkaline and acid phosphatase participate in the regulation of the nutrient economy [81].

The results observed in this study in regard to catalase activity show the significant stimulating effect of the treatments. Catalase is a well-documented enzyme in scientific literature and has the best-known chemical structure. It is an enzyme from the group of oxidoreductases (E.C.1.11.1.6), found in microorganisms, plants and animals. In the soil environment, it is present in the cells of all microorganisms that use oxygen for respiratory processes (aerobes, facultative anaerobes). Obligatory anaerobes show very little or no catalase activity. It is considered one of the main enzymes with antioxidant activity and works mainly to remove excess hydrogen peroxide (H_2O_2) by converting water and oxygen in all aerobic organisms [82]. Brzezińska et al. [72] showed a significant relationship between catalase activity and the oxygenation of soils. The stimulating effect of biostimulators on the activity of this enzyme was demonstrated by Niewiadomska et al. [29] in a study conducted during the cultivation of soybean and white lupin, while Stępniewska et al. [83] showed significant positive correlations between soil catalase activity and organic matter content, biomass, oxygen absorption, carbon dioxide secretion, as well as dehydrogenase, glucosidase, amidase, and phosphodiesterase activity.

As the results of our study on the activity of all enzymes tested in the experiment show, it is noteworthy that in all years of the experiment the highest dehydrogenase activity was observed during the flowering period of the plants, regardless of the biostimulators/fertilizers used. There were similar dependencies observed for the acid and alkaline phosphatases and catalase in 2016 and 2018. There were no such dependencies in 2017, when the highest phosphatase activity was noted during the period of full growth of the plants, i.e., at phases BBCH 35–40, whereas the highest catalase activity was observed after the plants had been harvested.

It is very likely that the significant increase in the activity of the enzymes tested in the experiment during the flowering of pea plants was caused by secretions of the root system. The substances contained in root secretions and in the dying cells of root tissues are rich sources of nutrients and energy for various groups of microorganisms. Hupe et al. [84] proved that the phase of pea development had significant influence on the dynamics of nutrients in the plant's root zone, and consequently, on the soil enzyme activity. The researchers observed strong rhizodeposition of carbon and nitrogen in the period ranging from the emergence of plants to their flowering. They noted significant inhibition of nitrogen rhizodeposition after the flowering period and attributed this effect to the displacement of nitrogen in the plants. The reduced amount of organic nitrogen substances in relation to carbon in the rhizosphere after the flowering of the plants explains the reduced metabolic level of the enzymes analyzed in the experiment.

5. Conclusions

The results of the study showed that the effect of biostimulators and foliar fertilizers was closely related to weather conditions, and so it was not possible to clearly indicate whether there was a positive effect from the treatments on pea seed yield in each year. Adob Zn IDHA was the only fertilizer that stimulated yields, especially under average moisture conditions and less so in drought conditions in two of three years of the study. Among tested biostimulants and fertilizers, a repeatable in years increase of the yield in both, dry and wet years, was obtained after the application of Adob Zn IDHA foliar fertilizer. In addition, this fertilizer stimulated vegetative development of the plants, i.e., plant height in the average year and plant dry mass in the wetter year.

Similarly, in regard to the enzymatic parameters tested, it cannot be clearly determined which biostimulator treatments was best for pea cultivation and improved soil biochemical activity. A significant relationship between the effect of applied biostimulators and the development phase of the plant, as well as the year of the study was indicated. Depending on the year, the positive effect of Bolero Mo application on dehydrogenase, acid phosphatase and catalase activity, and Adob B application on alkaline phosphatase activity was noted. The BNF level was best influenced by the Titanit biostimulator, but it should be noted that all the treatments used in this study were found to stimulate this parameter.

Supplementary Materials: The following are available online at http://www.mdpi.com/2073-4395/10/10/1558/s1, Table S1: Effects of experimental treatments on DHA; three-way ANOVA results, Table S2: Effects of experimental treatments on PAC; three-way ANOVA results, Table S3: Effects of experimental treatments on PAL; three-way ANOVA results, Table S4: Effects of experimental treatments on CAT; three-way ANOVA results, Table S5: Effects of experimental treatments on BIF; three-way ANOVA results, Table S6: Effects of experimental treatments on Y; two-way ANOVA results, Table S7: Effects of experimental treatments on M; two-way ANOVA results, Table S8: Effects of experimental treatments on TSW; two-way ANOVA results, Table S9: Effects of experimental treatments on SSPW; ANOVA results, Table S10: Effects of experimental treatments on H; two-way ANOVA results, Table S11: Effects of experimental treatments on SW/P; two-way ANOVA results, Table S12: Effects of experimental treatments on CCI; two-way ANOVA results, Table S13: Effects of experimental treatments on F_v/F_m; two-way ANOVA results, Table S14: Effects of experimental treatments on PDM; two-way ANOVA results, Table S15: Effects of experimental treatments on LAI; two-way ANOVA results, Table S16: Effects of experimental treatments on BNF; two-way ANOVA results.

Author Contributions: H.S., A.N., K.R., K.P., A.F. and A.W.-M. conceived and designed the experiments; H.S., A.N., K.R., K.P., A.F., A.W.-M., and L.D. performed the field experiments and analyzed the data; A.B. carried out the statistical analysis; H.S., A.N., wrote the paper; H.S., A.N., K.R. revised the manuscript. All authors have read and agreed to the published version of the manuscript.

Funding: This research was funded by Polish Ministry of Agriculture and Rural Development project number HOR 3.3/2011–2015 And The APC was funded by Ministry of Science and Higher Education programme Project No. 005/RID/2018/19.

Acknowledgments: This study was made possible by a grant from the Polish Ministry of Agriculture and Rural Development, Project: Improving domestic sources of plant protein, their production, trading and use in animal feed, project No. HOR 3.3/2011–2015 and within the framework of Ministry of Science and Higher Education programme as 'Regional Initiative Excellence' in years 2019–2022, Project No. 005/RID/2018/19.

Conflicts of Interest: The authors declare no conflict of interest.

References

1. Wang, N.; Daun, J.K. Effect of variety and crude protein content on nutrients and certain antinutrients in field peas (*Pisum sativum*). *J. Sci. Food Agric.* **2004**, *84*, 1021–1029. [CrossRef]
2. N'Dayegamiye, A.; Whalen, J.K.; Tremblay, G.; Nyiraneza, J.; Grenier, M.; Drapeau, A.; Bipfubusa, M. The benefits of legume crops on corn and wheat yield, nitrogen nutrition, and soil properties improvement. *Agron. J.* **2015**, *107*, 1653–1665. [CrossRef]
3. Michalk, K. The legal framework for short rotation coppice in Germany in the context of the 'greening'of the EU's common agricultural policy. In *Bioenergy from Dendromass for the Sustainable Development of Rural Areas;* Wiley-VCH: Hoboken, NJ, USA, 2015; pp. 367–374.
4. Hillocks, R.J. Farming with fewer pesticides: EU pesticide review and resulting challenges for UK agriculture. *Crop Prot.* **2012**, *31*, 85–93. [CrossRef]
5. Vermeulen, S.J.; Aggarwal, P.K.; Ainslie, A.; Angelone, C.; Campbell, B.M.; Challinor, A.J.; Lau, C. Options for support to agriculture and food security under climate change. *Environ. Sci. Policy* **2012**, *15*, 136–144. [CrossRef]
6. Parry, M.A.J.; Flexas, J.; Medrano, H. Prospects for crop production under drought: Research priorities and future directions. *Ann. App. Biol.* **2005**, *147*, 211–226. [CrossRef]
7. Drobek, M.; Frąc, M.; Cybulska, J. Plant Biostimulants: Importance of the Quality and Yield of Horticultural Crops and the Improvement of Plant Tolerance to Abiotic Stress—A Review. *Agronomy* **2019**, *9*, 335. [CrossRef]
8. Du Jardin, P. Plant biostimulants: Definition, concept, main categories and regulation. *Sci. Hortic.* **2015**, *196*, 3–14. [CrossRef]
9. Colla, G.; Rouphael, Y.; Canaguier, R.; Svecova, E.; Cardarelli, M. Biostimulant action of a plant-derived protein hydrolysate produced through enzymatic hydrolysis. *Front. Plant Sci.* **2014**, *5*, 448. [CrossRef]
10. Yakhin, O.I.; Lubyanov, A.A.; Yakhin, I.A.; Brown, P.H. Biostimulants in plant science: A global perspective. *Front. Plant Sci.* **2017**, *26*, 2049. [CrossRef]
11. Kunicki, E.; Grabowska, A.; Sekara, A.; Wojciechowska, R. The effect of cultivar type, time of cultivation, and biostimulant treatment on the yield of spinach (*Spinacia oleracea* L.). *Folia Hortic. Ann.* **2010**, *22*, 9–13. [CrossRef]
12. Ferreira, M.I.; Lourens, A.F. The efficacy of liquid seaweed extract on the field canola plants. *S. Afr. J. Plant Soil* **2002**, *19*, 159–161. [CrossRef]
13. Sultana, V.; Ehteshamul-Haque, S.; Ara, J.; Athar, M. Comparative efficacy of Brown, Green and red Seaweeds in the control of Root infesting fungi and okra. *Int. J. Environ. Sci. Technol.* **2005**, *21*, 75–81.
14. Piotrowska-Dlugosz, A.; Charzynski, P. The impact of the soil sealing degree on microbial biomass, enzymatic activity, and physicochemical properties in the ekranic technosols of Toruń (Poland). *J. Soil Sediment.* **2015**, *15*, 47–59. [CrossRef]
15. Ahemad, M.; Khan, S.M. Functional aspects of plant growth promoting *Rhizobacteria*: Recent Advancements. *Insight Microbiol.* **2011**, *1*, 39–54. [CrossRef]
16. Sulewska, H.; Ratajczak, K.; Panasiewicz, K.; Kalaji, H.M. Can pyraclostrobin and epoxiconazole protect conventional and stay-green maize varieties grown under drought stress? *PLoS ONE* **2019**, *14*, e0221116. [CrossRef]
17. Strażyński, P.; Mrówczyński, M. (Eds.) *Metodyka Integrowanej Ochrony i Produkcji Grochu dla Doradców;* Instytut Ochrony Roślin-Państwowy Instytut Badawczy: Poznań, Poland, 2016.
18. IUSS Working Group WRB. *World Reference Base for Soil Resources 2006, First Update 2007;* World Soil Resources Reports No. 103; FAO: Rome, Italy, 2007.
19. Molga, M. *Agrometeorology;* PWRiL: Warsaw, Poland, 1986. (In Polish)

20. Sawicka, A. The ecological aspects of dinitrogen fixation. In *Annals of the University of Agriculture in Poznan; Dissertations Scientific*: Poznań, Poland, 1983; Volume 134.
21. Thalmann, A. Zur Methodik der Bestimmung der Dehydrogenase aktivität im Boden mittels triphenytetrazoliumchlorid (TTC). *Landwirtsch Forsch* **1968**, *21*, 249–258.
22. Tabatabai, M.A.; Bremner, J.M. Use of p-nitrophenyl phosphate for assay of soil phosphatase activity. *Soil Biol. Biochem.* **1969**, *1*, 301–307. [CrossRef]
23. Johnson, J.L.; Temple, K.L. Some Variables Affecting the Measurement of "Catalase Activity" in Soil 1. *Soil Sci. Soc. Am.* **1964**, *28*, 207–209. [CrossRef]
24. Stefanic, F.; Ellade, G.; Chirnageanu, J. Researches concerning a biological index of soil fertility. In Proceedings of the 5th Symposium of Soil Biology, Bucharest, Romania, 28 June–3 July 1981; Nemes, M.P., Kiss, S., Papacostea, P., Stefanic, C., Rusan, M., Eds.; Romanian National Society of Soil Science: Bucharest, Romania, 1984; pp. 35–45.
25. COBORU. Available online: http://www.coboru.pl/PlikiWynikow/14_2016_WPDO_12_GRS.pdf (accessed on 27 May 2020).
26. Stevenson, F.J.; Cole, M.A. *Cycles of Soil: Carbon, Nitrogen, Phosphorus, Sulfur, Micronutrients*, 2nd ed.; John Wiley & Sons: Hoboken, NJ, USA, 1999.
27. Fageria, N.K.; Baligar, V.C. Growth components and zinc recovery efficiency of upland rice genotypes. *Pesq. Agropec. Bras.* **2005**, *40*, 1211–1215. [CrossRef]
28. Duffy, B. Zinc and plant disease. In *Mineral Nutrition and Plant Disease*; Datnoff, L.E., Elmer, W.H., Huber, D.M., Eds.; APS Press: St. Paul, MN, USA, 2007; pp. 155–175.
29. Niewiadomska, A.; Sulewska, H.; Wolna-Maruwka, A.; Ratajczak, K.; Waraczewska, Z.; Budka, A. The Influence of Bio-Stimulants and Foliar Fertilizers on Yield, Plant Features, and the Level of Soil Biochemical Activity in White Lupine (*Lupinus albus* L.) Cultivation. *Agronomy* **2020**, *10*, 150. [CrossRef]
30. Raj, A.B.; Raj, S.K. Zinc and boron nutrition in pulses: A review. *J. App. Nat. Sci.* **2019**, *11*, 673–679. [CrossRef]
31. Sulewska, H.; Ratajczak, K. Chemical composition of selected preparations supporting plant development and evaluation of their activity in soybean cultivation. *Chem. Indus.* **2017**, *96*, 1352–1355.
32. Kuniya, N.; Chaudhary, N.; Patel, S. Effect of sulphur and zinc application on growth, yield attributes, yield and quality of summer cluster bean [*Cyamopsis tetragonoloba* (L.)] in light textured soil. *IJCS* **2018**, *6*, 1529–1532.
33. Hassanein, M.S.; Shalaby, M.A.F.; Rashad, E.M. Improving growth and yield of some faba bean cultivars by using some plant growth promoters in newly cultivated land. *Ann. Agric. Sci. Moshtohor.* **2000**, *38*, 2141–2155.
34. Tobbal, Y.F.M. Physiological Studies on the Effect of Some Nutrients and Growth Regulators on Plant Growth and Metabolism. Ph.D. Thesis, Faculty of Science Al-Azhar University, Cairo, Egypt, 2006.
35. El-Sallami, I.H.; Gad, M.M. Growth and flowering responses of New York aster (*Aster novibelgii* L.) to a slow release fertilizer and foliar applied zinc. *Assuit J. Agric. Sci.* **2005**, *36*, 121–136.
36. Mostafavi, K. Grain yield and yield components of soybean upon application of different micronutrient foliar fertilizers at different growth stages. *Int. J. Agric. Res. Rev.* **2012**, *2*, 389–394.
37. Gomaa, A.O. Studies on the response of *Matthiola incana* plants to some growth conditions. 2-In Field: Effect of foliar spray with zinc sulphate, calcium and paclobutrazol on growth and flowering. *Egypt. J. Appl. Sci.* **2003**, *18*, 291–318.
38. Hansch, R.; Mendel, R.R. Physiological functions of mineral micronutrients (Cu, Zn, Mn, Fe, Ni, Mo, B, Cl). *Curr. Opin. Plant. Biol.* **2009**, *12*, 259–266. [CrossRef] [PubMed]
39. Brkić, S.; Milaković, Z.; Kristek, A.; Antunović, M. Pea yield and its quality depending on inoculation, nitrogen and molybdenum fertilization. *Plant Soil Environ.* **2004**, *50*, 39–45. [CrossRef]
40. Omer, F.A.; Abbas, D.N.; Khalaf, A.S. Effect of molybdenum and potassium application on nodulation, growth and yield of lentil (*Lens culinaris* MEDIC). *Pak. J. Bot.* **2016**, *48*, 2255–2259.
41. Madny, A.E.M. Response of Some Field Crops Grown under Newly Reclaimed Soil Conditions to Boron Fertilization. Ph.D. Thesis, Faculty of Science Al-Azhar University, Cairo, Egypt, 2004.
42. Shaaban, M.M.; Abdalla, F.E.; Abou El-Nour, E.A.A.; El-Saady, A.M. Boron nitrogen interaction effect on growth and yield of faba bean plants grown under sandy soil conditions. *Int. J. Agric. Res.* **2006**, *1*, 322–330.

43. Sulewska, H.; Ratajczak, K.; Niewiadomska, A.; Koziara, W.; Panasiewicz, K.; Faligowska, A. Preparaty zawierające tytan, krzem, bor, cynk i molibden w uprawie łubinu białego i grochu siewnego. *Przem chem.* **2018**, *97*, 1182–1185. [CrossRef]
44. Kumar, R.; Kumar, S.S.; Pandey, A.C. Effect of seed soaking in nitrogen, phosphorus, potassium and boron on growth yield of garden pea (*Pisum sativum* L.). *Ecol. Environ. Conserv.* **2006**, *14*, 383–386.
45. Moghazy, A.M.; Saed, S.M.E.; Awad, E.M.S. The influence of boron foliar spraying with compost and mineral fertilizers on growth, green pods and seed yield of pea. *Nat. Sci.* **2014**, *12*, 50–57.
46. Fageria, N.K.; Baligar, V.C.; Zobel, R.W. Yield, nutrient uptake and soil chemical properties as influenced by liming and boron application in common bean in a No–Tillage system. *Commun. Soil Sci. Plan.* **2007**, *38*, 1637–1653. [CrossRef]
47. Subasinghe, S.; Dayatilake, G.A.; Senaratne, R. Effect of B, Co and Mo on nodulation, growth and yield of cowpea (*Vigna unguiculata*). *Trop. Agric. Res. Ext.* **2003**, *6*, 108–112.
48. Bonilla, I.; Perez, H.; Cassab, G.; Lara, M.; Sanchez, F. The effect of boron deficiency on development in determinate nodules: Changes in cell wall pectin contents and nodule polypeptide expression. In *Boron in Soils and Plants. Developments in Plant and Soil Sciences*; Bell, R.W., Rerkasem, B., Eds.; Springer: Dordrecht, The Netherlands, 1997; Volume 76.
49. Fageria, N.K.; Baligar, V.C.; Clark, R.B. Micronutrients in crop production. *Adv. Agron.* **2002**, *77*, 185–268.
50. Ross, J.R.; Slaton, N.A.; Brye, K.R.; Delong, R.E. Boron fertilization influences on soybean yield and leaf and seed born concentration. *Agron. J.* **2006**, *98*, 198–205. [CrossRef]
51. Malinowska, E.; Kalembasa, S. The yield and content of Ti, Fe, Mn, Cu in celery leaves (*Apium graveolens* L. var. dulce mill. pers.) as a result of Tytanit application. *Acta Sci. Pol. Hortorum Cultus* **2012**, *11*, 69–80.
52. Grenda, A. Tytanit—An activator of metabolic processes. *Chem. Sustain. Agric.* **2003**, *4*, 263–269.
53. Sacała, E. Role of silicon in plant resistance to water stress. *J. Elementol.* **2009**, *14*, 619–630. [CrossRef]
54. Artyszak, A.; Gozdowski, D.; Kuchcińska, K. The effect of silicon foliar fertilization in sugar beet-*Beta vulgaris* (L.) ssp. vulgaris conv. crassa (Alef.) prov. altissima (Döll). *Turk. J. Field Crops* **2015**, *20*, 115–119. [CrossRef]
55. Liang, Y.; Si, J.; Römheld, V. Silicon uptake and transport is an active process in *Cucumis sativus*. *New Phytol.* **2005**, *167*, 797–804. [CrossRef] [PubMed]
56. Kowalczyk, K.; Zielony, T. Effect of Goteo treatment on yield and fruit quality of tomato grown on rock wool. In *Monographs Series: Biostimulators in Modern Agriculture: Solanaceous Crops*; Dąbrowski, Z.T., Ed.; Editorial House Wieś Jutra: Warsaw, Poland, 2008; pp. 21–26.
57. Gajc-Wolska, J. The Influence of Grafting and Biostimulators on the Yield and Fruit Quality of Greenhouse Tomato CV. (*Lycopersicon esculentum* Mill.) Grown in the Field. *Veg. Crops Res. Bull.* **2010**, *72*, 63–70. [CrossRef]
58. Gajewski, M.; Gos, K.; Bobruk, J. The influence of Goëmar Goteo biostimulator on yield and quality of two Chinese cabbage cultivars. In *Monographs Series: Biostimulators in Modern Agriculture: Vegetable Crops*; Dąbrowski, Z.T., Ed.; Editorial House Wieś Jutra: Warsaw, Poland, 2008; pp. 21–27.
59. Zboińska, M. W jaki sposób rośliny pobierają i asymilują azot? *Eduk. Biol. Środowiskowa* **2018**, *2*, 19–31.
60. Weisany, W.; Raei, Y.; Allahverdipoor, K.H. Role of some of mineral nutrients in biological nitrogen fixation. *Bull. Environ. Pharmacol. Life Sci.* **2013**, *2*, 77–84.
61. Sládková, A.; Száková, J.; Havelcová, M.; Najmanová, J.; Tlustoš, P. The contents of selected risk elements and organic pollutants in soil and vegetation within a former military area. *Soil Sediment. Contam. Int. J.* **2015**, *24*, 325–342. [CrossRef]
62. Kováčik, P.; Havrlentová, M.; Šimanský, V. Growth and Yield Stimulation of Winter Oilseed Rape (*Brasssica Napus*, L.) by Mg-Titanit Fertiliser. *Agriculture* **2014**, *1*, 132–141.
63. Rousk, K.; Sorensen, P.L.; Michelsen, A. What drives biological nitrogen fixation in high arctic tundra: Moisture or temperature? *Ecosphere* **2018**, *9*, e02117. [CrossRef]
64. Niewiadomska, A.; Majchrzak, L.; Borowiak, K.; Wolna-Maruwka, A.; Waraczewska, Z.; Budka, A.; Gaj, R. The Influence of Tillage and Cover Cropping on Soil Microbial Parameters and Spring Wheat Physiology. *Agronomy* **2020**, *10*, 200. [CrossRef]
65. Moeskops, B.; Buchan, D.; Sleutel, S.; Herawaty, L.; Husen, E.; Saraswati, R.; Setyorini, D.; De Neve, S. Soil microbial communities and activities under intensive organic and conventional vegetable farming in West Java, Indonesia. *Appl. Soil Ecol.* **2010**, *45*, 112–120. [CrossRef]

66. Wolińska, A.; Stępniewska, Z. Dehydrogenase Activity in the Soil Environment. In *Dehydrogenases*; Canuto, R.A., Ed.; Intech: Rijeka, Croatia, 2020; Chapter 8; pp. 183–210.
67. Januszek, K.; Błońska, E.; Długa, J.; Socha, J. Dehydrogenase activity of forest soils depends on the assay used. *Int. Agrophys.* **2014**, *29*, 47–59. [CrossRef]
68. Brzezińska, M.; Włodarczyk, T.; Stępniewski, W.; Przywara, G. Soil oxygen status and catalase activity. *Acta Agrophys.* **2005**, *5*, 555–565.
69. Bastida, F.; Kandeler, E.; Moreno, J.L.; Ros, M.; García, C.; Hernández, T. Application of Fresh and Composted Organic Wastes Modifies Structure, Size and Activity of Soil Microbial Community under Semiarid Climate. *Appl. Soil Ecol.* **2008**, *40*, 318–329. [CrossRef]
70. Salazar, S.; Sanchez, L.; Alvarez, J.; Valverde, A.; Galindo, P.; Igual, J.; Peix, A.; Santa-Regina, I. Correlation Among Soil Enzyme Activities Under Different Forest System Management Practices. *Ecol. Eng.* **2011**, *37*, 1123–1131. [CrossRef]
71. Gałązka, A.; Gawryjołek, K.; Perzyński, A.; Gałązka, R.; Księżak, J. Changes in Enzymatic Activities and Microbial Communities in Soil under Long-Term Maize Monoculture and Crop Rotation. *Pol. J. Environ. Stud.* **2017**, *26*, 39–46. [CrossRef]
72. Siwik-Ziomek, A.; Szczepanek, M. Soil Extracellular Enzyme Activities and Uptake of N by Oilseed Rape Depending on Fertilization and Seaweed Biostimulant Application. *Agronomy* **2019**, *9*, 480. [CrossRef]
73. Prashantha, G.M.; Prakash, S.S.; Umesha, S.; Chikkaramappa, T.; Subbarayappa, C.T.; Ramamurthy, V. Direct and residual effect of zinc and boron on soil enzyme activities at harvest in finger millet-groundnut cropping system. *J. Pharm. Phytochem.* **2019**, *8*, 2447–2451.
74. Dick, W.A.; Cheng, L.; Wang, P. Soil acid and alkaline phosphatase activity as pH adjustment indicators. *Soil Biol. Biochem.* **2000**, *32*, 1915–1919. [CrossRef]
75. Schneider, K.; Turrion, M.B.; Grierson, P.F.; Gallardo, J.F. Phosphatase activity, microbial phosphorus, and fine root growth in forest soils in the Sierra de Gata, western Spain. *Biol. Fert. Soils* **2001**, *34*, 151–155. [CrossRef]
76. Bielińska, E.; Mocek-Płóciniak, A. Impact of the tillage system on the soil enzymatic activity. *Arch. Environ. Prot.* **2012**, *38*, 75–82. [CrossRef]
77. Aon, M.A.; Colaneri, A.C. Temporal and spatial evolution of enzymatic activities and physical-chemical properties in an agricultural soil. *Appl. Soil Ecol.* **2001**, *18*, 155–270.
78. Margalef, O.; Sardans, J.; Fernández-Martínez, M.; Molowny-Horas, R.; Janssens, I.A.; Ciais, P.; Goll, D.; Richter, A.; Obersteiner, M.; Asensio, D.; et al. Global patterns of phosphatase activity in natural soils. *Sci. Rep.* **2017**, *2*, 1337. [CrossRef] [PubMed]
79. Niewiadomska, A.; Sulewska, H.; Wolna-Maruwka, A.; Ratajczak, K.; Głuchowska, K.; Waraczewska, Z.; Budka, A. An Assessment of the Influence of Co-Inoculation with Endophytic Bacteria and Rhizobia, and the Influence of PRP SOL and PRP EBV Fertilisers on the Microbial Parameters of Soil and Nitrogenase Activity in Yellow Lupine (*Lupinus luteus* L.) Cultivation. *Pol. J. Environ. Stud.* **2018**, *6*, 2687–2702. [CrossRef]
80. Fukuda, T.; Osaki, M.; Shinano, T.; Wasaki, J. *Cloning and Characterization of Two Secreted Acid Phosphatases from Rice Call, Plant Nutrition: Food Security and Sustainability of Agro-Ecosystems through Basic and Applied Research*; Kluwer Academic Publisher: New York, NY, USA, 2001; pp. 34–35.
81. Adetunji, A.T.; Lewu, F.B.; Mulidzi, R.; Ncube, B. The biological activities of β-glucosidase, phosphatase and urease as soil quality indicators: A review. *J. Soil Sci. Plant. Nutr.* **2017**, *17*, 794–807. [CrossRef]
82. Chelikani, P.; Ramana, T.; Radhakrishnan, T.M. Catalase: A repertoire of unusual features. *Indian J. Clin. Biochem.* **2005**, *20*, 131–135. [CrossRef] [PubMed]
83. Stępniewska, Z.; Wolińska, A.; Ziomek, J. Response of soil catalase activity to chromium contamination. *J. Environ. Sci.* **2009**, *21*, 1142–1147. [CrossRef]
84. Hupe, A.; Schulz, H.; Bruns, C.; Haase, T.; Heß, J.; Joergensen, R.G.; Wichern, F. Even flow? Changes of carbon and nitrogen release from pea roots over time. *Plant Soil* **2018**, *431*, 143–157. [CrossRef]

Article

Evaluation of Nitrogen Fertilization Systems Based on the in-Season Variability in the Nitrogenous Growth Factor and Soil Fertility Factors—A Case of Winter Oilseed Rape (*Brassica napus* L.)

Witold Grzebisz *, Remigiusz Łukowiak and Karol Kotnis

Department of Agricultural Chemistry and Environmental Biogeochemistry, Poznan University of Life Sciences, Wojska Polskiego 28, 60-637 Poznan, Poland; remigiusz.lukowiak@up.poznan.pl (R.Ł.); karol88@poczta.onet.pl (K.K.)

* Correspondence: witold.grzebisz@up.poznan.pl; Tel.: +48-618-487-788

Received: 8 September 2020; Accepted: 30 October 2020; Published: 3 November 2020

Abstract: Application of nitrogen (N) in contrastive chemical form changes availability of soil nutrients, affecting crop response. This hypothesis was evaluated based on field experiments conducted in 2015/16 and 2016/2017. The experiment consisted of three nitrogen fertilization systems: mineral-ammonium nitrate (AN) (M-NFS), organic-digestate (O-NFS), $^2/_3$ digestate + $^1/_3$ AN (OM-NFS), and N rates: 0, 80, 120, 160; 240 kg ha^{-1}. The content of nitrogen nitrate (N-NO_3) and available phosphorus (P), potassium (K), magnesium (Mg) and calcium (Ca) were determined at rosette, onset of flowering, and maturity of winter oilseed rape (WOSR) growth from three soil layers: 0.0–0.3, 0.3–0.6, 0.6–0.9 m. The optimum N rates were: 139, 171 and 210 kg ha^{-1} for the maximum yield of 3.616, 3.887, 4.195 t ha^{-1}, for M-NFS, O-NFS, OM-NFS. The N-NO_3 content at rosette of 150 kg ha^{-1} and its decrease to 48 kg ha^{-1} at the onset of flowering was the prerequisite of high yield. The key factor limiting yield in the M-NFS was the shortage of Ca, Mg, O-NFS—shortage of N-NO_3. Plants in the OM-NFS were well-balanced due to a positive impact of the subsoil Mg and Ca on the N-NO_3 content and productivity. The rosette stage was revealed as the cardinal for the correction of WOSR N nutritional status.

Keywords: soil; nitrate nitrogen content; contents of available phosphorus; potassium; magnesium; calcium; cardinal stages of WOSR growth; PCA

---

1. Introduction

In current agriculture, the amount of food production depends on the use of N fertilizers, both mineral and organic [1]. In modern agriculture, which is dominated by varieties of a huge yield potential, resulting from a high rate of biomass growth, the requirement for N is high [2]. This requirement can be fulfilled by taking into account the required rate of a crop plant growth and the uptake of the adequate amount of N, and synchronization of its supply with a plant requirement. N is taken up by plants in two chemical forms, i.e., as nitrogen nitrate (N-NO_3, NN), and ammonium (N-NH_4) [3,4]. This N form, as opposed to ammonium, does not undergo fixation by soil colloids therefore being easily available to a crop plant within a broad range of soil pH [3]. Nitrate N, in spite of a higher metabolic cost of assimilation, results in the production of carbohydrates, consequently leading to a higher rate of plant growth with respect to ammonium [5,6]. In the light of present knowledge, NN, in fact its soil resources, can be considered as the soil nitrogenous growth factor (NGF). This hypothesis is also supported by thousands of scientific papers in the area of agriculture that point to the strong yield increase in grown crops in response to the application of fertilizer N. The classic example is winter oilseed rape (WOSR) [7–11].

A plant requires about 16 (19) nutrients other than N to cover its life cycle. All these nutrients are responsible for N efficiency in metabolic and physiological processes in a crop plant, affecting both the rate of its growth, and final yield [5,12]. The primary source of these nutrients is the soil solution, being enriched by compounds incorporated into the soil through the application of organic or mineral fertilizers [13,14]. The rate of uptake of these nutrients by the growing plant, including ammonium, is significantly distinct from nitrate ions, being as a rule much slower [15]. Most of this set of nutrients, but especially phosphorus (P), potassium (K), magnesium (Mg), and calcium (Ca), undergo numerous agrochemical processes, temporally changing their concentration in the soil solution and consequently reducing/increasing their available pools to crop plants [16,17]. The shortage of a particular nutrient leads, as a rule, to an inefficient use of N, irrespective of its source (soil, manure, mineral fertilizers) [8,14,18]. It is well-recognized by farmers that the efficient use of N requires its balancing through an adequate supply of other nutrients. Therefore, the production potential of a given soil, including both the current content of available nutrients, and soil pH is, in general, recognized as soil fertility, and can be defined as the soil fertility factor (FF).

One of the most important characteristics of any crop plant is the concentration and consequently the amount of any given nutrient accumulated by the crop plant during the growing season, which affects the rate of biomass growth and expression of yield components [2,19]. The total vegetative season of a seed crop growth, based on the mode of development of its particular organs, can be divided into three major periods [20]. The period of crop foundation (PCF), extending from seed emergence to the onset of stem elongation (BBCH 30, STME, phenological stages of crop plants growth as proposed by Mayer [21] and Böttcher et al. [22], respectively), covers the WOSR growth stages related to root system growth and the rosette build-up. The second major period, termed as the period of yield foundation (PYF), ending at the end of inflorescence emergence (INFE), comprises stages responsible for the build-up of the primary generative yield components. The third major period, termed as the period of yield realization (PYR), extends from the onset of flowering (BBCH 60, flowering (FL)) up to physiological crop maturity (BBCH 89, the end of pod development-maturation (PDV-M phase). The borderline phases define the cardinal stages of WOSR growth. It can therefore be assumed that the plant nutritional status depends on both the status of the NGF and soil fertility factors (FFs) just at a borderline between the major phases, which can be named the cardinal stages. Thus, the status of NGF in a respective cardinal stage should be related to the content of NN, and the FF current availability status of other nutrients present in the soil occupied by WOSR roots. Accumulation of N by WOSR yields a high level (3.75 t ha^{-1}) continuous up to BBCH 79, i.e., to the end of pod growth (BBCH 79) [23]. The soil zone occupied by roots for WOSR is usually related to a soil depth down to 0.9 m [24,25]. So far, the analysis of mineral N has been limited to the onset of crop growth in spring [26,27]. For N budgeting, taking into account NN pressure on environment, its content should be also analyzed after harvest [28,29]. The content of available nutrients, defining the current status of FFs should also be measured in the same soil layer as is practiced for mineral N. To date, the in-season status and variability of both groups of production factors is still a classic *black box*.

The key sources of N in intensive crop production are mineral N fertilizers (N_f) [1,3]. Consumption of N_f has progressively increased during the last hundred years, and it is expected to have increased up to 182×10^6 t by 2050, i.e., 75% more as compared to 2010 [30]. Some of the projected N_f consumption can be potentially substituted by other N sources, mainly organic by-products (wastes) of human activity [31,32]. As a fast-developing source of renewable energy, biogas plants seem to be a great source of N, which could be used in crop production. The by-product of the anaerobic digestion is digestate, which is rich in mineral and organic N compounds. However, its concentration in raw biogas slurry is highly variable both in total content (0.1–0.5% fresh weight) and in ammonium (30 to 70% of total N) [33,34]. Digestate, depending on the substrate, also contains other nutrients (macro, and micronutrients), as well as enzymes and hormones. All these compounds significantly affect the level of soil fertility, including the content of soil available N, and in consequence affect plant growth and yield [35,36]. However, a key question remains respecting the efficiency of biogas N in crop production in comparison with the classical N source, i.e., to N_f. The latest study on maize response to digestate showed its higher efficiency, i.e., a net yield increase as compared to ammonium nitrate [37].

The objective of the study was to discriminate the three nitrogen fertilization systems, resulting from the application of two distinct nitrogen forms applied as mineral and organic, and their mixture on WOSR yield, based on the relation between the nitrogenous growth factor (NGF) and soil fertility factors (FFs) in the cardinal stages of winter oilseed rape growth.

2. Materials and Methods

2.1. Site Description

The primary sources of data are two series of field experiments with winter oilseed rape (*Brassica napus* L.), which were carried out during the 2015/2016 and 2016/2017 seasons in Baniewice (53°05′ N; 14°36′ E), Poland. The field experiment was conducted on soil with texture loamy sand in the top-soil over sandy loam, classified as Albic Luvisol (World Refernce Base -WRB, 2016) [38]. The content of the available nutrients was measured each year just after a fore-crop harvesting from three soil depths of 0.0–0.3, 0.3–0.6, and 0.3–0.9 m. The content of available nutrients was favorable for high-yielding WOSR (Table 1).

The local climate, classified as intermediate between atlantic and continental, is seasonally variable (Table 2). Precipitation during the period extending from January to July amounted to 332 mm in 2016 and to 622 mm in 2017 (237 in July). In June, a critical month with respect to pods and seed growth [39], the amount of rainfall was extremely low. Air temperatures were in both years higher in comparison to the long-term average.

Table 1. Soil characteristics of the experimental plots in the consecutive growing seasons [1], mg kg^{-1} soil.

Year/Soil Layer cm	P	K	Ca	Mg	P	K	Ca	Mg
	2015/2016				2016/2017			
0–30	80 ± 29 [2] H [3]	155 ± 38 H [3]	390 ± 61 L [4]	62 ± 6 H [3]	88 ± 72 H	105 ± 37 M	430 ± 156 VL	72 ± 12 M
30–60	30 ± 12 VL	96 ± 27 M	275 ± 100 VL	59 ± 11 H	53 ± 44 L	80 ± 22 M	369 ± 86 VL	65 ± 16 M
60–90	17 ± 6 VL	71 ± 16 M	309 ± 299 VL	73 ± 20 H	53 ± 34 L	70 ± 23 L	298 ± 44 VL	62 ± 12 H

[1] Mehlich 3 extraction solution [40]; [2] average ± standard deviation; [3] fertility class: VL—very low; L—low; M—medium; H—high; VH—very high [41].

2.2. Experimental Design

The field experiment was arranged as a two-factorial design. The first factor was the nitrogen fertilization system (NFS), related to the type of N fertilizer applied: (i) ammonium nitrate, 34-0-0, (acronym M-NFS), (ii) organic N: digestate (O-NFS), (iii) mixed: $^2/_3$ of digestate + $^1/_3$ of ammonium nitrate (AN) (OM-NFS). The rate of applied N was: 0 (N control), 60, 120, 180, and 240 kg ha^{-1}. Each year, digestate was applied at the end of November, and AN on the 1st of March of the following year. The applied digestate on average contained: DM: 72, N-NH_4: 7.2, P_2O_5: 5.45, K_2O: 2.85, MgO: 1.88, S: 1.99 kg m^{-3}. Winter barley was a fore-crop for WOSR. A hybrid WOSR variety *Impression Cl*, characterized by a high yielding potential for medium fertile soils, was used as a test crop. It was sown at the rate of 3.0 kg ha^{-1} at the end of August and harvested at the end of July from an area of 12 m^2. The seeds were harvested at maturity when the moisture content was 8% dry weight.

Table 2. Main characteristics of meteorological conditions during the winter oilseed rape (WOSR) growing season on the background of the long-term averages [1].

Growing Season	Consecutive Months during the WOSR Growing Season												
	VIII	IX	X	XI	XII	I	II	III	IV	V	VI	VII	Average
	Temperature, °C												
2015/2016	21.1	14.1	8.5	7.1	6.7	−0.9	3.7	4.3	8.8	15.7	18.5	19	10.6
2016/2017	17.8	16.8	8.6	9.3	3	−0.7	1.6	6.7	7.4	14	17.2	17.7	10.0
1981–2010	23.0	13.9	9.4	4.4	1.1	0.2	0.9	4.0	8.6	13.6	16.2	18.6	9.5
	Precipitation, mm												
2015/2016	26.5	40.9	61.0	57.4	45.7	38.6	56.6	33.3	33.5	17.4	73.9	78.5	563.3
2016/2017	69.0	0.0	54.0	49.0	57.8	66.7	58.0	51.2	47.0	56.7	106.0	236.6	852
1981–2010	80.6	62.0	52.2	53.0	50.1	57.5	30.2	35.6	36.9	65.6	57.9	93.8	675.3

[1] meteorological station at Szczecin.

2.3. Chemical Analysis and Indices

Soil samples were collected from each field three times a year, i.e., (i) at the onset of the stem elongation phase of WOSR growth (BBCH 30, STME), (ii) at the onset of flowering (BBCH 60, FL), and (iii) at the crop physiological maturity (BBCH 89). Soil samples at each stage of WOSR growth were taken from three soil depths as follows: 0–0.3, 0.3–0.6, and 0.6–0.9 m.

The soil content of NH_4-N and NO_3-N was determined in field-fresh (not air dried) soil samples within 24 h after sampling. Twenty-gram soil samples were shaken for 1 h with 100 mL of 0.01 M calcium chloride ($CaCl_2$) solution (soil/solution ratio 1:5; *m/v*). Concentrations of NO_3-N were determined by the colorimetric method using flow injection analyses (FIAstar5000, FOSS). The method of NO_3-N concentration analysis consists of two basic steps: reduction from nitrate to nitrite by using a cadmium column, followed by colorimetric determination of nitrite based on the Griess–Ilosvay reaction with N-(1-Naphtyl) ethylene-diamine dichloride as a diazotizing agent. The measurement was performed at a wavelength of 540 nm. The total content of nitrate nitrogen (NN) was expressed in kg ha^{-1}. The N_{min} content for a given soil layer was calculated using indices, which were constructed based on a soil textural class and soil bulk density [42].

Soil pH was measured in a 1 M potassium chloride (KCl) solution (soil/solution ratio 1:2.5; *m/v*). The content of plant available nutrients, including P and K, Mg, and Ca, was determined by the Mehlich 3 method [39]. The P concentration in the extract was determined by the colorimetric method, using the molybdenum blue method. Concentration of K in the collected extracts was determined by flame photometry. Concentration of Mg and Ca were determined by atomic absorption spectrometry (AAS)-flame type.

The partial productivity of NN was calculated based on yield (Y, kg ha^{-1}) and the amount of NN present in a soil layer of 0.0–0.9 m at BBCH 30 (kg ha^{-1}):

$$PFP_{N30} = Y/N_{30}, \text{ kg seeds kg}^{-1} \text{ of N} \quad (1)$$

2.4. Statistical Analyses

The effect of individual research factors (year, N fertilization system, and N rate) for each soil layer and the interaction between them was assessed by two-way ANOVA. Differences between the mean values were compared using the HSD – honestly significant difference test, according to the Tukey method, where the significance level was assessed at 0.05. STATISTICA 13® software was used to conduct all statistical analyses. In the second step of the diagnostic procedure, the relationships between variables representing soil properties were analyzed by principal component analysis (PCA) (StatSoft, Inc., Krakow, Poland 2013). In the third step of the diagnostic procedure, stepwise regression was applied to define an optimal set of variables for a given N characteristic. In the computational procedure, a consecutive variable was removed from the multiple linear regressions in a step-by-step manner. The best regression model was chosen based on the highest F-value for the model and the significance of all independent variables.

3. Results

3.1. In-Season Variability in the Nitrate Nitrogen (N-NO_3, NN) Content

A significant impact of the nitrogen fertilization system (NFS) was recorded in 2017 (Table 3). In this particular year, a significantly higher seed yield was recorded in the organic-mineral (OM-NFS) treatment, in which $^2/_3$ of the total N rate were applied in the form of digestate and the remaining one ($^1/_3$) as ammonium nitrate. In 2016, a year with a water shortage during WOSR flowering, no significant difference between NFSs was observed, but a higher yield was harvested from the OM-NFS (Tables 2 and 3). The effect of the progressively increased N rates was revealed each year. In both years, the effect of the applied N was only significant with respect to the absolute control (AC). In 2016, the relative yield increase on the plot fertilized with N applied at a rate of 60 kg ha^{-1}, compared to AC, was 63%, whereas in 2015 it doubled, reaching 115%.

Table 3. In-season variability of the nitrate nitrogen content variability.

Year	Factor	Level of	Yield	N_{30}	N_{60}	N_{89}	$N_{60\text{-}30}$	PFP_{N30}
(Y)		Factor	t ha^{-1}		kg ha^{-1}			kg Seeds kg^{-1} N
2015/2016	Fertilization	M	3.072	107.8 ^{a}	45.2 ^{a,b}	39.4 ^{a}	−62.6 ^{b}	38.4
	system	O	3.068	77.3 ^{c}	40.6 ^{b}	29.8 ^{c}	−36.8 ^{a}	42.0
	(FS)	OM	3.363	88.7 ^{b}	49.0 ^{a}	34.6 ^{b}	−39.7 ^{a}	45.0
	F-value		0.11	28.4 ***	9.45 ***	13.0 ***	17.1 ***	2.32
	Nitrogen	0	1.937 ^{b}	39.4 ^{c}	33.6 ^{c}	18.5 ^{c}	−5.7 ^{a}	51.2 ^{a}
	rate	60	3.161	54.0 ^{c}	42.3 ^{b}	25.4 ^{c}	−11.7 ^{a}	59.7 ^{a}
	(N)	120	3.438 ^{a}	97.8 ^{b}	53.3 ^{a}	36.8 ^{b}	−44.5 ^{b}	36.6 ^{b}
	kg ha^{-1}	180	3.643 ^{a}	104.2 ^{b}	43.3 ^{b}	38.1 ^{b}	−60.9 ^{b}	35.3 ^{b}
		240	3.660 ^{a}	161.0 ^{a}	52.1 ^{a}	54.1 ^{a}	−108.9 ^{c}	26.3 ^{a}
	F-value		24.0 ***	164 ***	20.8 ***	61.5 ***	89.6 ***	23.1 ***
2016/2017	Fertilization	M	3.062 ^{b}	112.4	35.1 ^{b}	41.8 ^{b}	−77.3	34.3 ^{b}
	system	O	3.332 ^{a,b}	105.0	28.1 ^{a}	35.1 ^{a}	−76.8	37.0 ^{b}
	(FS)	OM	3.357 ^{a}	113.6	30.4 ^{a}	31.2 ^{a}	−83.2	42.8 ^{a}
	F-value		3.91 *	1.12	10.4 ***	19.9 ***	0.58	8.56 ***
	Nitrogen	0	1.634 ^{b}	40.6 ^{d}	13.4 ^{c}	24.9 ^{a}	−27.2 ^{a}	41.4 ^{b}
	rate	60	3.517 ^{a}	63.6 ^{c}	18.3 ^{c}	29.9 ^{a,b}	−45.3 ^{a}	63.0 ^{a}
	(N)	120	3.675 ^{a}	108.1 ^{b}	36.5 ^{b}	41.9 ^{c,d}	−71.6 ^{b}	34.8 ^{b,c}
	kg ha^{-1}	180	3.908 ^{a}	128.1 ^{b}	37.2 ^{c}	35.7 ^{b,c}	−90.9 ^{b}	32.8 ^{c}
		240	3.517 ^{a}	211.3 ^{a}	50.8 ^{a}	47.8 ^{d}	−160.5 ^{c}	18.1 ^{d}
	F-value		73.7 ***	134 ***	115 ***	34.3 ***	73.2 ***	71.7 ***
	Y × FS		1.26	5.63 **	6.20 **	6.20 **	7.34 **	0.31
	Y × N		2.53 *	7.97 ***	18.0 ***	5.62 ***	2.34	2.14
	FS × N		2.09 *	29.3 ***	5.37 ***	18.6 ***	27.8 ***	9.73 ***
	Y × FS × N		0.32	5.56 ***	3.15 **	5.13 ***	6.04 ***	5.87 ***

***, **, * significant at $p < 0.001$; < 0.01; < 0.05, respectively; n.s.—non significant; $^{a, b, c, d}$ significance letters, a—the highest, d—the lowest; ^{a} means within a column followed by the same letter indicate a lack of significant difference between the treatments. N30, N60, N89—nitrate N content at BBCH 30, 60, and 89, kg ha^{-1} respectively; N30–N60—nitrate N balance at BBCH 60 vs. BBCH 30; PFP_{N30}-partial factor productivity of nitrogen nitrate (N-NO_3) at WOSR stage of BBCH 30 (rosette).

Yield, taking into account both experimental factors and year, significantly depended on the interaction of N rates (N) and year (Y × N) and on NFS and N rates (NFS × N). Due to its significance for all the other studied N characteristics, the second interaction was considered as the most representative for this study (Table 3). As shown in Figure 1, an N rate of 60 kg ha^{-1}, irrespective of the NFS, resulted in a significant yield increase. The further yield response to increasing N rates was NFS specific. For the M-NFS, an N rate of 120 and 180 kg ha^{-1} did not result in any yield increase. An N rate of 240 kg ha^{-1} resulted in a yield drop as compared to the N rate of 60 kg ha^{-1}. A similar trend was observed for the O-NFS, but without any drop with respect to the treatment with the N rate of 60 kg ha^{-1}. A significantly different trend was observed for the OM-NFS, in which yield increased significantly in accordance with the progressively applied N rate. The regression models of WOSR yield response to the N_f rate are as follows:

$$\text{Y-M-NFS} = -0.00009\text{N}^2 + 0.025\text{N} + 1.88 \text{ for } n = 5, R^2 = 0.93, \text{P} \leq 0.05 \quad (2)$$

$$\text{Y-O-NFS} = -0.00007\text{N}^2 + 0.024\text{N} + 1.83 \text{ for } n = 5, R^2 = 0.94, \text{P} \leq 0.05 \quad (3)$$

$$\text{Y-OM-NFS} = -0.00005\text{N}^2 + 0.021\text{N} + 1.99 \text{ for } n = 5, R^2 = 0.97, \text{P} \leq 0.05 \quad (4)$$

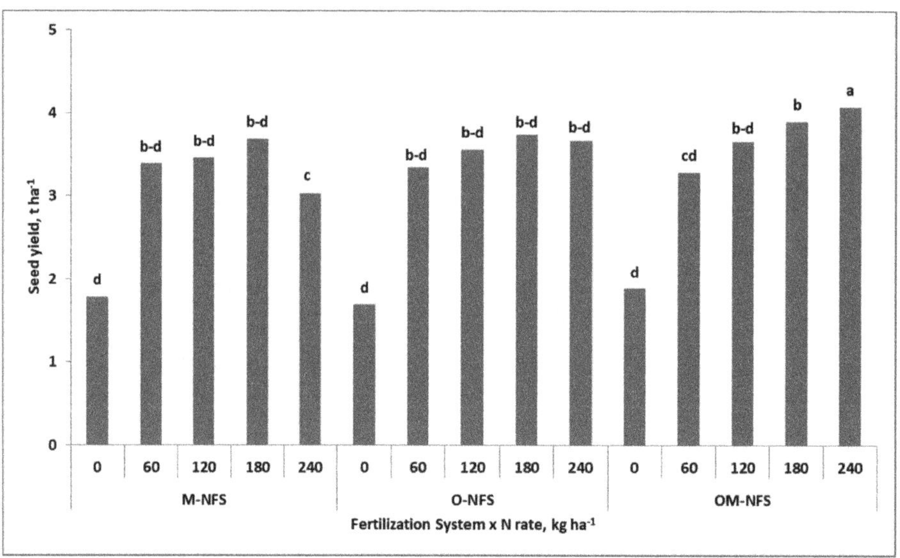

Figure 1. Yield of winter oilseed rape as affected by N rates on the background of nitrogen fertilization systems (NFSs). M-NFS, O-NFS, OM-NFS—mineral, organic, organic-mineral, respectively. [a, b, c, d] significance letters, a—the highest, d—the lowest; [a] means within a column followed by the same letter indicate a lack of significant difference between the treatments.

The key attributes of quadratic models, such as the optimum N rate (N_{op}) and the respective maximum yield (Y_{max}), were used to distinguish the tested NFSs. The lowest N_{op} of 138.9 kg ha^{-1} was the attribute of the M-NFS. The respective Y_{max} amounted to 3.616 t ha^{-1}. The parameters of the O-NFS were 171.4 kg ha^{-1} of applied N, and the Y_{max} of 3.887 t ha^{-1}. The highest values of both parameters were found for the OM-NFS. A N_{op} of 210 kg ha^{-1} resulted in a theoretical Y_{max} of 4.195 t ha^{-1}.

The NN content at the rosette stage of WOSR growth significantly responded to NFSs in 2016 (Table 3). In this particular year, the highest NN content was recorded in the M-NFS, followed by OM-NFS, and O-NFS. In 2017, the NN content in the M-NFS was almost at the same level as in 2016, but no significant differences between NFS treatments were recorded. As shown in Figure 2, the NN content followed different regression models in response to increasing N rates. The cubic regression model fitted the M-NFS and O-NFS data the best. The course of the model for the M-NFS showed a significant response to the N rate of 60 kg ha^{-1}, next stabilizing up to the N rate of 180 kg ha^{-1}. A sudden increase in the NN content was recorded following the N rate of 180 kg ha^{-1}. The course of this model suggests huge NN resources during STME in the plot fertilized with 240 kg ha^{-1} of N. The course of NN on the O-NFS was completely different. Its content increased exponentially up to an N rate of 145.7 kg ha^{-1}, then significantly decreased. This course, in contrast to the M-NFS, indicates a full exploitation of NN resources in the plot fertilized with 240 kg ha^{-1} of N. The NN course for the OM-NFS fitted the quadratic function the best, and was characterized by an exponential increase in the content of NN from the N rate of 6.9 kg ha^{-1}, thus indicating huge NN resources during the STME in plots fertilized with the highest N rates.

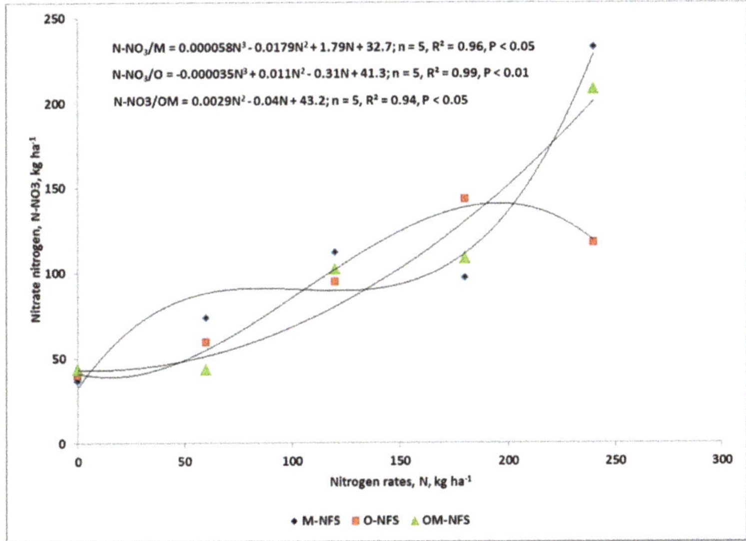

Figure 2. Trends in the nitrate N content at the onset of WOSR stem elongation stage of growth for three nitrogen fertilization systems (NFSs). M-NFS, O-NFS, OM-NFS—mineral, organic, organic-mineral, respectively.

The NN content at BBCH 60, i.e., at the onset of WOSR flowering (FL), was significantly driven by both experimental factors and the interaction between them. It is necessary to stress that the NN content at end of the STME was significantly correlated with its status at the beginning of this phase, i.e., BBCH 30 (Table A1). In 2016, the impact of NFS on the NN content, evaluated independently of N rates, was the same as at BBCH 30. In 2017, a significantly higher amount of NN was recorded in the M-NFS. The effect of the progressively increased N rates was significantly different in both years. In 2016, the NN content increased up to the N rate of 120 kg ha^{-1} and then fell significantly, whereas in 2017 it increased in accordance with the increased N rates. As shown in Figure 3, the course of the NN content was significantly affected by the interaction of NFS and N rates. For the M-NFS, the data obtained fitted the quadratic regression model the best, characterized by an N optimum of 176 kg ha^{-1} and the respective maximum N nitrate content at BBCH 30 (N_{30max}) of 46.7 kg ha^{-1}. This model clearly demonstrates that at the onset of FL, resources of NN were exploited more strongly, exceeding the N rate of 176 kg ha^{-1}. A quite different course of NN was observed for the O-NFS and OM-NFS treatments. The NN content increased progressively with the increased N rate. As compared to the quadratic model, characterizing the M-NFS, the linear model as obtained for the O-NFS shows nitrate N resources to be still present in the soil at the onset of FL in plots fertilized with highest N rates.

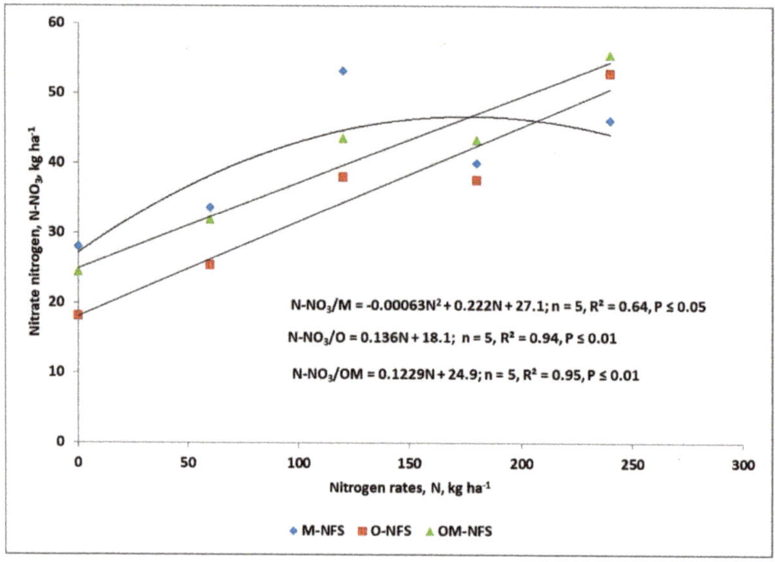

Figure 3. Trends in the N nitrate content at the onset of WOSR flowering for three nitrogen fertilization systems (NFSs). M-NFS, O-NFS, OM-NFS—mineral, organic, organic-mineral NFS, respectively.

The difference in the course of the NN content during STME is best described by its quantitative change. As a rule, in both years, the NN content during this period significantly decreased. The effect of N rates was significant in both years, and the NN content decreased progressively with the increasing N rates.

The dominant trends were evaluated for the NFS x N interaction (Figure 4). For the M-NFS, the cubic regression model fitted the obtained data the best. An initial drop in the NN content of 55 kg ha^{-1} on the AC plot underwent stabilization at this level on plots fertilized with an N rate extending from 60 to 180 kg ha^{-1}. A marked decrease in the NN content was only recorded on the plot with 240 kg ha^{-1} of applied N. The course of the cubic model obtained for the O-NFS was completely different and can be divided into three stages. The first phase covers the stabilized range of the NN content. It amounted to −20 kg ha^{-1} of NN. In the next phase, NN content showed a progressive drop, which lasted up to the N rate of 175 kg ha^{-1}, reaching the lowest value of −106 kg ha^{-1}. In the third stage, resulting from the application of an N rate of 240 kg ha^{-1}, the NN content increased again, and the final decrease in its content with respect to onset of the STME was only −65 kg ha^{-1}. The trend obtained fully explains the progressive response of the NN content at the onset of FL to the progressively increased N fertilizer rates. The trend in the NN content for the OM-FS followed the quadratic regression model. The stability phase at the level of 16 kg ha^{-1} lasted only to the theoretical N rate of 29.3 kg ha^{-1}. Following this N value, an exponential decrease in the NN content along with increasing N rates was recorded.

The third studied cardinal phase of WOSR growth with respect to the NN content was ripening (RE). The NN status at the RE phase was significantly correlated with its status at both the rosette and the onset of flowering stages, respectively (Table A1). It can be therefore concluded that the general pattern of NN management by WOSR did not change significantly from the onset of the STME phase. The course of NN content with respect to the increasing N rates was described by different regression models (Figure 5). For the M-NFS, the linear regression model showing a progressive increase in the NN content fitted the data obtained the best. The cubic regression model best described treatments fertilized with organic N, but its course was treatment specific. For the O-NFS, three distinct phases of NN content were distinguished. The first phase, stabilized at the level of 16 kg ha^{-1} of NN, was related to an N rate of 15 kg ha^{-1}. The second phase, with a progressive increase in NN content, extended up to an N rate of 190 kg ha^{-1}. The third phase, characterized by an NN content decrease, covered the part of the curve with the highest N fertilizer rates. An opposite trend was found for the OM-NFS. The NN content, in spite of a slight variability, showed up to the N rate of 150 kg ha^{-1} a stabilization at the level of 30 kg ha^{-1}. A sudden increase in NN content was observed on the plot with the highest N rate.

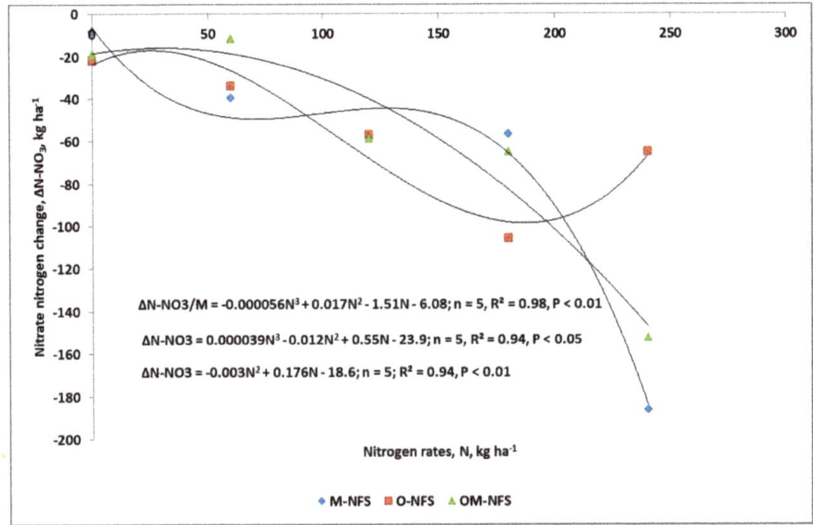

Figure 4. Trends in the N nitrate content change during the stem elongation phase of WOSR growth for three nitrogen fertilization systems (NFSs). M-NFS, O-NFS, OM-NFS—mineral, organic, organic-mineral, respectively.

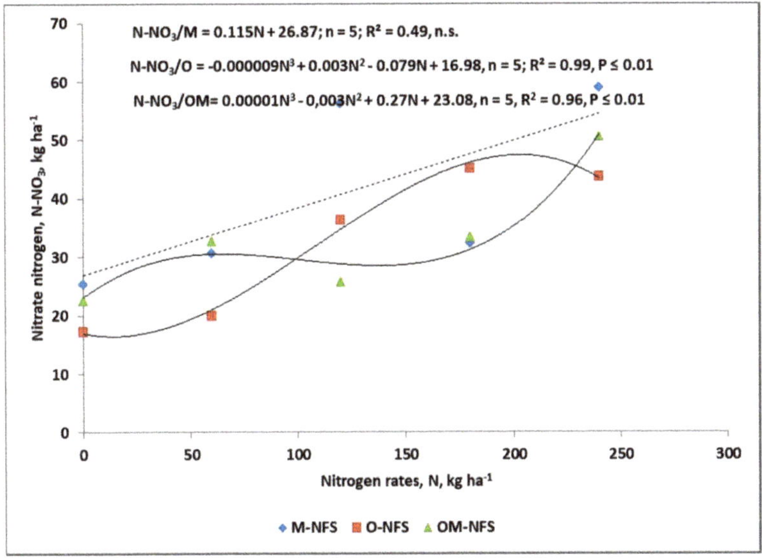

Figure 5. Trends in the N nitrate content at WOSR ripening for three nitrogen fertilization systems (NFSs). M-NFS, O-NFS, OM-NFS—mineral, organic, organic-mineral NFS, respectively.

The observed differences, both in seed yield and NN content at BBCH 30 and during the STME in response to the tested treatments, can be explained by analysis of the partial factor productivity of N. In this study, it was developed based on the amount of NN in a soil layer of 0.9 m at the onset of STME. In both years, the index values averaged over N rates were slightly higher for O-NFS and OM-NFS treatments. This dependence was significant for the OM-NFS treatment in 2017. The impact of N rates, averaged over NFSs, was significant for both years. The impact of NFS on the partial factor productivity of N-NO_3 (PFP_{N30}) indices is shown in Figure 6. The key differences between NFSs result from the index value on the plot fertilized with an N rate of 60 kg ha^{-1}. In the M-NFS, the index was stable on this plot with respect to the N control. On treatments fertilized with organic N, the index increased by 31% for the O-NFS, and by 71% for the OM-FS. Following an N rate of 60 kg ha^{-1}, PFP_{N30} indices fell in response to increasing N rates, showing strong differences between NFSs, as proved by the regression models developed:

$$\text{M-NFS } PFP_{N30} = -0.16N + 57.7 \text{ for } n = 4, R^2 = 0.73, P \leq 0.05 \quad (5)$$

$$\text{O-NFS: } PFP_{N30} = 0.0017N^2 - 0.63N + 89 \text{ for } n = 4, R^2 = 0.99, P \leq 0.01 \quad (6)$$

$$\text{OM-NFS: } PFP_{N30} = 0.0019N^2 - 0.85N + 120.7 \text{ for } n = 4, R^2 = 0.92, P \leq 0.01 \quad (7)$$

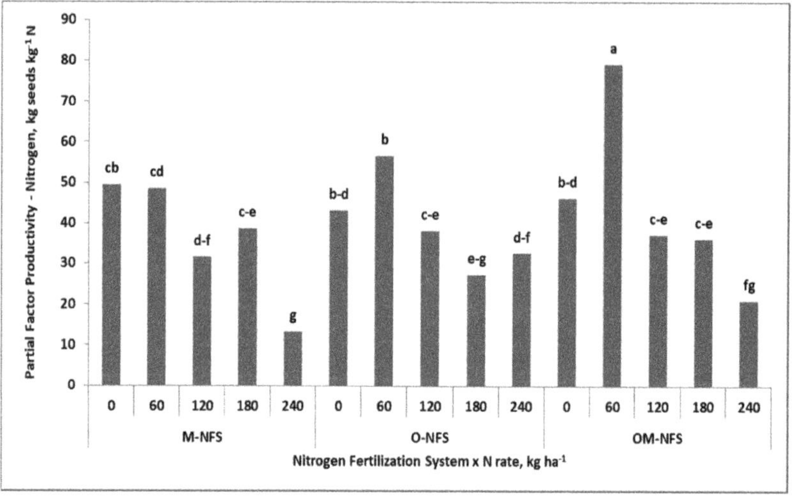

Figure 6. Partial factor productivity of nitrate N at BBCH 30 on the background of nitrogen fertilization systems (NFSs). M-NFS, O-NFS, OM-NFS—mineral, organic, organic-mineral, respectively. [a, b, c, d, e, f, g] significance letters, a—the highest, g—the lowest; [a] means within a column followed by the same letter indicate a lack of significant difference between the treatments.

The PFP_{N30} models for treatments fertilized with organic N, in contrast to M-NFS, showed a stabilization phase, as indicated by the fixed PFP_{N30} minimum. For the O-NFS, its lowest value of 30.6 kg seeds kg^{-1} N was achieved for an N rate of 185.3 kg ha^{-1}. For the OM-NFS, these characteristics were 25.6 kg seeds per kg^{-1} N and 223.7 kg ha^{-1}, respectively. The PFP_{N30} index was, as a rule, negatively correlated with the NN content in all stages of WOSR growth (Table A1).

3.2. *In-Season Variability in Contents of Available Nutrients*

3.2.1. The Onset of Stem Elongation (STME)

The general rule governing the content of available P and K was its decline with soil depth. The content of available P in the topsoil at the onset of STME was, in general, in the suitable class (Table 4, [41]). In both years, a significant impact of NFSs on the P content was recorded in the first subsoil layer (0.3–0.6 m). Significantly lower values were recorded in treatments fertilized with organic N (O-NFS, OM-NFS). The effect of N rates was significant for the third soil layer in 2016, and for both subsoil layers in 2017. The lowest content of P was, as a rule, recorded in treatments fertilized with moderate N rates (120 and 180 kg N ha^{-1} in 2017 and 120 kg N ha^{-1} in 2017), and the highest in the N control plot. The content of available K in the topsoil was in the suitable class in 2016 and in the low in 2017. In the first year of the study, a significant impact of NFSs was observed for the third soil layer, and in the second in the whole subsoil. The general rule was the same as that observed for P, i.e., a significantly higher content of K was recorded in the M-NFS, and the effect of N rates was insignificant.

The content of available Mg in the topsoil was in the low class in 2016 and did not show any response to the experimental treatments. It is necessary to stress that its content, in contrast to P, increased with soil depth. In 2017, a significant impact of NFSs was recorded, but only for the third soil layer. A significantly higher Mg content was recorded in treatments fertilized with organic N, moving these two treatments to the suitable class of Mg availability. The effect of N rates was insignificant. The content of available calcium (Ca) was, in general, low (extremely low) and responded to applied treatments, but only in the wet 2017. The highest impact of NFSs was recorded in the topsoil. A significant increase with respect to the M-NFS was recorded on plots fertilized with organic N. The effect of N rates was observed in each soil layer. As a rule, the highest Ca content was recorded in soil fertilized with an N rate of 120 and 180 kg ha^{-1}.

For the whole N fertilization system, analyzed irrespectively of the N source, the first three Principal Components (PCs) explained 68.6% of the total variance. PC1 was significantly associated with four of fifteen variables. The highest positive loadings were recorded for Ca and Mg in the subsoil (Ca_b, Ca_c, Mg_b). A significant but negative loading was found for the K content in the second subsoil layer (K_c) (Table A2). PC2 had high, but negative loadings with P and Mg contents in the topsoil (P_a, Mg_a) and with P in the first subsoil layer (P_b). PC3 was significantly but negatively associated with the nitrate N content at BBCH 30 N_{30} and yield (Y). A significant impact on yield was exerted by K_c and N_{30} (Table S1a; Figure S1a). The applied stepwise analysis showed however a significant dependence of Y on Mg_c, and N_{30}:

$$Y = 1.852 - 0.025\ Mg_c^{*} + 0.012 N_{30}^{**}\ \text{for}\ n = 30, R^2 = 86 \quad (8)$$

The four PCs explained 90.6% of the total variance in the M-NFS. PC1 was significantly correlated with six of fifteen analyzed variables (Table A2). Positive coefficients of correlation (for coefficient of determination - $R^2 > 0.50$) were recorded for $Ca_{a,b,c}$, yield, and negative for P_c, K_c. PC2 was significantly but negatively correlated with P_a, P_b, Mg_a. PC3 was significantly and positively associated with N_{30}, and PC4 with Mg_c. The strongest negative correlation with PC1 and Y was exerted by K in the third soil layer (K_c). The third group of variables associated with PC2 were contents of available P, Mg_a, and N_{30}. The fourth group of variables was represented by Mg_c and PFP_{N30} showed another direction with respect to the previous group (Table S1b; Figure S1b).

The three PCs accounted for 82.2% of the total variance for the O-NFS. PC1 had significant loadings with six of fifteen variables. The contents of Ca_b and Mg (Mg_a, Mg_b) followed the same direction. PC2 had positive loadings with Mg_c and PFP_{N30}, and negative with N_{30}. PC3 was associated with Y, showing a significant relationship with N_{30}, but at the same time negative with P_c (Table S1c; Figure S1c).

Table 4. The content of available nutrients along the soil layers at the onset of stem elongation of winter oilseed rape (BBCH 30), mg kg^{-1} soil.

Year (Y)	Factor	Level	Phosphorus (P)			Potassium (K)			Magnesium (Mg)			Calcium (Ca)		
		of Factor	a	b	c	a	b	c	a	b	c	a	b	c
2015/	Fertilization	M	51.5	31.0 ^{a}	20.4	111.8	68.8	47.4 ^{a}	56.9	58.3	74.7	385.1	231.7	329.3
2016	system	O	50.7	25.4 ^{b}	19.0	109.0	65.4	42.3 ^{a,b}	56.2	55.2	79.8	385.7	226.9	338.4
	(FS)	OM	48.6	22.4 ^{b}	18.4	108.3	60.8	39.6 ^{b}	58.0	56.3	76.0	395.9	222.5	312.4
	F-value		0.39	7.19 **	0.93	0.20	1.79	4.11 *	0.26	0.33	0.42	0.04	0.07	0.28
	Nitrogen	0	48.9	28.0	23.2 ^{a}	107.5	66.5	49.5	56.1	55.9	83.1	384.5	221.4	319.8
	rate	60	50.0	26.2	19.4 ^{a,b}	106.8	64.9	41.9	57.5	57.3	86.2	400.0	234.9	364.0
	(N)	120	51.2	21.4	16.7 ^{b}	109.3	64.6	41.5	56.3	58.7	78.5	356.5	242.4	313.0
	kg ha^{-1}	180	50.1	25.2	16.1 ^{b}	108.8	63.2	41.1	57.4	58.8	69.8	436.6	253.9	369.4
		240	51.2	30.5	20.7 ^{a,b}	116.0	65.7	41.5	57.9	52.4	66.6	367.0	182.6	267.3
	F-value		0.10	2.53	4.69 **	0.48	0.10	1.98	0.12	0.54	2.52	0.64	1.38	1.67
2016/	Fertilization	M	62.1	39.2 ^{a}	27.5 ^{a}	98.8	74.6 ^{a}	48.3 ^{a}	70.0	64.8	71.1 ^{b}	357.5 ^{b}	302.8	320.0
2017	system	O	57.0	32.4 ^{b}	22.4 ^{a,b}	94.9	65.5 ^{a,b}	35.5 ^{b}	68.4	62.9	80.6 ^{a,b}	558.3 ^{a}	331.3	378.0
	(FS)	OM	56.5	31.5 ^{b}	21.2 ^{b}	84.5	55.1 ^{b}	37.2 ^{b}	67.1	62.5	93.3 ^{a}	489.7 ^{a}	263.0	422.1
	F-value		0.48	5.22 **	4.35 *	2.26	4.67 *	6.73 **	0.34	0.10	3.52 *	8.92 ***	2.1	2.66
	Nitrogen	0	63.0	32.9 ^{a,b}	28.3 ^{a}	91.0	66.3	48.1	63.6	58.1	80.8	385.8 ^{b,c}	264.2 ^{b,c}	351.8 $^{a\ b}$
	rate	60	56.3	36.9 $^{a\ b}$	25.7 ^{a,b}	88.9	65.5	37.8	68.9	64.4	76.8	417.3 ^{b,c}	264.7 ^{b,c}	343.8 $^{a\ b}$
	(N)	120	54.5	27.9 ^{a}	18.4 ^{b}	95.2	73.8	41.0	70.1	70.7	94.0	559.0 ^{a,b}	390.0 ^{a}	406.7 $^{a\ b}$
	kg ha^{-1}	180	57.9	32.6 ^{a,b}	20.6 ^{a,b}	90.8	56.3	33.4	72.8	66.4	83.6	614.7 ^{a}	386.2 ^{a,b}	501.2 ^{a}
		240	60.9	41.6 ^{b}	25.4 ^{a,b}	97.8	63.5	41.3	66.9	57.4	73.1	365.8 ^{c}	190.2 ^{c}	263.2 ^{c}
	F-value		0.36	4.59 **	3.74 *	0.33	1.16	2.41	1.13	1.21	1.08	6.38 ***	8.07 ***	4.72 **
F value for FS × N			0.25	3.09 **	7.20 ***	0.97	0.66	2.62*	0.97	0.77	2.72 **	3.44 **	4.40 ***	10.6 ***

***, **, * significant at $p < 0.001$; < 0.01; < 0.05, respectively; n.s.—non significant; ^{a,b,c} significance letters, a—the highest, c—the lowest; ^{a} means within a column followed by the same letter indicate a lack of significant difference between the treatments, a, b, c—soil layers of 0.0–0.30, 0.30–0.60, 0.60–0.90 cm, respectively.

The three PCs explained 85.9% of the OM-NFS total variance. PC1 was significantly and positively associated with Mg_b, Mg_c, and Ca for all soil layers (Table S1d; Figure S1d). Negative loadings were exerted by K_b, K_c. PC2 had high loadings with P_a, P_b and Mg_a. PC3 was associated with N_{30} and Y, but they had negative loadings. The highest positive impact on Y was exerted by N_{30}.

3.2.2. The Onset of Flowering—Change in Nutrient Availability during STME

The analysis of soil fertility at the onset of FL comprises two components: (i) current status of the content of available nutrient, (ii) the net content of available nutrient change during STME (Tables 4 and 5). In both years, the content of available P in the topsoil was not affected by the studied NFSs. As compared to the onset of STME, the P content increased. The effect of NFSs was observed in both years, and a significant increase due to application of organic N was recorded in 2017. The effect of N rates was significant in 2017. The P content in the subsoil underwent a great change during STME. In 2016, it was depleted, being the strongest in the M-NFS. In 2017, an opposite trend was recorded. As a result, in 2016, a significantly higher P content at the onset of FL was recorded in treatments with organic N, whereas in 2017, with mineral N. In 2016, the lowest P depletion with respect to the N rates was recorded on the plot fertilized with 120 kg ha^{-1} of N. In 2017, the net P content increase was variable, but the highest was recorded in the plot with 240 kg ha^{-1} of N, leading to its highest content at the onset of FL.

The content of available K showed a much lower variability during STME as compared to P. In both years, its highest content was recorded in the M-NFS. The significant differences between NFSs as observed in 2016 were due to an extremely high increase in the K content in soil fertilized with mineral N. In 2017, a net K content increase with soil depth was recorded only in the O-NFS treatment. The application of mineral N resulted in K depletion in the topsoil, which was compensated by its increase in the subsoil, especially in the OM-NFS. The effect of N rates on both the K content and its change during STME was not significant. However, in 2017, a year with a significantly lower initial K content, a decreasing trend in accordance with the increasing N rates was observed in the topsoil. The K status at the onset of FL was a result of the decreasing trend in the net K content increase during STME, which underwent depletion in plots fertilized with the highest N rate.

At the onset of FL, the content of available Mg showed a great change with respect to the onset of STME and responded significantly to NFSs in the subsoil. In 2016, as in the case of K, a significantly higher Mg content was recorded in the M-NFS. The key reason for the observed Mg status was a net increase in its content, especially in the deepest soil layer. In 2017, an opposite trend was observed, i.e., the content of available Mg was depleted, being the strongest in the M-NFS. The increasing N rates significantly affected the content of available Mg in the third soil layer in 2016, and in both subsoil layers in 2017. In both years, the content of available Ca showed an extremely high variability in response to the experimental factors. In 2016, in the topsoil and in the first subsoil layer, a significantly higher content of Ca was recorded in the M-NFS, whereas in 2017, in the OM-NFS. The effect of N rates on the content of Ca in both years was recorded for the topsoil. In 2016, the observed difference between N plots was due to significant differences in the net Ca content increase, as occurred in plots fertilized with 60, 120, and especially with 240 kg ha^{-1} of N. In 2017, the Ca content at the onset of FL resulted from a great variability in its content change during STME. A net increase was the attribute of plots fertilized with an N rate of 60 and 240 kg ha^{-1}. In the other plots, the Ca content underwent depletion. In both years, these trends were also observed in the deeper soil layers.

The pooled analysis of NFSs show that three PCs accounted for 58.64% of the total variance (Table A3). PC1 was associated with five of fifteen variables, of which Mg had positive loadings, irrespective of soil layer. P showed the opposite direction on the PC1 axis in the subsoil. PC2 was significantly associated with Ca_c, and PC3 with Y. The impact of soil variables on Y was insignificant, and the highest positive was exerted only by the nitrate N content at BBCH 60 (N_{60}), which in turn was positively affected by Mg (Table S2a; Figure S2a).

Table 5. The content of available nutrients along the soil layers at the onset of winter oilseed rape flowering (BBCH 60), mg kg^{-1} soil.

Year	Factor	Level	Phosphorus (P)			Potassium (K)			Magnesium (Mg)			Calcium (Ca)		
(Y)		of Factor	a	b	c	a	b	c	a	b	c	a	b	c
2015/2016	Fertilization	M	57.7	18.4 ^{b}	16.0	142.2 ^{a}	77.0	58.2 ^{a}	85.3	95.4	136.3 ^{a}	568.5 ^{a}	323.9 ^{a}	378.0
	system	O	56.7	25.0 ^{a}	15.3	128.9 ^{a,b}	72.6	47.4 ^{b}	82.8	74.5	107.5 ^{b}	458.4 ^{b}	327.1 ^{a}	408.0
	(FS)	OM	59.3	24.1 ^{a}	16.2	113.7 ^{b}	70.7	52.3 ^{a,b}	78.1	73.9	118.3 ^{a,b}	406.2 ^{b}	222.4 ^{b}	365.7
	F-value		0.32	7.77 **	0.28	3.47 *	0.77	4.43 *	0.78	7.74 **	3.22 *	7.01 ***	10.13 ***	0.66
	Nitrogen	0	59.3	25.9	15.2	130.8	75.1	48.3	82.7	90.8	99.7 ^{b}	430.7 ^{b}	353.5 ^{a,b}	438.3 ^{a,b}
	rate	60	56.8	22.0	14.6	117.3	76.2	57.8	84.2	84.5	161.9 ^{a}	507.9 ^{a,b}	357.8 ^{a}	522.4 ^{a}
	(N)	120	59.6	21.4	15.6	126.5	72.2	52.5	81.6	81.5	123.7 $^{a\ b}$	477.0 ^{a,b}	258.8 ^{b,c}	358.0 ^{b,c}
	kg ha^{-1}	180	54.4	22.2	15.7	131.1	72.8	52.5	78.8	75.1	122.5 ^{a,b}	377.7 ^{b}	223.6 ^{c}	362.6 ^{b,c}
		240	59.5	21.0	18.0	135.7	70.8	52.2	83.0	74.4	95.9 ^{b}	595.1 ^{a}	262.1 ^{a,b,c}	238.1 ^{c}
	F-value		0.59	1.38	1.09	0.50	0.22	1.04	0.15	1.45	6.28 ***	4.12 **	6.33 ***	9.30 ***
2016/2017	Fertilization	M	68.8	50.4 ^{a}	38.1 ^{a}	96.9 ^{a,b}	90.7 ^{a}	47.9	57.4	51.2 ^{b}	58.1 ^{b}	414.5 ^{a,b}	239.5 ^{b}	332.8
	system	O	69.3	39.8 ^{b}	26.0 ^{b}	113.6 ^{a}	72.3 ^{b}	43.5	62.8	56.7 ^{a,b}	71.9 ^{a,b}	379.9 ^{b}	307.0 ^{a,b}	415.8
	(FS)	OM	56.7	32.2 ^{b}	29.3 ^{b}	82.9 ^{b}	64.1 ^{b}	47.3	57.3	66.6 ^{a}	82.0 ^{a}	515.2 ^{a}	376.0 ^{a}	437.6
	F-value		2.64	9.05 ***	7.84 **	6.49 **	9.31 *	0.67	0.96	3.95 *	6.15 **	3.83 *	5.00 *	2.51
	Nitrogen	0	67.7 ^{a,b}	41.9 ^{a,b}	28.7 ^{b}	102.5	74.4	48.1	57.3	53.5	72.5	270.0 ^{b}	227.3	329.1
	rate	60	67.3 ^{a,b}	40.9 ^{a,b}	26.0 ^{b}	102.0	76.3	43.9	59.5	60.9	76.6	445.9 ^{a,b}	312.3	446.2
	(N)	120	52.0 ^{b}	30.4 ^{b}	32.2 ^{a,b}	105.2	86.4	51.2	62.6	52.9	68.3	503.6 ^{a}	316.3	386.2
	kg ha^{-1}	180	60.4 ^{a,b}	39.5 ^{a,b}	26.4 ^{b}	91.8	65.6	43.8	59.2	63.9	67.8	459.9 ^{a}	369.5	403.9
		240	77.4 ^{a}	51.3 ^{a}	42.5 ^{a}	87.3	75.8	44.1	57.4	59.8	68.1	503.1 ^{a}	312.1	411.6
	F-value		2.78 *	3.68 *	5.63 ***	0.99	1.64	0.77	0.27	0.88	0.38	4.32 **	1.67	0.91
	FS × N		1.38	3.87 ***	6.09 ***	2.31*	1.31	1.29	0.43	2.91 **	4.43 ***	4.35 ***	4.82 ***	5.08 ***

***, **, * significant at $p < 0.001$; < 0.01; < 0.05, respectively; n.s.—non significant; $^{a, b, c}$ significance letters, a—the highest, c—the lowest; ^{a} means within a column followed by the same letter indicate a lack of significant difference between the treatments, a, b, c—soil layers of 0.0–0.30, 0.30–0.60, 0.60–0.90 cm, respectively.

For the M-NFS, two PCs were extracted that cumulatively explained 59.35% of the total variance. PC1 accounted for 45.2% of the total variance and had high positive loadings for Mg, including all soil layers, and for K, but only for the topsoil (K_a). Negative loadings were recorded for P, also including all soil layers. PC2 was not associated with any variable with high loading ($R^2 > 0.50$). Yield was weakly related to the soil variables but N_{60} showed a significant and positive response to Mg in the subsoil and Ca in the second subsoil layer (Table S2b; Figure S2b).

Three PCs explained 73.24% of the total variance for the O-NFS (Table A3). PC1 was dominated by six of fifteen variables that had high loadings. A positive impact on PC1 was exerted only by Y, and a negative one by PFP_{N30}, and also by Mg and Ca in the subsoil. The second set of variables changed in the opposite direction to Y. PC2 was associated with P (P_a, P_b) that had high and positive loadings. PC3 grouped two variables, i.e., K_a with a negative and Ca_a with a positive loading. The efficiency of N_{30} was significantly affected by Ca and Mg contents in the second subsoil layer. Yield was positively, but not significantly, dependent on N_{60}, but negatively on the contents of Ca and Mg in the first subsoil layer (Table S2c; Figure S2c).

Four PCs accounted for 85.07% of the total variance for the OM-NFS. PC1 was associated with six of fifteen variables, such as K_a, K_c, and Mg_a, which had high and positive loadings. A negative loading was exerted by Ca_b. PC2 showed high and positive loadings with Mg_b and P_b. PC3 had negative loadings with Ca_a and Y. PC4 was significantly, but negatively, associated with P_a. The NN content at the onset of FL was significantly correlated with Mg_a, and negatively with P_b. The content of Ca in the topsoil and N_{60} showed the highest and at the same time positive relationships with yield (Table S2d; Figure S2d).

3.2.3. Physiological Maturity

The content of available P at the physiological maturity of WOSR as compared to the onset of FL, as a rule, was slightly higher in 2016, and considerably lower in 2017. A significant effect of NFSs was observed only in 2016 (Table 6). A higher P content was recorded for treatments fertilized with organic N, being significant in the subsoil. The effect of N rates was non-significant, although a decreasing trend in accordance with the increasing N rates was observed. The content of available K was only affected by NFS. In 2016, as compared to FL, much lower K values were recorded in the topsoil, but an opposite trend was observed in the subsoil. In 2017, an increase in K content was noted for the subsoil.

A depletion of the available Mg content, as compared to FL, was recorded in 2016 for all soil layers. In 2017, it appeared only in the topsoil. In the subsoil, a net increase in Mg content was recorded. The impact of NFS was observed only in 2016 for the subsoil, in which a significantly higher content of Mg was recorded for the M-NFS and O-NFS. The effect of N rates was observed only for the second subsoil layer in 2016. The content of available Ca showed the highest variability. Ca status was significantly affected by both experimental factors and their interaction with a particular year. In both years, the impact of NFS was recorded for all three soil layers. As a rule, the lowest content of available Ca was found for the OM-NFS. The highest Ca values were the attribute of M-NFS. The effect of the increasing N rates was highly variable between soil layers. For the topsoil, the lowest content of Ca was generally recorded for the plot fertilized with 180 kg ha^{-1} of N. In the subsoil, no consistent rule governing the content of available Ca was observed.

The four PCs accounted for 69.50% of the total variance in the pooled NFSs (Table A4). PC1 had the highest and most positive loadings for the Ca content in the subsoil. PC2 was associated with K_c that exerted a negative effect in its value. PC3 explained 14.16% of the total variance, and no significant loading was recorded. PC4 was controlled by the nitrate N content at BBCH 89 (N_{89}), having a negative loading. Yield showed a positive relationship with N_{89} and P_a, but the latter changed in the opposite direction to that of N_{89} (Table S3a; Figure S3a).

Table 6. The content of available nutrients along the soil layers at WOSR ripening (BBCH 89), mg kg^{-1} soil.

Year	Factor	Level	Phosphorus (P)			Potassium (K)			Magnesium (Mg)			Calcium (Ca)		
(Y)		of Factor	a	b	c	a	b	c	a	b	c	a	b	c
2016	Fertilization	M	57.2	21.8 ^{b}	13.5 ^{b}	125.2 ^{a}	82.3 ^{b}	61.2	67.3	86.1 ^{a}	115.1 ^{a}	335.8 ^{b}	346.4 ^{a}	423.5 ^{a}
	system	O	62.6	25.6 ^{a,b}	17.5 ^{a}	115.9 ^{a,b}	78.8 ^{a,b}	54.7	72.1	75.9 ^{a}	100.3 ^{a}	581.1 ^{a}	314.8 ^{a}	368.7 ^{a}
	(FS)	OM	60.8	27.9 ^{a}	16.8 ^{a}	99.2 ^{b}	65.9 ^{b}	58.4	59.6	53.9 ^{b}	71.4 ^{b}	332.2 ^{b}	176.4 ^{b}	225.7 ^{b}
	F-value		0.63	5.13 **	4.17 *	4.77 *	3.71 *	0.89	2.90	6.66 **	10.1 ***	21.8 ***	23.1	21.3 ***
	Nitrogen	0	62.8	26.0	16.0	120.6	72.1	56.7	65.4	62.9	76.8 ^{a}	396.4 ^{a,b}	258.3	272.6 ^{b}
	rate	60	60.9	25.5	17.8	123.3	80.3	50.7	65.0	78.6	79.8 ^{a}	473.9 ^{a,b}	323.6	307.4 ^{a,b}
	(N)	120	60.4	25.0	14.7	108.8	78.2	61.6	61.4	70.0	111.6 ^{b}	379.5 ^{a,b}	293.8	349.4 ^{a,b}
	kg ha^{-1}	180	59.6	23.4	15.8	113.7	80.4	61.2	65.6	83.6	106.5ab	336.3 ^{b}	258.9	398.5 ^{a}
		240	57.4	25.7	15.2	100.8	67.4	60.1	74.3	64.9	103.2ab	495.9 ^{a}	261.2	368.5 ^{ab}
	F-value		0.20	0.34	0.75	1.37	0.97	1.06	0.98	1.16	3.17 *	2.86 *	1.42	3.04 *
2017	Fertilization	M	64.8	31.2	19.5	111.6	82.2 ^{a}	67.7 ^{a}	69.5	61.5	70.3	484.9 ^{a}	243.3 ^{a}	235.3 ^{a}
	system	O	54.3	29.3	23.0	110.3	78.6 ^{a,b}	52.2 ^{a,b}	76.3	67.7	86.3	382.2 ^{a,b}	187.4 ^{b}	210.0 ^{a,b}
	(FS)	OM	52.1	32.3	22.6	103.7	63.9 ^{b}	47.6 ^{b}	69.0	71.2	93.1	319.5 ^{b}	118.5 ^{c}	136.4 ^{b}
	F-value		1.99	0.08	0.33	0.41	3.37 *	5.0 *	1.27	0.56	1.69	5.12 **	16.2 ***	5.32 **
	Nitrogen	0	63.0	34.7	25.7	105.4	78.6	63.8	66.8	64.6	74.6	379.7b	167.1b	192.9 ^{a,b}
	rate	60	63.1	38.2	24.6	105.3	64.5	51.1	73.8	75.4	87.4	316.4b	180.2ab	152.4 ^{b}
	(N)	120	48.8	26.2	19.2	104.2	90.0	53.5	79.7	73.1	111.3	595.2a	260.3a	280.5 ^{a}
	kg ha^{-1}	180	54.8	24.2	18.3	110.4	74.6	52.1	71.4	64.8	76.9	316.8b	185.1ab	226.4 ^{a,b}
		240	55.7	31.5	20.5	117.3	66.6	58.8	66.3	56.2	66.0	369.6b	122.6b	117.1 ^{b}
	F-value		0.97	0.64	0.59	0.41	2.24	0.77	1.37	0.83	2.26	5.87 ***	6.13 ***	4.89 **
	FS × N		0.31	1.49	0.53	1.98	0.54	0.76	0.40	0.57	1.37	2.99 **	8.42 ***	6.41 ***

***, **, * significant at $p < 0.001$; < 0.01; < 0.05, respectively; n.s.—non significant; $^{a, b, c}$ significance letters, a—the highest, c—the lowest; ^{a} means within a column followed by the same letter indicate a lack of significant difference between the treatments; a, b, c—soil layers of 0.0–0.30. 0.30–0.60. 0.60–0.90 cm, respectively.

For the M-NFS, three PCs accounted for 75% of the total variance. PC1 was associated with seven of fourteen variables. Positive loadings were identified for P in the subsoil and Ca_a, but negative values for Mg and Ca for the subsoil. Variables within each subgroup were significantly and positively correlated with each other. Yield exerted a high positive loading on PC2. PC3 had the highest and most positive loading with K_a, and K_b. Yield was weakly related to other soil variables but showed an opposite direction to P_a and K_c. The latter variable was significantly and negatively correlated with Y (Table S3b; Figure S3b).

The O-NFS was associated with three PCs, which explained 73.5% of the total variance. PC1 showed high loadings of P_b, which were positive, and of Mg_b, Ca_b, Ca_c, which were negative. PC2 showed high positive loading with N_{89} and moderate with Y (0.65) and Kc (0.63). These three variables were significantly correlated with each other, although only N_{89} significantly affected Y. The opposite direction to that set of variables was exerted by P variables, of which P_a and P_b were significantly and negatively correlated with Y. PC3 showed a high but negative loading with K_a (Table S3c; Figure S3c).

In the OM-NFS, four PCs accounted for 81.88% of the total variance. PC1 had positive loadings for Ca, but negative for P and Mg_b in the subsoil. PC2 was significantly affected by Mg_a and Mg_c, which had high and positive loadings, and P_a with a negative effect. PC3 was dominated with K_a. PC4 was associated with N_{89} due to its high loading and moderately affected by Y (0.61). Yield was closely, but not significantly, related to K_a and N_{89} (Table S3d; Figure S3d).

4. Discussion

4.1. In-Season Variability in the Growth Factor—Nitrate Nitrogen

The yield variability of winter oilseed rape (WOSR) was evaluated based on the in-season variability in the N-NO_3 content, i.e., the nitrogenous growth factor (NGF), directly impacting the rate of WOSR plant growth during the growing season. Yield responded to the applied N, irrespective of its chemical form, but significantly only to its lowest rate of 60 kg ha^{-1}. This N rate resulted in a yield increase of 60% in 2016, a year with drought in June, and a 115% increase in 2017. The recorded yield increase fully corroborates the opinion on the high sensitivity of WOSR to drought during the end of inflorescence emergence and the onset of flowering [18,39]. The strong response of WOSR to the lowest N rate is typical for soil naturally poor in inorganic N, or soils strongly depleted of this N form [43]). The second problem, which became apparent during the study, refers to the lack of WOSR response to increasing N rates, applied as ammonium nitrate (AN) or digestate, which in the basic form is ammonium [44]. The application of 240 kg ha^{-1} of N as AN resulted in a yield reduction. The observed tendencies are broadly explained by assuming an imbalance of applied nutrients or referring to the low nitrogen use efficiency (NUE) of WOSR plants [10,45–49]. In the studied case, this observation was fully corroborated by the trend of partial factor productivity of N indices, which were calculated based on the N-NO_3 content in the soil at BBBCH 30. The observed PFP_{N30} stabilization on treatments with the highest N rates was due to the lower rate of the NN content decrease during STME, subsequently leading to the better N utilization by WOSR during the seed filling period (SFP). The net N increase during the SFP, as has been documented recently by Łukowiak and Grzebisz [43], is the prerequisite of the high yield of WOSR.

One of the most important targets of soil or plant N monitoring during the growing season is to make a reliable yield prognosis [3,50]. In the study, the N-NO_3 content in the soil profile during two of the cardinal stages of WOSR growth, i.e., rosette (BBCH 30) and the onset of flowering (BBCH 60), significantly limited seed yield, and therefore can be used as a yield predictor (Table A1). The relationship obtained was best described using the quadratic regression model:

$$\text{BBCH 30: Y-N}_{30} = -0.00012\text{N}_{30}^2 + 0.042\text{N}_{30} + 0.76 \text{ for } n = 15, R^2 = 0.72, \text{P} \le 0.05 \quad (9)$$

$$\text{BBCH 60: Y-N}_{60} = -0.022\text{N}_{60}^2 + 0.21\text{N}_{60} - 1.56 \text{ for } n = 15, R^2 = 0.66, \text{P} \le 0.05 \quad (10)$$

As results from equations developed, the N-NO_3 content of 150 kg ha^{-1}, as recorded at BBCH 30, resulted in a theoretical Y_{max} of 3.91 t ha^{-1}. The calculated yield is very close to that achieved in the OM-NFS on the plot with the N rate of 180 kg ha^{-1}. The yield prognosis at the onset of flowering was significantly lower when compared to BBCH 30, as indicated by the Y_{max} of 3.451 t ha^{-1}. This yield could be achieved provided the N-NO_3 amounted to 47.7 kg ha^{-1}. Based on these two data sets, it can be concluded that a typical feature of N management by WOSR during the stem elongation phase is a significant decline in the N-NO_3 content. The average N-NO_3 content decrease, based on the respective N optima, was 102.3 kg ha^{-1} (= 150 kg ha^{-1} at BBCH 30 minus 47.7 kg ha^{-1} at BBCH 60). The study showed that the change in the N-NO_3 content (ΔN-NO3) within the STME can be determined based on the N-NO_3 content precisely at BBCH 30:

$$\Delta \text{N-NO}_3 = -0.85\text{N}_{30} + 23.4 \text{ for } n = 15, R^2 = 0.98, \text{P} \le 0.01 \quad (11)$$

The direction coefficient of the linear regression model obtained clearly indicates a high efficiency of N-NO_3 during STME, which reached 85% of the N-NO_3 content in the whole analyzed soil profile. This value can be used as an index of nitrate N uptake efficiency by WOSR.

The analysis of the N-NO_3 content at BBCH 30 and its rate of change during STME can be used as a basis for the revision of the current hypothesis about the critical stages of WOSR yield development. In fact, it is well-documented that two phases, i.e., inflorescence emergence and flowering, are crucial for the establishment of basic yield components. Any disturbance in plant growth during these two phases, either abiotic or nutritional, leads to yield reduction [18,39,51]. The observed dependence of the N-NO_3 decrease during STME on its content in the soil profile at BBCH 30, as documented in this study, explicitly indicates the rosette stage as the cardinal for exploiting WOSR yielding potential [52]. In the well-established WOSR canopy, as in this study, the NN content underwent significant depletion, irrespective on its amount and N form (Figure 7).

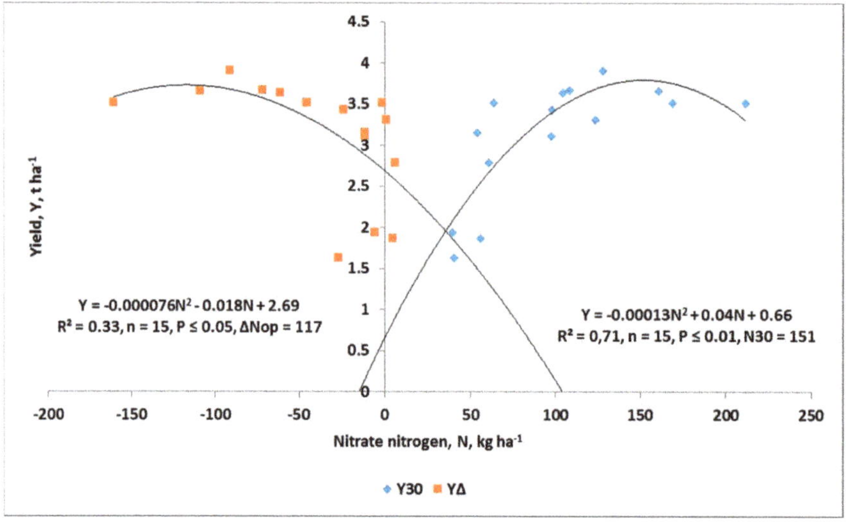

Figure 7. Yield prediction based on the N-NO_3 management during stem elongation phase of WOSR growth.

4.2. In-Season Variability in Soil Fertility Factor—Available Nutrients

The yield of WOSR as discussed in the previous section responded significantly to the applied fertilizer N, irrespective of its chemical form. The key question remaining is to what the extent the soil fertility factor (FF), as defined by the in-season status of four basic nutrients (P, K, Mg, Ca), was related to the content of N-NO_3, defining the NGF status, and in consequence, seed yield. The applied PCA was revealed as a useful tool to evaluate the nitrogen fertilization system based on the above described factors. It was also a useful tool to define the production role of particular subsoil layers with respect to the availability of a given nutrient, or set of nutrients, as yield limiting factors. So far, knowledge about nutrient availability from deeper soil layers, especially during the growing season, is low [53].

The analysis of the NGF, i.e., variability in the N-NO_3 content during the growing season, clearly indicates the rosette and the onset of flowering stages as cardinal for determining WOSR yield. Based on the PCA, it can be concluded that at the onset of STME, the primary set of soil nutrients, including the content of available Ca and Mg, representing the FF significantly limited yield, irrespective of the studied NFS. It is necessary to stress that the importance of both nutrients for NFS productivity resulted more strongly from their contents in the subsoil than in the topsoil. This conclusion stresses the importance of Ca as the yield nutritional factor. In fact, the yield forming function of Ca, in spite of the high requirements of some crops, such as oilseed rape, is weakly recognized [23,53]. In contrast to Ca, the yield forming functions of Mg are well recognized, but are mostly related to the onset of flowering and during the SFP [18,49].

The detailed analysis of the impact of FFs on yield for each of the studied NFSs showed a significantly different response to the applied N form. The general pattern of the FF relation with Y, as presented above, was to a great extent typical for the M-NFS. The shortage of Ca and Mg in connection with a shortage of NN resulted in the insufficient exploitation of both P and especially K from the second subsoil layer. The content of available K in this layer in the M-NFS can be used as a single yield (Y) predictor at BBCH 30:

$$Y = -0.68K_c + 6.32 \text{ for } n = 10, R^2 = 0.85, P \leq 0.01 \quad (12)$$

The negative sign of the direction coefficient implicitly indicates that the low exploitation of K from the subsoil, just at the rosette stage of WOSR, is the reason for the yield decline. The OM-NFS yielded higher, because N was better balanced with basic nutrients, such as P and K, which were more strongly exploited by WOSR plants from the subsoil as compared to the M-NFS (Table S1b and Table 3d). It was found that a higher content of available K in the subsoil resulted in favorable conditions for the uptake of Ca and Mg. This hypothesis was fully corroborated at the onset of flowering, when available Ca positively affected N availability, finally resulting in a yield increase. The same effect in the OM-NFS on N availability was exerted by P and Mg. It is well-documented that all these three nutrients are strongly exploited by WOSR from subsoil [54–57].

A significantly different pattern of FF impact on the content of NN and yield was observed in the O-NFS. In this particular NFS, yield significantly depended on the content of N-NO_3, i.e., the NGF was found to be the direct yield driver (Table S1c; Figure S1c). The productivity of this system was also dependent on the content of Mg and Ca in the deep subsoil layer (c). The higher availability of both nutrients in this layer significantly affected the unit productivity of N (PFP_{N30}). This was the key reason for the much higher productivity of the O-NFS system as compared to the M-NFS.

This study explicitly shows that a reasonable, but significant, decrease in the NN content during STME is the prerequisite of high WOSR yield (Figure 7). A similar decrease in the content of other nutrients can therefore be assumed. K requires special attention, because during this period it reaches both the maximum rate of uptake and the maximum value of its accumulation just at the end of the inflorescence emergence (INFE) [23,56]. In 2016, a net K content increase was recorded in all three soil layers. The same trend was observed in 2017 for treatments fertilized fully or partly with organic N. The observed phenomenon was not, however, related to the amount of K applied in digestate. Therefore, it cannot be explained by the direct impact of this fertilizer on K availability through K input, or indirectly by ammonium oxidation and its subsequent impact on the cation exchange [58]. The only reasonable explanation for the K content increase with soil depth is the WOSR activity in the rhizosphere, including its acidification [59].

At the onset of flowering, being at the same time the end phase of intensive uptake of N by WOSR, the key limiting nutrient for the whole system became the content of available Mg, which controlled the amount of NN, the direct yield driver. The result obtained corroborates the importance of Mg for stabilizing the nutritional status of WOSR plants at this particular phase [18]. The insufficient availability of Mg on the one hand and excess of available P in the subsoil on the other points to FFs as critical factors for WOSR yield performance, irrespective of the NFS. Magnesium content and to a lesser extent Ca in the subsoil exerted a positive impact on the content of NN, a direct growth factor. Therefore, it can be concluded that the higher the content of available Mg in the subsoil, the better the supply of N to WOSR plants during the INFE, the phase responsible for seed set [49]. In the O-NFS, the impact of the key FF variables on the NFS functioning revealed a positive impact of K_c on the content of NN at the end of the INFE phase, concomitant with a significant improvement in N productivity, which was exerted by the Mg and Ca present in the second subsoil layer (Table S2c). The positive impact of Mg content on the NN content was fully corroborated in the OM-NFS. The higher content of both nutrients at the onset of flowering resulted in a positive yield response. The yield increase also resulted from the concomitant positive impact of available Ca on N productivity. In both years, the highest net increase in the content of available Ca during the STME was recorded on plots fertilized with the highest N rate, which resulted in the highest yield. These three simultaneously occurring processes led to a higher seed yield and the advantage of the OM-NFS over other tested NFSs.

The dominant impact of FFs as limiting production factors was fully corroborated at WOSR physiological maturity. The entire NFS production efficiency was governed by the content of available Ca and P in the subsoil. It is well documented that in soil of a neutral pH range, the content of available P depends on the content of Ca and vice versa (Table S3a) [17]. The negative relationship between the content of available K in the second subsoil layer with yield clearly indicates the stability of the soil/plant system during the spring growing season:

$$Y = -0.0015K_c^2 + 0.15K\text{-}0.334 \text{ for } n = 10, R^2 = 0.55, P \leq 0.05 \quad (13)$$

The presented equation shows that the depletion of the K content in the second subsoil layer at WOSR physiological maturity to 50.3 mg kg^{-1} is the prerequisite of the maximum yield of 3.466 t ha^{-1}. A quite different set of variables were responsible for O-NFS. In the final phase of WOSR growth, the two decisive variables for yield were the amount of NN, K, and also P. The importance of this set of nutrients for exploiting WOSR yielding potential corroborates the latest study by Grzebisz et al. [49]. The key importance of K during the SFP results from its physiological function, as related to the transport of assimilates to the growing pods and seeds [60]. The better the supply of assimilates to the growing pods and seeds, the higher the number of seeds, subsequently resulting in a higher yield. The shortage of NN at this stage led to the weak exploitation of P from the whole soil profile. This observation is in agreement with Grzebisz et al. [61], who showed that the exploitation of P from its soil resources depends on the WOSR sink strength, which is related to the number of seeds per unit area.

5. Conclusions

The rosette stage of WOSR growth (BBCH 30) was revealed as the cardinal for seed yield performance. The highest yield was obtained, provided two basic conditions were fulfilled. The first refers to the N-NO_3 content at this particular stage, amounting to 150 kg ha^{-1}. The second was the considerable decrease in the N-NO_3 content during the stem elongation phase of WOSR growth. The yield was significantly dependent on the nitrogen fertilization system and increased in the order of NFS: M, O, OM. The maximum yields of 3.616, 3.887, 4.195 t ha^{-1} were achieved given the optimum N rates of 138.9, 171.4, 210 kg ha^{-1}, respectively. The key factors limiting yield were significantly related to the particular NFS. In the case of M-NFS, yield was limited by the availability of Ca and Mg, but especially of K in the deepest subsoil layer (0.6–0.9 m). In the O-NFS, the key limiting factor was the shortage of N-NO_3. The highest yield obtained in the OM-NFS was due to the positive impact of Ca and Mg on both the amount of N-NO_3 and its productivity. During the second cardinal stage of WOSR yield performance, i.e., at the onset of flowering, Mg and to a lesser extent the contents of Ca in the subsoil both resulted in an N-NO_3 content increase—a direct growth factor. In can be therefore concluded that the higher the content of available Mg in the subsoil at the onset of WOSR flowering, the better the supply and utilization of N by plants, resulting in the highest yield. PCA analysis was found to be an effective tool in discriminating against nitrogen fertilization systems differing in the source of N (mineral vs. organic).

Supplementary Materials: The following Tebles and Figures are available online at http://www.mdpi.com/2073-4395/10/11/1701/s1, Table S1a. Spearman's matrix of correlation coefficients for the pooled nitrogen fertilization systems at BBCH 30 of WOSR growth, $n = 30$; Table S1b. Spearman's matrix of correlation coefficients for the mineral nitrogen fertilization systems at BBCH 30 of WOSR growth, $n = 10$; Table S1c. Spearman's matrix of correlation coefficients for organic nitrogen fertilization systems at BBCH 30 of WOSR growth, $n = 10$; Table S1d. Spearman's matrix of correlation coefficients for the organic-mineral nitrogen fertilization systems at BBCH 30 of WOSR growth, $n = 10$; Table S2a. Spearman's matrix of correlation coefficients for the pooled nitrogen fertilization systems at BBCH 60 of WOSR growth, $n = 30$; Table S2b. Spearman's matrix of correlation coefficients for the mineral nitrogen fertilization systems at BBCH 60 of WOSR growth, $n = 10$; Table S2c. Spearman's matrix of correlation coefficients for organic nitrogen fertilization systems at BBCH 60 of WOSR growth, $n = 10$; Table S2d. Spearman's matrix of correlation coefficients for the organic-mineral nitrogen fertilization systems at BBCH 60 of WOSR growth, $n = 10$; Table S3a. Spearman's matrix of correlation coefficients for the pooled nitrogen fertilization systems at BBCH 89 of WOSR growth, $n = 30$; Table S3b. Spearman's matrix of correlation coefficients for the mineral nitrogen fertilization systems at BBCH 89 of WOSR growth, $n = 10$; Table S3c. Spearman's matrix of correlation coefficients for organic nitrogen fertilization systems at BBCH 89 of WOSR growth, $n = 10$; Table S3d. Spearman's matrix of correlation coefficients for the organic-mineral nitrogen fertilization systems at BBCH 89 of WOSR growth, $n = 10$. Figure S1. Principal components analysis (PCA) of nitrate N, available nutrients content and yield indices at the onset of WOSR stem elongation for (a) total NFS, (b) M-NFS, (c) O-NFS, (d) OM-NFS; NFS—nitrogen fertilization system; N_{30}—nitrate N content at BBCH 30 (rosette); PFP_{N30}—partial factor productivity of N-NO_3 at BBCH 30; P—phosphorus, K—potassium, Mg—magnesium, Ca—calcium; a, b, c—soil layers of 0.0–0.30, 0.30–0.60, 0.60–0.90 cm, respectively, Y—yield. Figure S2. Principal components analysis (PCA) of nitrate N, available nutrients content and yield indices at the onset of WOSR flowering for (a) total NFS, (b) M-NFS, (c) O-NFS, (d) OM-NFS; NFS—nitrogen fertilization system; N_{60}—nitrate N content at BBCH 60; PFP_{N30}—partial factor productivity of N-NO_3 at BBCH 30; P—phosphorus, K—potassium, Mg—magnesium, Ca—calcium; a, b, c—soil layers of 0.0–0.30, 0.30–0.60, 0.60–0.90 cm, respectively; Y—yield. Figure S3. Principal components analysis (PCA) of nitrate N, available nutrients content and yield indices at WOSR ripening for (a) total NFS, (b) M-NFS, (c) O-NFS, (d) OM-NFS; NFS—nitrogen fertilization system; N89—nitrate N content at BBCH 89; P—phosphorus, K—potassium, Mg—magnesium, Ca—calcium; a, b, c—soil layers of 0.0–0.30, 0.30–0.60, 0.60–0.90 cm, respectively, Y—yield.

Author Contributions: Conceptualization, W.G., and R.Ł.; methodology, W.G., R.Ł.; software, K.K.; validation, W.G.; R.Ł.; formal analysis, R.Ł.; investigation, K.K., R.Ł.; resources, R.Ł.; K.K.; writing—original draft preparation, R.Ł.; writing—review and editing, W.G.; visualization, K.K.; supervision, R.Ł.; project administration, R.Ł. All authors have read and agreed to the published version of the manuscript.

Funding: This research received no external funding.

Conflicts of Interest: The authors declare no conflict of interest.

Appendix A

Table A1. Spearman's coefficients of correlation between nitrate N contents at cardinal stages of WOSR growth and yield, $n = 15$.

Variables	N_{60}	N_{89}	$N_{30–60}$	Yield
N_{30}	0.44	0.48	0.67 **	0.67 **
N_{60}	1.00	0.94 ***	−0.37	0.22
N_{89}		1.00	−0.28	0.22
N_{30-60}			1.00	0.51 *

***, **, * significant at $p < 0.001$; < 0.01; < 0.05, respectively. N30, N60, N89—nitrate N content at BBCH 30, 60, and 89, kg ha^{-1}, respectively; N30–N60—nitrate N balance at BBCH 60 vs. BBCH 30; PFP_{N30}—partial factor productivity of N-NO_3 at WOSR stage of BBCH 30 (rosette).

Appendix B

Table A2. Principal component analysis of the content of available nutrients and nitrate nitrogen, and yield at the onset of WOSR stem elongation (BBCH 30).

Available Nutrients Other	Soil Layer Variables	Nitrogen Fertilization System, NFS												
		Total			Mineral				Organic			Organic–Mineral		
		PC1	PC2	PC3	PC1	PC2	PC3	PC4	PC1	PC2	PC3	PC1	PC2	PC3
Phosphorus	a	−0.08	**−0.80**	0.17	−0.22	**−0.88**	−0.14	−0.12	−0.65	−0.51	−0.16	0.20	**0.78**	0.12
	b	−0.32	**−0.84**	0.07	0.07	**−0.91**	0.00	0.06	−0.43	−0.41	**−0.71**	−0.36	**0.85**	−0.15
	c	−0.52	−0.57	0.44	**−0.73**	−0.58	−0.30	0.17	−0.27	−0.21	**−0.93**	−0.47	0.57	0.23
Potassium	a	−0.47	0.42	−0.32	−0.50	0.25	0.63	0.52	**0.88**	0.27	0.12	−0.31	**−0.89**	−0.29
	b	−0.48	−0.32	0.25	−0.51	−0.17	−0.66	0.38	−0.59	0.36	0.04	**−0.82**	−0.17	−0.29
	c	**−0.72**	−0.03	0.35	**−0.85**	0.11	−0.21	0.15	0.41	0.65	−0.27	**−0.88**	0.10	0.05
Magnesium	a	0.49	**−0.82**	0.06	0.43	**−0.84**	−0.29	−0.01	**−0.89**	−0.32	0.15	0.60	**0.70**	−0.20
	b	**0.75**	−0.28	0.29	0.67	−0.11	−0.50	0.48	**−0.91**	−0.06	0.25	**0.85**	−0.09	0.13
	c	0.65	0.28	0.28	−0.13	0.36	−0.05	**0.89**	−0.40	**0.76**	0.32	**0.86**	0.01	0.24
Calcium	a	0.68	−0.18	−0.23	**0.82**	0.13	0.22	0.08	−0.64	−0.20	0.20	**0.74**	0.22	−0.36
	b	**0.81**	−0.23	0.23	**0.79**	−0.06	−0.54	0.23	**−0.94**	0.14	0.15	**0.94**	−0.26	−0.13
	c	**0.85**	0.09	0.20	**0.86**	0.14	−0.04	0.30	−0.65	**0.70**	0.12	**0.97**	−0.15	0.02
N–NO_3	N_{30}	−0.10	−0.50	**−0.79**	0.06	−0.61	**0.73**	0.29	0.20	**−0.89**	0.37	−0.09	0.29	**−0.94**
PFP_{N30}	PFP_{N30}	0.08	0.41	0.68	0.02	0.62	−0.67	−0.29	−0.21	**0.91**	0.04	−0.15	−0.01	**0.83**
Yield	Y	0.38	−0.08	**−0.71**	**0.92**	−0.11	0.28	−0.05	0.17	−0.46	**0.84**	0.05	−0.09	**−0.72**

Bold—correlation coefficient > 0.70.

Appendix C

Table A3. Principal component analysis of the content of available nutrients and nitrate nitrogen, and yield at the onset of flowering (BBCH 60).

Available Nutrients	Soil Layer	Nitrogen Fertilization System—NFS												
		Total			Mineral			Organic			Organic–Mineral			
Other Variables		PC1	PC2	PC3	PC1	PC2	PC3	PC1	PC2	PC3	PC1	PC2	PC3	PC4
Phosphorus	a	−0.56	−0.04	0.10	**−0.72**	0.08	−0.31	0.02	**0.87**	−0.08	0.29	−0.07	−0.21	**−0.83**
	b	**−0.88**	0.08	0.02	**−0.89**	0.16	0.01	0.02	**0.95**	−0.06	−0.45	**−0.74**	0.10	−0.35
	c	**−0.80**	−0.15	−0.23	**−0.82**	0.20	−0.07	0.52	0.65	0.16	−0.49	−0.59	−0.51	0.10
Potassium	a	0.53	−0.33	0.24	**0.77**	0.07	0.15	−0.38	−0.23	**−0.70**	**0.83**	−0.19	−0.17	0.22
	b	−0.20	−0.27	0.20	−0.38	0.31	0.43	−0.45	0.20	−0.26	0.59	−0.50	0.33	−0.22
	c	0.41	−0.58	0.17	0.61	−0.37	−0.21	0.39	−0.58	−0.46	**0.75**	−0.46	−0.18	0.05
Magnesium	a	**0.83**	−0.21	0.22	**0.90**	−0.16	−0.06	−0.34	**−0.90**	0.00	**0.90**	0.32	0.17	−0.02
	b	**0.82**	0.22	−0.07	**0.94**	0.23	−0.18	**−0.83**	−0.23	−0.04	−0.06	**0.90**	0.18	0.08
	c	**0.79**	0.20	0.11	**0.87**	0.17	−0.44	**−0.72**	−0.31	0.00	0.41	0.58	0.35	−0.32
Calcium	a	0.32	−0.21	−0.61	0.64	−0.01	0.54	−0.09	−0.50	**0.76**	−0.08	−0.05	**−0.92**	−0.34
	b	0.27	0.68	−0.47	0.69	0.15	0.61	**−0.85**	0.16	0.42	**−0.73**	0.50	−0.27	−0.14
	c	0.20	**0.85**	−0.11	0.29	0.57	−0.37	**−0.83**	0.33	0.20	−0.55	0.59	0.35	−0.28
N–NO_3	N_{60}	0.43	−0.46	−0.34	0.38	**0.70**	−0.47	0.50	−0.66	0.10	0.66	0.52	−0.51	0.02
PFP_{N30}	N_{30}	0.16	0.40	0.55	0.31	**−0.70**	−0.15	**−0.75**	0.04	−0.03	0.07	−0.28	0.61	−0.59
Yield	Y	0.11	−0.20	**−0.74**	0.20	0.65	0.42	**0.72**	−0.18	0.28	0.19	0.41	**−0.73**	−0.34

Bold—correlation coefficient > 0.70.

Appendix D

Table A4. Principal component analysis of the content of available nutrients and nitrate nitrogen, and yield at physiological maturity (BBCH 89).

Available Nutrients	Soil Layer	Nitrogen Fertilization System—NFS													
		Total				Mineral			Organic			Organic–Mineral			
Other Variables		PC1	PC2	PC3	PC4	PC1	PC2	PC3	PC1	PC2	PC3	PC1	PC2	PC3	PC4
Phosphorus	a	−0.25	−0.69	0.34	−0.08	0.60	−0.60	−0.09	0.50	−0.56	−0.21	0.14	−0.70	0.14	0.51
	b	−0.62	0.05	0.59	−0.35	**0.82**	0.24	0.14	**0.78**	−0.49	0.27	**−0.80**	−0.21	0.21	0.41
	c	−0.60	0.22	0.58	−0.10	**0.78**	−0.22	0.15	0.63	−0.19	0.37	**−0.89**	−0.20	0.09	0.23
Potassium	a	0.36	−0.30	−0.09	0.27	−0.50	−0.42	**0.70**	0.18	0.26	**−0.81**	0.05	0.24	**−0.92**	−0.02
	b	0.48	−0.32	0.37	−0.26	−0.01	−0.07	**0.72**	−0.42	−0.37	0.39	−0.08	−0.38	0.53	0.05
	c	0.12	**−0.73**	−0.19	−0.49	0.49	−0.61	0.11	−0.29	0.63	0.15	0.55	−0.55	0.37	0.30
Magnesium	a	0.35	0.52	0.45	−0.30	0.29	0.67	0.48	−0.37	−0.14	0.49	−0.46	**0.86**	0.06	0.11
	b	0.66	0.23	0.40	0.25	**−0.84**	0.04	0.28	**−0.88**	−0.28	0.19	**−0.80**	0.39	0.25	0.11
	c	0.59	0.36	0.29	−0.08	**−0.72**	0.29	−0.19	−0.66	−0.41	0.41	−0.27	**0.73**	0.41	0.29
Calcium	a	0.51	−0.05	0.30	−0.34	**0.73**	0.29	0.59	−0.67	−0.44	−0.54	0.41	0.54	0.52	0.00
	b	**0.88**	−0.24	0.11	0.10	**−0.82**	−0.20	0.35	**−0.76**	−0.40	−0.42	**0.73**	0.43	0.28	−0.11
	c	**0.92**	−0.12	0.02	0.05	**−0.94**	0.09	0.15	**−0.88**	−0.32	−0.06	**0.75**	0.38	0.38	0.26
$N-NO_3$	N_{89}	0.16	0.11	−0.44	**−0.82**	0.42	0.67	0.15	−0.38	**0.84**	0.27	0.42	0.10	−0.36	**0.70**
Yield	Y	0.21	0.61	−0.50	−0.23	−0.11	**0.84**	−0.16	−0.56	0.65	−0.01	0.17	0.34	−0.51	0.61

Bold—correlation coefficient > 0.70.

References

1. Dowson, C.J.; Hilton, J. Fertiliser availability in a resource-limited world: Production and recycling of nitrogen and phosphorus. *Food Policy* **2011**, *36*, 14–22. [CrossRef]
2. Benton Jones, J., Jr. *Agronomic Handbook*; CRC Press: Boca Raton, FL, USA, 2003; p. 450.
3. Luce, M.S.; Whalen, J.K.; Ziadi, N.; Zebarth, B.J. Nitrogen dynamics and indices to predict soil nitrogen supply in humid temperate soils. *Adv. Agron.* **2011**, *112*, 55–102.
4. Li, S.X.; Wang, Z.; Stewart, B. Responses of crop plants to ammonium and nitrate N. *Adv. Agron.* **2013**, *118*, 205–397.
5. Marschner, P. (Ed.) *Marchnerr's Mineral Nutrition of Higher Plants*, 3rd ed.; Academic Press: London, UK, 2012; p. 672.
6. Tegeder, M.; Masclaux-Daubresse, C. Source and sink mechanisms of nitrogen transport and use. *New Phytol.* **2018**, *217*, 35–53. [CrossRef] [PubMed]
7. Berry, P.M.; Spink, J.; Foulkes, M.J.; White, P.J. The physiological basis of genotypic differences in nitrogen use efficiency in oilseed rape (*Brassica napus* L.). *Field Crops Res.* **2010**, *119*, 365–373. [CrossRef]
8. Rathke, G.W.; Behrens, T.; Diepenbrock, W. Integrated nitrogen management strategies to improve seed yield, oil content and nitrogen efficiency of winter oilseed rape (*Brassica napus* L.): A review. *Agric. Ecosyst. Environ.* **2016**, *117*, 80–108. [CrossRef]
9. Sieling, K.; Kage, H. N balance as an indicator of N leaching in an oilseed rape–winter wheat–winter barley rotation. *Agric. Ecosyst. Environ.* **2006**, *115*, 261–269. [CrossRef]
10. Szczepaniak, W.; Barłóg, P.; Łukowiak, R.; Przygocka-Cyna, K. Effect of balanced nitrogen fertilization in four-year rotation on plant productivity. *J. Cent. Eur. Agric.* **2013**, *14*, 64–77. [CrossRef]
11. Szczepaniak, W. A mineral profile of oilseed rape in critical stages of growth-nitrogen. *J. Elem.* **2014**, *19*, 759–778. [CrossRef]
12. Marschner, H.; Kirkby, E.A.; Cackmak, J. Effect of mineral nutritional status on shoot-root partitioning of Photo-assimilates and cycling of mineral nutrients. *J. Exp. Bot.* **1996**, *47*, 1255–1263. [CrossRef]
13. Gutser, R.; Ebertseder, T.; Weber, A.; Schraml, M.; Schmidhalter, U. Nitrogen release from organic fertilizers after short and long-term application to arable land. *J. Plant Nutr. Soil Sci.* **2005**, *168*, 439–446. [CrossRef]
14. Sharma, L.K.; Bali, S.K. A review of methods to improve nitrogen use effciency in agriculture. *Sustainability* **2018**, *10*, 51. [CrossRef]
15. Raynaud, X.; Leadley, P. Soil characteristics play a key role in modeling nutrient competition in plant communities. *Ecology* **2004**, *85*, 2200–2214. [CrossRef]
16. Wallace, A.; Wallace, G.A. *Closing the Crop-Yield Gap through Better Soil and Better Management*; Wallace Laboratories: Los Angeles, CA, USA, 2003; p. 162.
17. Penn, C.J.; Camberato, J.J. A critical review on soil chemical processes that control how soil pH affects phosphorus availability to plants. *Agriculture* **2019**, *9*, 120. [CrossRef]
18. Szczepaniak, W.; Grzebisz, W.; Potarzycki, J.; Łukowiak, R.; Przygocka-Cyna, K. Nutritional status of winter oilseed rape in cardinal stages of growth as yield indicator. *Plant Soil Environ.* **2015**, *61*, 291–296. [CrossRef]
19. Fageria, N.K.; Bailigar, V.C.; Clark, R.B. *Physiology of Crop Production*; CRC Press: New York, NY, USA, 2006; p. 356.
20. Sylvester-Bradley, R.; Lunn, G.; Foulkes, J.; Shearman, V.; Spink, J.; Ingram, J. Management Strategies for Yield of Cereals and Oilseed Rape. HGCA Conference: Agronomic Intelligence: The Basis for Profitable Production. HGCA, 18. Available online: www.hgca.com/publications (accessed on 14 August 2020).
21. Mayer, U. BBCH Monograph. In *Growth Stages of Mono- and Dicotyledonous Plants*, 2nd ed.; Federal Biological Research Center for Agriculture and Forestry: Berlin, Germany, 2001; Available online: http://www.jki.bund.de/fileadmin/dam_uploads/_veroeff/bbch/BBCH-Skala_Englisch.pdf (accessed on 10 August 2020).
22. Böttcher, U.; Rampin, E.; Hartmann, K.; Zanetti, F.; Flenet, F.; Morison, M.; Kage, H. A phenological model of winter oilseed rape according to the BBCH scale. *Crop Pasture Sci.* **2016**, *67*, 345–358. [CrossRef]
23. Barłóg, P.; Grzebisz, W.; Diatta, J. Effect of timing and nitrogen fertilizers on nutrients content and uptake of winter oilseed rape. Part II. Dynamics of nutrients uptake. In *Chemistry for Agriculture*; Górecki, H., Dobrzański, Z., Kafarski, P., Eds.; Czech-Pol Trade: Prague, Czech Republic, 2005; Volume 6, pp. 113–123.
24. White, C.; Sylvester-Bradley, R.; Berry, P.M. Root length densities of UK wheat and oilseed rape crops with implications for water capture and yield. *J. Exp. Bot.* **2015**, *66*, 2293–2303. [CrossRef]

25. Ulas, A.; Schulte auf'm Erley, G.; Kamh, M.; Wiesler, F.; Horst, W.J. Root-growth characteristics contributing to genetic variation in nitrogen efficiency of oilseed rape. *J. Plant Nutr.* **2012**, *175*, 489–498. [CrossRef]
26. Schmidhalter, U. Development of a quick on-farm test to determine nitrate levels in soils. *J. Plant Nutr. Soil Sci.* **2005**, *168*, 432–438. [CrossRef]
27. Olfs, H.W.; Blankenau, K.; Brentrup, F.; Jasper, J.; Link, A.; Lammel, J. Soil-and plant-based nitrogen-fertilizer recommendations in arable farming. *J. Plant Nutr. Soil Sci.* **2005**, *168*, 414–431. [CrossRef]
28. Grzebisz, W.; Łukowiak, R.; Sassenrath, G. Virtual nitrogen as a tool for assessment of nitrogen at the field scale. *Field Crops Res.* **2018**, *218*, 182–184. [CrossRef]
29. Parris, K. Agricultural nutrient balances as agri-environmental indicators: An OECD perspective. *Environ. Pollut.* **1998**, *102*, 219–225. [CrossRef]
30. Conijn, J.G.; Bindraban, P.S.; Schroder, J.J.; Jongschaap, R.E.E. Can our global food system meet food demand within planetary boundaries? *Agric. Ecosyst. Environ.* **2018**, *252*, 244–256. [CrossRef]
31. Odlare, M.; Lindmark, J.; Ericsson, A.; Pell, M. Use of organic wastes in agriculture. *Energy Procedia* **2015**, *75*, 2472–2476. [CrossRef]
32. Vaneeckhaute, C.; Meers, E.; Michels, E.; Buysse, J.; Tack, F.M.G. Ecological and economic benefits of the application of bio-based mineral fertilizers in modern agriculture. *Biomass Bioenergy* **2013**, *49*, 239–248. [CrossRef]
33. Nkoa, R. Agricultural benefits and environmental risks of soil fertilization with anaerobic digestates: A review. *Agron. Sustain. Dev.* **2014**, *34*, 473–492. [CrossRef]
34. Tambone, F.; Orzi, V.; D'Imporzano, G.; Adani, F. Solid and liquid fractionantion of digestate: Mass balance, Chemical characterization, and agronomic and environmental value. *Bioresour. Technol.* **2017**, *243*, 1251–1256. [CrossRef] [PubMed]
35. Makadi, M.; Tomocsik, A.; Orosz, V. Digestate: A new nutrient source—A review. In *Biogas*; Kumar, S., Ed.; InTech: Rijeka, Croatia, 2012; pp. 295–312.
36. Barłóg, P.; Hlisnikovský, L.; Kunzová, E. Effect of digestate on soil organic carbon and plant-available nutrient content compared to cattle slurry and mineral fertilization. *Agronomy* **2020**, *10*, 379. [CrossRef]
37. Przygocka-Cyna, K.; Grzebisz, W. The multifactorial effect of digestate on the availability of soil elements and grain yield and its mineral profile—The case of maize. *Agronomy* **2020**, *10*, 275. [CrossRef]
38. Weymann, W.; Bottcher, U.; Sieling, K.; Kage, H. Effects of weather conditions during different growth phases on yield formation of winter oilseed rape. *Field Crops Res.* **2015**, *173*, 41–48. [CrossRef]
39. WRB, I.W.G. *World Reference Base for Soil Resources 2014. World Soil Resources Reports*; Food and Agriculture organization of the United Nations: Rome, Italy, 2019; Volume 106.
40. Mehlich, A. Mehlich 3 soil test extractant: A modification of Mehlich 2 extractant. *Commun. Soil Sci. Plant Anal.* **1984**, *15*, 1409–1416. [CrossRef]
41. Trávník, K.; Zbíral, J.; Němec, P. *Agrochemical soil testing—Mehlich III*; Central Institute for Supervising and Testing in Agriculture: Brno, Czech Republic, 1999. (In Czech)
42. Fotyma, E.; Fotyma, M.; Pietruch, C. The content of mineral N in arable soils in Poland. *Fertil. Fertil.* **2004**, *3*, 11–54.
43. Łukowiak, R.; Grzebisz, W. Effect of site specific nitrogen management on seed nitrogen—A driving factor of winter oilseed rape (*Brassica napus* L.) yield. *Agronomy* **2020**, in press.
44. Möller, K.; Stinner, W. Effects of different manuring systems with and without biogas digestion on soil mineral nitrogen content and on gaseous nitrogen losses (ammonia, nitrous oxides). *Eur. J. Agron.* **2009**, *30*, 1–16. [CrossRef]
45. Grzebisz, W.; Łukowiak, R.; Biber, M.; Przygocka-Cyna, K. Effect of multi-micronutrient fertilizers applied to foliage on nutritional status of winter oilseed rape and development of yield forming elements. *J. Elem.* **2010**, *15*, 477–491. [CrossRef]
46. Wang, Y.; Liu, T.; Li, X.; Ren, T.; Cong, R.; Lu, J. Nutrient deficiency limits population development, yield formation and nutrient uptake of direct sown winter oilseed rape. *J. Integr. Agric.* **2015**, *14*, 670–680. [CrossRef]
47. Bouchet, A.S.; Laperche, A.; Bissuel-Belaygue, C.; Snowdon, R.; Nesi, N.; Stahl, A. Nitrogen use efficiency in rapeseed. A review. *Agron. Sustain. Dev.* **2016**, *36*, 38. [CrossRef]
48. Szczepaniak, W.; Grzebisz, W.; Barłóg, P.; Przygocka-Cyna, K. Mineral composition of winter oilseed rape (*Brassica napus* L.) seeds as a tool for oil seed prognosis. *J. Cent. Eur. Agric.* **2017**, *18*, 196–213. [CrossRef]

49. Grzebisz, W.; Szczepaniak, W.; Grześ, S. Sources of nutrients for high-yielding winter oilseed rape (*Brassica napus* L.) during post-flowering growth. *Agronomy* **2020**, *10*, 626. [CrossRef]
50. Barłóg, P.; Grzebisz, W. Effect of timing and nitrogen fertilizer application on winter oilseed rape (*Brassica napus* L.). II. Nitrogen uptake dynamics and fertilizer efficiency. *J. Agron. Crop Sci.* **2004**, *190*, 314–323. [CrossRef]
51. Wang, X.; Mathieu, A.; Cournede, P.H.; Allirand, J.M.; Jullien, A.; de Reffye, P.; Zhang, B.G. Variability and regulation of the number of ovules, seeds, and pods according to assimilate availability in winter oilseed rape (*Brassica napus* L.). *Field Crops Res.* **2011**, *122*, 60–69. [CrossRef]
52. Zhang, Y.; Lu, P.; Ren, T.; Lu, J.; Wang, L. Dynamics of growth and nitrogen capture in winter oilseed rape hybrid and line cultivars under contrasting N supply. *Agronomy* **2020**, *19*, 1183. [CrossRef]
53. Kautz, T.; Walter, A.; Kemna, A.; Vetterlein, D.; Ewert, F.; Wiesenberg, G.L.B.; Schneider, H.; Scherer, H.W.; Vanderborght, J.; Munch, J.; et al. Nutrient acquisition from arable subsoils in temperate climates: A review. *Soil Biol. Biochem.* **2013**, *57*, 1003–1022. [CrossRef]
54. White, P.J.; Broadley, M.R. Calcium in plants. *Ann. Bot.* **2003**, *92*, 487–511. [CrossRef]
55. Łukowiak, R.; Barłóg, P.; Grzebisz, W. Soil mineral nitrogen and the rating of $CaCl_2$ extractable nutrients. *Plant Soil Environ.* **2017**, *63*, 177–183.
56. Barraclough, P.B. Root growth, macro-nutrient uptake dynamics and soil fertility requirements of a high-yeilding winter oilseed rape crop. *Plant Soil* **1989**, *119*, 59–70. [CrossRef]
57. Łukowiak, R.; Grzebisz, W.; Sassenrath, G.F. New insights into phosphorus management in agriculture. *Sci. Total Environ.* **2016**, *542*, 1062–1077. [CrossRef]
58. Nieder, R.; Dinesh, K.B.; Schere, H.W. Fixation and defixation of ammonium in soils: A review. *Biol. Fertil. Soil* **2011**, *47*, 1–14. [CrossRef]
59. Grinsted, M.J.; Hedley, M.J.; White, R.E.; Nye, P.H. Plant induced changes in the rhizospehere of rape (*Brassica napus* var. *Emerald) seedlings. New Phytol.* **1982**, *91*, 19–29. [CrossRef]
60. Pan, Y.; Lu, Z.; Lu, J.; Li, X.; Cong, R. Effects of low sink demand on leaf photosynthesis under potassium deficiency. *Plant Physiol. Bioch.* **2017**, *113*, 110–121. [CrossRef]
61. Grzebisz, W.; Szczepaniak, W.; Barłóg, P.; Przygocka-Cyna, K.; Potarzycki, J. Phosphorus sources for winter oilseed rape (*Brassica napus* L.) during reproductive growth–magnesium sulfate management impact on P use efficiency. *Arch. Agron. Soil Sci.* **2018**, *64*, 1646–1662. [CrossRef]

Article

The Effect of Farmyard Manure and Mineral Fertilizers on Sugar Beet Beetroot and Top Yield and Soil Chemical Parameters

Lukáš Hlisnikovský *, Ladislav Menšík, Kateřina Křížová and Eva Kunzová

Crop Research Institute in Prague, Drnovská 507/73, Prague 6–Ruzyně, 161 06 Prague, Czech Republic; ladislav.mensik@vurv.cz (L.M.); krizovak@vurv.cz (K.K.); kunzova@vurv.cz (E.K.)
* Correspondence: l.hlisnik@vurv.cz; Tel.: +420-773-636-546

Abstract: In order to recommend the dose of fertilization for sugar beet under currently unstable weather conditions, we analysed beetroot and top yields, sugar content (SC), and the effect of fertilization on soil chemistry over a three-year period (2016–2018). All three years were characterized by different weather conditions. The year 2016 was very warm and very dry. The year 2017 was warm with normal precipitation. The year 2018 was extraordinary warm and very dry. We compared the following ten fertilization treatments: unfertilized control, farmyard manure (FYM), mineral fertilizers NPK1–4, and FYM + NPK1–4. The applications of FYM, NPK, and FYM + NPK resulted in significantly higher yields of beetroots and tops as compared with the control, while no significant differences were recorded among FYM, NPK, and FYM + NPK treatments. The SC was not affected by the fertilization. The application of NPK resulted in a lower pH value, while the highest values were recorded for the control and FYM treatments. The application of FYM + NPK increased the content of organic carbon (Corg) in the soil, the total content of nitrogen (Ntot), and P and K concentrations. According to the results of the linear-plateau model, the recommended dose of N is 112 kg ha^{-1}, corresponding to a beetroot yield of 66 t ha^{-1}.

Keywords: *Beta vulgaris* L.; organic manure; weather conditions; soil chemistry; sugar concentration

Citation: Hlisnikovský, L.; Menšík, L.; Křížová, K.; Kunzová, E. The Effect of Farmyard Manure and Mineral Fertilizers on Sugar Beet Beetroot and Top Yield and Soil Chemical Parameters. *Agronomy* **2021**, *11*, 133. https://doi.org/10.3390/agronomy11010133

Received: 14 December 2020
Accepted: 8 January 2021
Published: 12 January 2021

1. Introduction

Sugar beet is one of the most important crops in the EU, as it is the only raw material for sugar extraction. Sugar beet acts as a good breaker of cereal crop rotations in the field and is also a good pre-crop for cereals (except for spring barley [1]), which are the most abundant arable crops in the EU. During most of the 20th century, sugar beet was a strategic crop in the Czech Republic. With the change from a centrally planned economy to a market economy in 1989, followed by the application of EU quotas restricting beetroot yields, sugar beet has undergone significant changes both in regards to the size of sown areas and in yields per hectare. Today, sugar beet is grown on an average area of 61,000 ha in the Czech Republic.

Beetroot and top yield and the quality of sugar beet are affected by a wide range of factors. Some of these factors are controllable by the farmers, such as crop rotation [2], tillage practices [3–6], or fertilization, however, some of them are not, such as weather conditions [7]. Fertilization represents a crucial factor influencing the final yield and quality, especially fertilization with nitrogen (N). The under application of N leads to a lower yield of beetroots and lower sucrose yield, while an over–application of N leads to imbalanced partitioning of assimilates, decreased sucrose content, and increased concentrations of impurities, resulting in reduced sucrose extraction [3,8–12] due to higher water retention by the beetroots and a lower amount of dry matter. An over–application of N also increases the concentration of soluble N compounds in the beetroots and this prevents subsequent extraction of sugar.

The determination of the optimal nitrogen dose varies from site to site, and therefore is site-specific dose. According to Chatterjee et al. [12], a single dose of 146 kg ha^{-1} of N was recommended in North Dakota and Minnesota for sugar beet, irrespective of soil type and soil organic matter content, but this recommendation should be lowered to 112 kg ha^{-1} of N, based on their two years of research. According toDeBruyn et al. [13], a dose of 157 kg ha^{-1} of N was associated with the highest beetroot yield, while 136 kg ha^{-1} of N was associated with the highest profit, in their three year experiment in Canada. In Europe, much attention is being paid to sugar beet nutrition experiments. Islamgulov et al. [14] experimented with the hybrid Hercules and found that 160 kg ha^{-1} of N provided the highest economic efficiency under the conditions of the middle Cis-Ural region. According to Malnou et al. [15], who analysed the response of sugar beet to N fertilization at five sites within the UK, a dose of 100 kg of N per ha, in the absence of organic manure, should be applied for maximum yield. Similar results (100–110 kg ha^{-1} of N) were published by Jaggard et al. [16], who analysed 161 experiments from England in their meta-analysis. The optimal dose can be determined by modelling. There are several models applicable depending on the crop evaluated, the data obtained, and the answer to the question being asked [17]. The quadratic model offers an answer to the maximum yield depending on the dose of nutrients. This model is very suitable for winter wheat because, with an increasing dose of nitrogen, wheat yields initially increase and begin to decline after reaching a critical value [18]. However, determining the dose of nutrients, in this way, may not be statistically significantly different from the lower dose of applied nutrients. Not every crop follows a parabolic course for the dependence of yields and doses of applied nutrients. For example, the reaction of sugar beet yields on doses of nitrogen may be linear [19], even the differences between the analysed fertilizer treatments are not significant. In that case, a linear-plateau model can provide useable answers [12,17,19].

Previously, the sugar beet crop, in the Czech Republic, was commonly fertilized with organic manures. We deliberately state "previously", because today's situation is completely different. There is a shortage of organic manure due to a reduction in animal production and there has been a significant split between animal and plant production, manifested by an insufficient amount of organic matter incorporated into the soil. The common doses of farmyard manure applied directly to potatoes and sugar beet range from 20 to 40 tons per hectare in the Czech Republic. As compared with mineral fertilizers, the content of nutrients in organic manures is non-standardized. Thus, the nutrient content may vary, depending on the animals from which it came, their diet, and other aspects. The mineralization process is also strongly dependent on weather conditions [20], and therefore farmers may not know exactly how much nutrients they applied to the soil, which may explain the recommendation to not use farmyard manure for sugar beet fertilization [21,22]. However, rising prices for mineral fertilizers [21] and the practice of growing sugar beet for the organic market [23] have increased the interest in the application of organic manures to sugar beet, especially in the USA, because the application of manures directly to sugar beet has a long tradition in Europe. Organic manures work in two ways. The first way represents direct releasing of nutrients into the soil environment through the process of mineralization. The second way represents the beneficial influence on the soil's physical, chemical, and biological properties [24–29], especially maintaining and increasing soil organic carbon (SOC) content. This indirect positive effect of livestock manure on crop yields was evidenced by Hlisnikovský et al. [18].

Concerning the issues discussed above, we analysed a three-year sequence in a long-term field experiment, and focused on how mineral fertilizers (different doses, NPK1–4), farmyard manure (FYM), and combinations of FYM and NPK (FYM + NPK1–4) affected the yield and quality of sugar beet beetroots and tops. In this paper, we also recommend the dose of fertilizers according to the linear-plateau regression model. The evaluation included three years (2016, 2017, and 2018). All three years were characterized by different weather conditions. The year 2016 was very warm and very dry, but with relatively good conditions for sugar beet. The year 2017 was warm, with normal precipitation. The year 2018 was extraordinary warm and very dry, significantly affecting sugar beet beetroot and top yields, therefore, in our experiment, we covered the unfavourable conditions that occurred more frequently and were connected with global weather change. Finally, an analysis of soil properties affected by the fertilizer treatments is also provided.

2. Materials and Methods

2.1. Site Description

The long-term field trial was located on the western border of the city of Prague (the Czech Republic, Central Europe, temperate climate zone, 50°05′15″ N, 14°17′28″ E). The trial was established to study the effect of different fertilizer treatments and crop rotations on yield and quality of arable crops and soil chemical properties. The year the trial was established was 1954. The annual mean precipitation and mean temperature from the establishment of the trial is shown in Figure 1. The standard climatological long-term average (1954–2019) precipitation and temperature was 490.4 mm and 8.65 °C, respectively. The standard climatological normal (1961–1990) of the precipitation and temperature was 472.8 mm and 7.97 °C, respectively. The average annual precipitation for the years 2016, 2017, and 2018 was 382.1, 470.0, and 345.3 mm, respectively. The average annual temperature for the same years was 10.0, 9.9, and 11.1 °C, respectively. The average temperature at the site had an increasing trend, and total precipitation also increased slightly (Figure 1). According to Kožnarová and Klabzuba [30], all three years were characterized by different weather conditions. The year 2016 was very warm and very dry, with conditions relatively good for sugar beet. The year 2017 was warm with normal precipitation, providing optimal conditions for sugar beet. The year 2018 was extraordinary warm and very dry, significantly reducing beetroot and top yields. The altitude of the trial site is 370 m a.s.l. According to the World Reference Base [31], the soil type is haplic Luvisol.

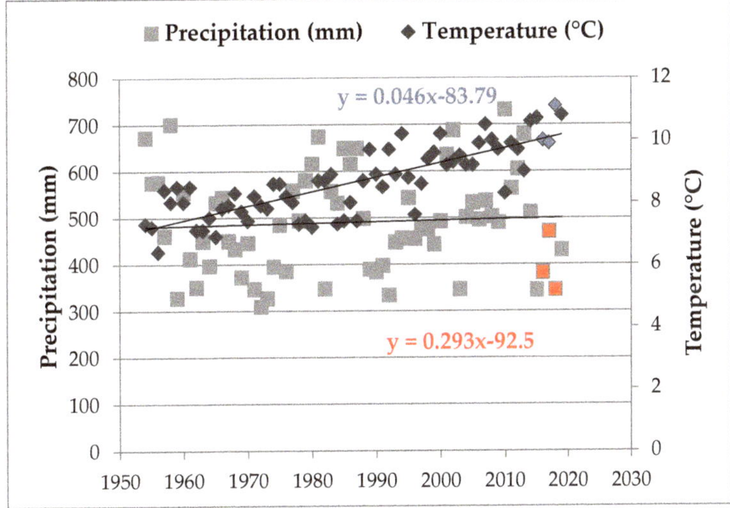

Figure 1. The mean annual precipitation (mm) and temperature (°C) at the experimental site in Prague (1954–2019). Blue and red squares indicate analysed years, blue is the linear regression equation for temperature and red is thelinear regression equation for precipitation.

2.2. Experimental Design Description

The long-term field trial consisted of five fields, marked as I, II, III, IV, and B. Each field consisted of 96 experimental plots (12 × 12 m), where 24 different fertilizer treatments were applied in four replications (24 × 4 = 96). Each field was arranged in a completely randomized block design. The results used, in this paper, analysed the yield and quality of sugar beet from fields IV (2016), III (2017), and II (2018). The crop rotation in these fields was equal, consisting of red clover, red clover, winter wheat, sugar beet, spring barley, potatoes, winter wheat, sugar beet, and spring barley. In this paper, we analysed the sugar beet following the red clover-winter wheat sequence.Among the 24 fertilizer treatments, the following 10 fertilizer treatments were analysed in this paper: (1) the control (unfertilized since 1954), (2) NPK1, (3) NPK2, (4) NPK3, (5) NPK4, (6) the farmyard manure (FYM), (7) FYM + NPK1, (8) FYM + NPK2, (9) FYM + NPK3, and (10) FYM + NPK4. The FYM was applied in October before moderate deep tillage (0.2 m) at a dose of 21 t ha^{-1}. The content of nutrients in the applied FYM was approximately 105 kg, 39 kg, and 124 kg of N, P, and K ha^{-1}, respectively. The doses of mineral N, P, and K are shown in Table 1.

Table 1. Doses of applied N, P, and K in the analysed fertilizer treatments.

Fertilizer Treatment	N (kg ha^{-1})	P (kg ha^{-1})	K (kg ha^{-1})
NPK1	80	64	150
NPK2	120	64	150
NPK3	160	80	200
NPK4	200	80	200
FYM + NPK1	(105) + 80	(39) + 64	(124) + 150
FYM + NPK2	(105) + 120	(39) + 64	(124) + 150
FYM + NPK3	(105) + 160	(39) + 80	(124) + 200
FYM + NPK4	(105) + 200	(39) + 80	(124) + 200

Note: Values in parentheses represent the expected amount of nutrients provided by the FYM.

Mineral N was applied as lime ammonium nitrate (27% N), mineral P as the superphosphate (8.3% P), and mineral K as potassium chloride (49.8% K). Mineral P and K fertilizers were applied in autumn and were incorporated into the soil by moderate deep tillage (0.2 m). Mineral N was applied in the spring, before the beet planting. The harvest of the sugar beet was in October. The sugar beet tops from each experimental plot were separated from the beetroots by hand trimmers and weighed in a net using a mobile digital scale. The harvest of sugar beet beetroots was done using a root crops digger. The beetroots were, then, weighted in the same way as the sugar beet tops, using the nets and mobile digital scale.

2.3. *Sugar Beet Analyses*

2.3.1. Nitrogen and Phosphorus Content in Plant Materials

Nitrogen and phosphorus contained in plant tissues were determined by mineralization with a mixture of sulfuric acid, hydrogen peroxide, and selenium. A portion of the analysed sample was oxidized with hydrogen peroxide in concentrated sulfuric acid. After decomposition of the hydrogen peroxide and distillation of the water, the mineralization was completed by boiling with sulfuric acid under the catalytic action of selenium. The resulting solution was analysed using a San plus System SKALAR analyser (Skalar Analytical B.V., Breda, The Netherlands).

2.3.2. The Contents of K, Ca, Mg, and Na in Plant Materials

The contents of K, Ca, Mg, and Na in plant tissues were determined by oxidation with hydrogen peroxide in a concentrated nitric acid medium in a closed system with a controlled temperature rise, using a Milestone microwave digestion system (Milestone Inc., Sorisole, Italy). The final analysis was carried out using a ICP–OES Trace Scan device (Thermo Jarrel Ash, Trace Scan, Franklin, TN, USA).

2.3.3. Sugar Content Analysis

Sampling and determination of sugar content were performed following the ČSN 46 2110 standard. Laboratory processing of whole sugar beet beetroots bean with its cleaning and subsequent mechanical processing. Beetroots were cut into slices to represent its entire profile. In this way, a sample was taken from each beetroot and further grated on a mechanical grater. The grated material was thoroughly mixed to be sufficiently homogeneous. From the sample, thus prepared, 26 g was again weighed, put into a beaker, and circumfused with the extraction solution. This was followed by heating in a water bath heated to 80 °C for 30 min. After this time, the samples were cooled to room temperature, filtered through a filter, and the filtrate was poured through a tube of an automatic polarimeter (Polarimeter MCP 200, Anton Paar, Graz, Austria).

2.4. Soil Analysis

The samples of the soil (Ap horizon, 0–30 cm) were taken using a soil probe. Totally, four soil samples were taken from each experimental plot. The pH value was analysed potentiometrically (inoLab pH 730, WTW, Xylem Analytics, Weilheim, Germany). The content of soil organic carbon (Corg) was analysed according to [2,3]. The content of nitrogen (Ntot) was done using sulfuric acid in the heating block (Tecator, Sweden), and following the Kjeldahl method [32]. The contents of soil P, K, Ca, and Mg were analysed via the Mehlich III solution [33], followed by ICP-OES analysis (Thermo Scientific iCAP 7400 Duo, Thermo Fisher Scientific, Cambridge, UK).

2.5. Data Analyses

The analysis of variance (ANOVA) was used to evaluate the effect of fertilizer treatment in one season. For the evaluation of fertilizer treatment, season, and their interaction, the multivariate analysis of variance (MANOVA) was used. Both analyses were followed by Tukey's HSD post hoc test to select the treatments and seasons that differentiated significantly. To perform all analyses, we used STATISTICA 13.3 software (TIBCO Software, Palo Alto, CA, USA). The linear-plateau model was calculated using the R software (R: A language and environment for statistical computing. R Foundation for Statistical Computing, Vienna, Austria, 2020), together with the three R packages [34–36].

3. Results

3.1. The Effect of Farmyard Manure (FYM) on Sugar Beet Beetroot and Top Yield

If we compare the effect of manure application, we find that the beetroot yield in the observed period (2016–2018) was significantly affected by both the fertilization treatment (d.f. = 1, F = 13.58, $p < 0.001$) and especially the weather conditions (d.f. = 2, F = 73.48, $p < 0.002$). The effect of the interaction between the treatment and year was also significant (d.f. = 2, F = 4.29; $p < 0.03$). The conditions of the year had the highest impact on beetroot yield (80%), followed by the fertilizer treatment (15%), and their interaction (5%).

The application of the FYM provided comparable results as the control. Significantly higher yields were recorded only in 2016 (Table 2). The average beetroot yield was 52.9 t ha^{-1} in the control, and 61.2 t ha^{-1} in the FYM treatment (2016–2018, Table 2). Comparing the years, the average yield was 66.2 t ha^{-1} and 67.2 t ha^{-1} in 2016 and 2017, respectively (without a statistical difference), while the significantly lower yield was recorded in 2018 (37.8 t ha^{-1}) (Table 2).

Table 2. The beetroot and top yield as affected by the fertilizer treatment (control and farmyard manure (FYM)) and year (2016–2018).

	Beetroot Yield (t ha^{-1})				Top Yield (t ha^{-1})			
	2016	2017	2018	$\bar{X}$	2016	2017	2018	$\bar{X}$
Control	57.4 ± 4.3A	65.7 ± 2.0A	35.6 ± 4.0A	52.9 ± 4.3A	20.4 ± 1.5A	22.8 ± 0.6A	9.0 ± 0.1A	17.4 ± 1.9A
FYM	75.0 ± 1.1B	68.7 ± 1.0A	39.9 ± 2.3A	61.2 ± 4.7B	23.8 ± 0.6A	24.8 ± 1.7A	9.6 ± 0.9A	19.4 ± 2.2B
	66.2 ± 3.9Bb	67.2 ± 1.2b	37.8 ± 2.3a		22.1 ± 1.0b	23.8 ± 0.9b	9.3 ± 0.4a	

Note: The mean values with the standard error of the mean followed by the same letter (small letters "a", horizontally; and big letters "A", vertically) are not significantly different (p, 0.05).

In the individual years, the top yield was not affected by the FYM application (Table 2). However, for the entire evaluated period (2016–2018), the differences among the compared treatments (d.f. = 1, F = 5.5, $p < 0.03$) and years (d.f. = 2, F = 113.0, $p < 0.001$) were significant.

While the effect of the year was 95%, the effect of fertilization was only 5%. As in the case of beetroots, this means that the differences between the compared fertilization treatments were very low, while the fluctuation between the years was very high (caused mainly by the severe drought in 2018). The average top yield was 17.4 t ha^{-1} in the control, while it was 19.4 t ha^{-1} in the FYM treatment. Comparing the years, the highest yields were recorded in 2017 (23.8 t ha^{-1}), followed by 2016 (22.1 t ha^{-1}), and 2018 (9.3 t ha^{-1}) (Table 2).

3.2. The Effect of Mineral NPK on Sugar Beet Beetroot and Top Yield

If we compare the entire period (2016–2018), the application of mineral NPK fertilizers generally increased the beetroot yield significantly (Table 3). According to MANOVA, the beetroot yield was mainly affected by the year (d.f. = 2, F = 146.3, $p < 0.0001$, 92%), showing a very high fluctuation among the years. The highest average yield was recorded in 2017 (72.2 t ha^{-1}), followed by 2016 (68.2 t ha^{-1}), and 2018 (44.4 t ha^{-1}). The effect of the fertilizer treatment was also significant (d.f. = 4, F = 11.4, $p < 0.001$), but the only significant difference was recorded between the control and NPK treatments. However, no significant differences among NPK1–4 treatments were recorded over the entire period (Table 3). The average beetroot yield was 52.9 t ha^{-1} (control), 61.3 t ha^{-1} (NPK1), 62.7 t ha^{-1} (NPK3), 63.3 t ha^{-1} (NPK2), and 67.7 t ha^{-1} (NPK4). When only NPK treatments were considered, yield response to N rates across three years plateaued at 112 kg ha^{-1} N with a corresponding beetroot yield of 66 t ha^{-1} (Figure 2, left).

Table 3. The beetroot and top yield as affected by the fertilizer treatment (control, NPK1–4) and years (2016–2018).

	Beetroot Yield (t ha^{-1})				Top Yield (t ha^{-1})			
	2016	2017	2018	$\overline{X}$	2016	2017	2018	$\overline{X}$
Control	57.4 ± 4.3A	65.7 ± 2.0A	35.6 ± 4.0A	52.9 ± 4.3A	20.4 ± 1.5A	22.8 ± 0.6A	9.0 ± 0.1A	17.4 ± 1.9A
NPK1	70.3 ± 1.3AB	72.9 ± 3.1A	40.6 ± 1.9AB	61.3 ± 4.6B	27.8 ± 1.3B	28.2 ± 0.7B	11.8 ± 0.6B	22.6 ± 2.3B
NPK2	70.7 ± 3.8AB	73.5 ± 2.5A	45.7 ± 0.7ABC	63.3 ± 4.0B	29.3 ± 1.5B	28.8 ± 1.1B	13.0 ± 0.5B	23.7 ± 2.4B
NPK3	68.3 ± 2.7AB	73.8 ± 2.3A	46.1 ± 0.8BC	62.7 ± 3.8B	32.6 ± 1.3B	30.5 ± 1.3BC	12.0 ± 0.4B	25.0 ± 2.8BC
NPK4	74.5 ± 4.3B	75.0 ± 2.0A	53.8 ± 2.6C	67.7 ± 3.4B	31.4 ± 0.7B	35.1 ± 1.9C	13.8 ± 0.6B	26.8 ± 2.9C
	68.2 ± 1.9b	72.2 ± 1.2b	44.4 ± 1.7a		28.3 ± 1.1b	29.1 ± 1.0b	11.9 ± 0.4a	

Note: The mean values with the standard error of the mean followed by the same letter (small letters "a", horizontally and big letters "A", vertically) are not significantly different (*p*, 0.05).

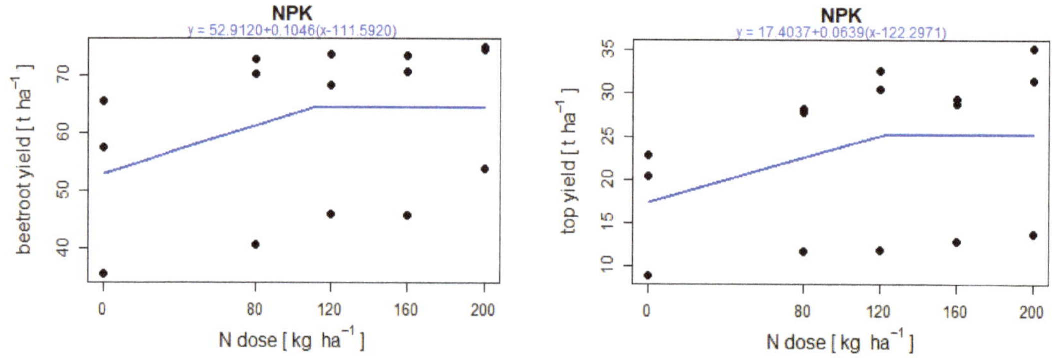

Figure 2. Means (black dots) of sugar beet beetroot yield (**left**) and top yield (**right**) at different N rates of NPK treatments in 2016, 2017, and 2018 combined and their linear-plateau regression (blue line).

In the case of the sugar beet top yield, the effect of the year (d.f. = 2, F = 425.9, $p < 0.0001$), fertilizer treatments (d.f. = 4, F = 34.3, $p < 0.001$), and their interactions (d.f. = 8, F = 3.8, $p < 0.002$) was significant. The lowest average top yield over the evaluated period was provided by the control treatment (17.4 t ha^{-1}). Significantly higher top yields were recorded in NPK treatments, with the highest top yield in the NPK4 treatment (26.8 t ha^{-1}) (Table 3). The year again had the greatest impact on the top yield (92%), followed by the fertilization treatment (7%). Comparable top yields were recorded in the years 2016 (28.3 t ha^{-1}) and 2017 (29.1 t ha^{-1}), while a significantly lower top yield was recorded in the dry year 2018 (11.9 t ha^{-1}) (Table 3). According to the linear-plateau model, the mean top yield response to N rates across three years plateaued at 122 kg ha^{-1} N, with a corresponding top yield of 25 t ha^{-1} (Figure 2, right).

3.3. Comparison of the FYM and FYM + NPK Treatments

Over the entire evalutated period (2016–2018), the combined application of the FYM with mineral NPK fertilizers significantly increased the beetroot yields (d.f. = 5, F = 19.6, $p < 0.001$) (Table 4). The lowest yield was recorded in the control (52.9 t ha^{-1}), followed by the FYM treatment (61.2 t ha^{-1}). The addition of mineral NPK fertilizers significantly increased the beetroot yields as compared with the control and FYM treatments (Table 4), ranging from 65.5 t ha^{-1} (FYM + NPK3) to 66.3 t ha^{-1} (FYM + NPK1). The differences among all FYM + NPK treatments were insignificant. The effect of the year was also significant (d.f. = 2, F = 333.7, $p < 0.0001$), as well as the year*treatment interaction (d.f. = 10, F = 2.5, $p = 0.014$). The comparison of years indicated the same results as the previous evaluation. While in the years with relatively favourable conditions (2016 and 2017) the differences were not significant (the average yields were 71.4 t ha^{-1} in 2016 and 72.4 t ha^{-1} in 2017), the conditions of the year 2018 sharply reduced the beetroot yield to an average value of 45.1 t ha^{-1}. The beetroot yield response to different rates of FYM and NPK fertilizers plateaued at 165 kg ha^{-1} N, with a corresponding beet yield 66 t ha^{-1} (Figure 3, left).

A similar effect of mineral fertilizers was found for the top yields. The top yield was significantly affected by the year (d.f. = 2, F = 493.8, $p < 0.0001$), fertilization treatment (d.f. = 5, F = 41.8, $p < 0.0001$), and their interaction (d.f. = 10, F = 4.5, $p < 0.001$). The lowest yields were provided by the control and FYM treatments (17.4 and 19.4 t ha^{-1}, respectively) (Table 4). The addition of mineral fertilizers increased the top yields significantly, ranging from 25.2 t ha^{-1} (FYM + NPK2) to 26.9 t ha^{-1} (FYM + NPK3). The differences between the FYM + NPK treatments were insignificant. Comparing the years, dry conditions during 2018 resulted in the lowest yield of the tops (12.1 t ha^{-1}), while significantly higher yields were recorded in 2016 and 2017 (28.8 and 29.7 t ha^{-1}, respectively). According to the linear-plateau model, the response of the sugar beet tops plateaued at 181 kg ha^{-1} N, with a corresponding yield of 24 t ha^{-1} (Figure 3, right).

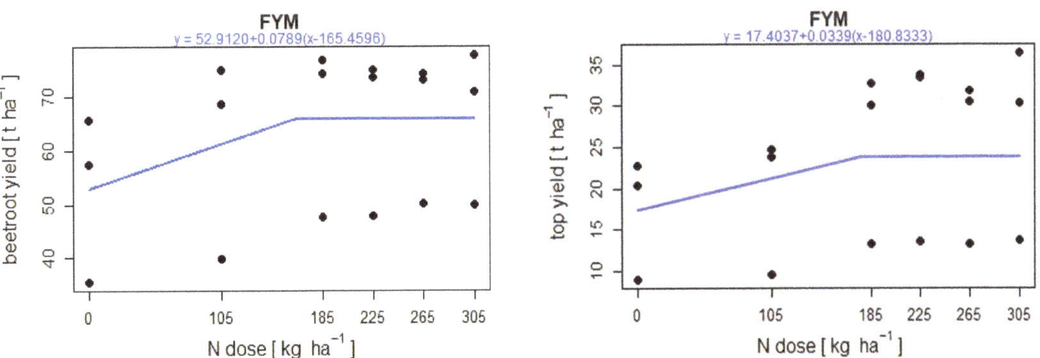

Figure 3. Means (black dots) of sugar beet beetroot yield (**left**) and top yield (**right**) at different N rates applied with the FYM and FYM + NPK treatments in 2016, 2017, and 2018 combined and their linear-plateau regression (blue line).

Table 4. The beetroot and top yield as affected by the fertilizer treatment (control, FYM, FYM + NPK1–4) and years (2016–2018).

	Beetroot Yield (t ha^{-1})				Top Yield (t ha^{-1})			
	2016	**2017**	**2018**	$\overline{X}$	**2016**	**2017**	**2018**	$\overline{X}$
Control	57.4 ± 4.3A	65.7 ± 2.0A	35.6 ± 4.0A	52.9 ± 4.3A	20.4 ± 1.5A	22.8 ± 0.6A	9.0 ± 0.1A	17.4 ± 1.9A
FYM	75.0 ± 1.1B	68.7 ± 1.0AB	39.9 ± 2.3AB	61.2 ± 4.7B	23.8 ± 0.6A	24.8 ± 1.7AB	9.6 ± 0.9A	19.4 ± 2.2A
FYM+ NPK1	76.9 ± 1.5B	74.4 ± 1.7BC	47.6 ± 0.9BC	66.3 ± 4.1C	32.8 ± 1.2B	30.1 ± 1.2BC	13.2 ± 0.3B	25.4 ± 2.7B
FYM + NPK2	73.3 ± 1.5B	74.3 ± 1.5BC	50.0 ± 1.4C	65.9 ± 3.5BC	31.9 ± 1.1B	30.5 ± 1.9BC	13.3 ± 0.4B	25.2 ± 2.6B
FYM + NPK3	75.1 ± 1.0B	73.6 ± 0.6BC	47.8 ± 1.8BC	65.5 ± 3.8BC	33.7 ± 0.2B	33.4 ± 0.9C	13.5 ± 0.3B	26.9 ± 2.9B
FYM + NPK4	70.9 ± 3.0B	77.8 ± 2.1C	50.0 ± 1.3C	66.2 ± 3.8C	30.3 ± 1.0B	36.4 ± 2.2C	13.8 ± 0.2B	26.8 ± 3.0B
	71.4 ± 1.6b	72.4 ± 1.0b	45.1 ± 1.4a		28.8 ± 1.1b	29.7 ± 1.1b	12.1 ± 0.4a	

Note: The mean values with the standard error of the mean followed by the same letter (small letters "a", horizontally; and big letters "A", vertically) are not significantly different (p, 0.05).

3.4. The Effect of Fertilization on Sugar Content (SC) and Chemical Elements Concentration

We must admit that due to limited funds, analyses of sugar beet in reduced quantities were performed over the years 2016–2018. This means that no repeated measurements were performed from each fertilizer treatment every single year. Therefore, the results of the statistical analysis presented here represent the average results for the entire analysed period. It is, therefore, necessary to take the results with a grain of salt.

According to the statistical analysis, no significant differences were recorded between the fertilizer treatments for any analysed parameter (the SC and the concentration of N, P, K, Ca, Mg, and Na) of the sugar beetroots. The SC varied from 19.7% (NPK4) to 21.9% (NPK1) (Table 5). The concentration of N, P, K, Ca, Mg, and Na was not affected by the fertilizer treatment (Table 5).

Table 5. The sugar content (%) and concentrations of N, P, K, Ca, Mg, and Na (%) in sugar beet beetroots as affected by the fertilizer treatment and over the years 2016–2018.

	SC (%)	N (%)	P (%)	K (%)	Ca (%)	Mg (%)	Na (%)
Control	19.9 ± 0.8	0.19 ± 0.01	0.02 ± 0.01	0.16 ± 0.03	0.07 ± 0.01	0.05 ± 0.01	0.006 ± 0.002
NPK1	21.9 ± 1.3	0.20 ± 0.02	0.02 ± 0.01	0.17 ± 0.03	0.05 ± 0.01	0.05 ± 0.01	0.005 ± 0.001
NPK2	20.5 ± 0.9	0.20 ± 0.01	0.08 ± 0.06	0.17 ± 0.02	0.16 ± 0.11	0.05 ± 0.01	0.006 ± 0.001
NPK3	19.8 ± 0.3	0.19 ± 0.02	0.02 ± 0.01	0.16 ± 0.03	0.05 ± 0.01	0.05 ± 0.01	0.006 ± 0.001
NPK4	19.7 ± 0.9	0.20 ± 0.02	0.02 ± 0.01	0.18 ± 0.04	0.04 ± 0.01	0.05 ± 0.01	0.007 ± 0.001
FYM	20.1 ± 0.5	0.19 ± 0.01	0.02 ± 0.01	0.15 ± 0.02	0.05 ± 0.01	0.05 ± 0.01	0.007 ± 0.001
FYM + NPK1	21.1 ± 1.1	0.20 ± 0.02	0.02 ± 0.01	0.16 ± 0.02	0.05 ± 0.01	0.05 ± 0.01	0.008 ± 0.001
FYM + NPK2	21.2 ± 0.8	0.22 ± 0.02	0.02 ± 0.01	0.20 ± 0.03	0.05 ± 0.01	0.05 ± 0.01	0.010 ± 0.003
FYM + NPK3	19.8 ± 0.7	0.20 ± 0.02	0.02 ± 0.01	0.16 ± 0.03	0.05 ± 0.01	0.05 ± 0.01	0.007 ± 0.001
FYM + NPK4	20.7 ± 0.6	0.21 ± 0.01	0.02 ± 0.01	0.17 ± 0.03	0.04 ± 0.01	0.05 ± 0.01	0.007 ± 0.001

Note: The mean values without letters were not significantly different.

Similar results were recorded in the case of the sugar beet tops, where the concentrations of N, P, K, Ca, and Mg were analysed. Except for P, the effect of the fertilizer treatment was insignificant. All results are shown in Table 6. In the case of P, the mean concentration varied from 0.15% (control) to 0.23% (NPK4 and FYM + NPK1 treatments). Higher concentrations of the P were found in the FYM + NPK treatments as compared with the control, FYM, and NPK treatments (Table 6).

Table 6. The concentrations of N, P, K, Ca, and Mg in sugar beet tops as affected by the fertilizer treatment and over the years 2016–2018.

	N (%)	P (%)	K (%)	Ca (%)	Mg (%)
Control	2.54 ± 0.21	0.15 ± 0.01A	3.66 ± 0.17	1.38 ± 0.12	0.85 ± 0.10
NPK1	2.39 ± 0.23	0.19 ± 0.01 AB	4.35 ± 0.21	1.26 ± 0.15	0.84 ± 0.10
NPK2	2.63 ± 0.15	0.20 ± 0.01 AB	4.43 ± 0.24	1.15 ± 0.18	0.79 ± 0.11
NPK3	2.67 ± 0.17	0.19 ± 0.01 AB	4.11 ± 0.11	1.25 ± 0.17	0.95 ± 0.10
NPK4	3.03 ± 0.16	0.23 ± 0.02B	4.20 ± 0.09	1.14 ± 0.17	0.84 ± 0.13
FYM	2.51 ± 0.22	0.18 ± 0.01AB	3.36 ± 0.56	1.25 ± 0.06	0.83 ± 0.02
FYM + NPK1	2.71 ± 0.29	0.23 ± 0.03B	3.85 ± 0.18	1.04 ± 0.11	0.76 ± 0.06
FYM + NPK2	2.96 ± 0.24	0.22 ± 0.01B	4.14 ± 0.06	1.10 ± 0.10	0.82 ± 0.06
FYM + NPK3	3.07 ± 0.12	0.22 ± 0.01B	3.55 ± 0.08	1.09 ± 0.06	0.89 ± 0.05
FYM + NPK4	2.89 ± 0.11	0.22 ± 0.01B	3.94 ± 0.19	1.08 ± 0.08	0.84 ± 0.01

Note: The mean values with the standard error of the mean followed by the same letter are not significantly different (p, 0.05). Mean values without letters were not significantly different.

3.5. The Effect of the Fertilizer Treatments on the Soil Properties

The application of different combinations and doses of fertilizers did not affect the value of the soil pH. The average values ranged from 6.08 (NPK3) to 6.60 (FYM). The concentration of N was slightly affected by the fertilizer treatment. The lowest concentrations were recorded in the control and FYM treatments (0.13%), while the highest concentrations were recored in the FYM+NPK4 treatment (0.16%). All other treatments provided results fitting within these extreme limits. In the case of soil carbon content, the distribution of the fertilizer treatments is clearer. The lowest C concentration was recorded in the control treatment (0.99%). All FYM + NPK treatments differed significantly from this value, and ranged from 1.26% to 1.35%, while the FYM and all NPK treatments filled the space between the control and FYM + NPK treatments. The concentration of soil P significantly varied among the treatments with lowest concentration in the control (20 mg kg^{-1}) and FYM (29 mg kg^{-1}) treatments and highest concentrations in NPK4 (70 mg kg^{-1}) and FYM + NPK4 (93 mg kg^{-1}) treatments. A similar pattern was recorded in the case of K (lowest concentrations were in the control and FYM treatments, while the highest concentrations were in the NPK4 and FYM + NPK2 treatments) (Table 7). The concentrations of Ca and Mg were not affected by the fertilizer treatment (Table 7).

Table 7. The basic soil chemical properties as affected by the fertilizer treatment and over the years 2016–2018.

	pH	Ntot (%)	Corg (%)	P (mg kg^{-1})	K (mg kg^{-1})	Ca (mg kg^{-1})	Mg (mg kg^{-1})
Control	6.44 ± 0.16A	0.13 ± 0.01A	0.99 ± 0.04A	20 ± 4A	150 ± 7A	3097 ± 104	180 ± 9
NPK1	6.26 ± 0.18B	0.14 ± 0.01AB	1.10 ± 0.03AB	53 ± 10BC	182 ± 8ABC	2813 ± 121	150 ± 10
NPK2	6.20 ± 0.29B	0.14 ± 0.01AB	1.15 ± 0.05AB	59 ± 5CD	198 ± 11ABC	2951 ± 231	149 ± 3
NPK3	6.08 ± 0.08C	0.13 ± 0.01A	1.16 ± 0.05AB	52 ± 2BC	173 ± 14AB	2790 ± 99	147 ± 16
NPK4	6.20 ± 0.12B	0.14 ± 0.01AB	1.22 ± 0.04AB	70 ± 1CDE	195 ± 10ABC	2800 ± 139	141 ± 7
FYM	6.60 ± 0.17A	0.14 ± 0.01AB	1.19 ± 0.05AB	29 ± 5AB	166 ± 15A	3240 ± 214	187 ± 1
FYM + NPK1	6.46 ± 0.21A	0.15 ± 0.01AB	1.27 ± 0.07B	64 ± 1CDE	199 ± 23ABC	3154 ± 308	176 ± 15
FYM + NPK2	6.34 ± 0.20AB	0.15 ± 0.01AB	1.31 ± 0.10B	87 ± 6DE	246 ± 11CD	3031 ± 244	172 ± 10
FYM + NPK3	6.19 ± 0.11B	0.15 ± 0.01AB	1.26 ± 0.02AB	73 ± 3CDE	202 ± 13ABC	2891 ± 142	170 ± 5
FYM + NPK4	6.26 ± 0.20B	0.16 ± 0.01B	1.35 ± 0.07B	93 ± 12E	254 ± 24D	3022 ± 129	177 ± 3

Note: The mean values with the standard error of the mean followed by the same letter are not significantly different (*p*, 0.05). Mean values without letters were not significantly different.

4. Discussion

As compared with the control, statistically higher beetroot yields in the FYM treatment were recorded only in 2016 (+17.6 t ha^{-1}). In the following years, the application of FYM resulted in comparable yields. If we compare the entire evaluated period (2016–2018), the application of FYM increased the average sugar beet beetroot yield by about 8 t ha^{-1}. In the case of the tops, no differences between the control and the FYM treatments were recorded in individual years. A comparison of the entire analysed period showed that the average top yield was significantly higher in the FYM treatment (+2.0 t ha^{-1}) (Table 2). As mentioned above, the mineralization of manure in the soil strongly depends on the weather and other soil parameters [20]. The years 2016 and 2017 represent seasons with relatively good (2016) and good (2017) conditions, resulting in very high yields in the control (especially in 2017, these yields are very high for the unfertilized control treatment and we assume that they are the result of exceptionally good climatic conditions during the season), and visible effect of the FYM (especially in 2016). The extremely unfavourable weather conditions in 2018 reduced beetroot yields by 43% and top yields by 69%. The explanation for the higher yields in the FYM treatment lies both in the direct supply of nutrients through mineralization and the course of mineralization. According to Barłóg et al. [8], three main periods of beetroot yield formation can be distinguished, i.e., early, midseason, and final period, with N requirements dominating in the first two stages. The FYM is a fertilizer with a high C/N ratio (in comparison with slurries), and contains a high amount of organic N that is not directly available to plants [20], therefore, FYM releases its nutrients slowly and over a longer period, covering critical periods of beet formation (if weather conditions allow it). A similar effect of FYM on sugar beet yield was published in [21]. According to their results, the yield response to different manure ratios across two years plateaued at 23 t ha^{-1}, with a corresponding beet yield of 62.2 t ha^{-1}. The sugar content (SC, %) was not affected by the FYM application. Both treatments (control and FYM) varied from 19.9% to 20.1%. The same situation happened in the case of other chemical elements in the beetroots and tops (Tables 5 and 6), therefore, the application of the FYM provided higher yields and, consequently, a higher amount of sugar harvested from the field, without significant changes in sugar beet chemical composition. This also applies to the comparison of all other fertilization treatments (Tables 5 and 6), except P concentration in sugar beet tops (Table 6), where P concentration slightly increases in NPK4 and all FYM + NPK treatments.

Application of mineral fertilizers significantly increased beetroot and top yield (Table 3). This result is expected and is in line with other published results [6,9,14,21], as N is the most important element for sugar beet and mineral fertilizers provide readily available N in precisely definable amounts. In our research, it was rather crucial to recommend an average dose of nitrogen that provided the best results during the years including both, the standard and the dry weather conditions. In the case of mineral fertilizers applied without organic manures, it is relatively simple and our results are comparable with other recommendations [12,15,16], while lower than recommendations of [13,14], but every experiment provides site-specific recommendations concerning soil and climate conditions of the site. In our case, a dose of 122 kg ha^{-1} N represents a breakpoint between the linear and plateau functions of the developed model, with the corresponding beetroot yield of 66 t ha^{-1}. Application of N above this value does not increase the beetroot yield significantly.

The combined application of FYM and mineral fertilizers (FYM + NPK treatments) had a different course each year. In years with good climatic conditions, the beetroot yield fertilized with the FYM was comparable (2016) or slightly lower (2017, only FYM + NPK4 treatment provided significantly higher yields as compared with the FYM treatment) than in the FYM + NPK treatments (Table 4). A significant difference only became apparent with the advent of drought in 2018, when treatments fertilized with FYM + NPK provided higher yields than treatments fertilized only with FYM. In years with a normal course, manure could cover the demands of beets during the season and provided very good yields. However, in the event of a drought, the efficiency of manure decreased as it responded more sensitively to unsuitable climatic conditions. The positive benefit of mineral fertilizers was also manifested in the case of the tops. The combined application of FYM and mineral fertilizers provided, on average, higher yields than the application of FYM alone. These results were predictable. For this article, it was more important to analyse the response of beetroots and tops to the dose of nutrients and to determine the recommended dose. From this point of view, it is interesting that the breaking point of the linear-plateau model occurred at the same value as in the NPK treatments, i.e., 66 t ha^{-1}, but the amount of nitrogen increased to 165 kg ha^{-1} (+53 kg ha^{-1} N as compared with the NPK treatments). The same situation occurred in the case of the beet tops, where the break occurred at a yield of 24 t ha^{-1}, and at a dose of 181 kg ha^{-1} N (+59 kg ha^{-1} N as compared with the NPK treatments). According to the data, the combined application of FYM and NPK did not bring any massive improvement in yields as compared with NPK or FYM applied alone, showing that maximum yielding potential of the sugar beet was reached under local soil-climate conditions. According to [8], the maximum yield potential of sugar beet in Europe is between 110 and 150 t ha^{-1} (calculations based on [7]), and around 80 t ha^{-1} in Poland, but the farmers' share of the actual yields is only 50 or 60% of that value.

According to the MANOVA results, both, beetroot and top yields were mainly affected by the weather conditions, while the effect of the fertilizer treatment was minor. This was mainly due to the extraordinary dry year in 2018. There is an increasing number of dry years and their frequent occurrence and weather instability, generally, are associated with the current global change in climate conditions. These extreme years should not be surprising in the coming period. The farmers in Europe are already adapting their approaches to this fact by selecting other crop varieties and species and adjusting the timing of cultivation [37].

A slightly different situation is found in the case of soil parameters. Application of NPK without organic manures resulted in generally lower pH values as compared with the control and FYM treatments (Table 7). The applications of FYM + NPK treatments resulted in between these two groups, which mean that FYM reduces the negative impact of NPK on soil pH. The same results were published by [24,38]. By affecting the value of the soil pH, organic manures also modify the environment for and activity of the microbial community in the soil [24], and the availability of nutrients. The concentration of soil N was not affected significantly by the fertilizer treatment, only high doses of applied mineral N (FYM + NPK4) resulted in significantly higher N concentration as compared with the control. In the case of P and K, the highest concentrations of both elements were recorded in FYM + NPK treatments, The combination of FYM with NPK significantly increased the soil C content. This result is in agreement with the results published by [24,39,40]. On the one hand, application of mineral fertilizers without manures can decrease soil carbon content when the C inputs to the soil from arable crops (including straw, roots, and post-harvest residues) are lower than the C decomposed by the soil microbial community. On the other hand, organic manures contain organic matter that directly affects the physiological, chemical, and biological properties of the soil. From this point of view, the combined application of FYM and mineral fertilizers results in maintaining soil fertility and is a sustainable approach to soil care [24,26,39].

5. Conclusions

The decisive factors determining sugar beet beetroot and top yield were weather conditions. During the years with relatively good (2016) and good (2017) conditions, the beetroot and top yield was on average 70 t ha^{-1} and 28.7 t ha^{-1}, respectively. In the extraordinary dry year of 2018, the average beetroot and top yield decreased to 44 t ha^{-1} and 12 t ha^{-1}, respectively.

The application of FYM at a dose of 21 t ha^{-1} significantly increased the beetroot and top yield, if we evaluate the entire period (2016–2018). In individual years, we recorded a significant difference only in the case of beetroots, in 2016. In general, the yields with FYM treatment were always higher than in the case of the non-fertilized control.

The application of mineral fertilizers significantly increased beetroot and top yield as compared with the unfertilized control. According to the results of the linear-plateau model, a suitable N dose is 112 kg ha^{-1} with a corresponding yield of 66 t ha^{-1}. This model took into account the average yields of years including standard and unsuitable climatic conditions.

In the case of the joint application of FYM and NPK, we did not record a significant increase in yield as compared with NPK applied alone.

The sugar content and the concentration of chemical elements in the beetroots and the tops were not significantly affected by the fertilization treatment, except for slightly higher concentrations of P in the tops.

The application of NPK in the soil resulted in lower pH values than we observed in the control and FYM treatments. The combined application of FYM and NPK slightly reduced the negative impact of NPK on soil pH. The application of NPK, FYM, and especially the combination of NPK and FYM significantly increased the content of Corg, P, and K in the soil.

Author Contributions: Conceptualization, L.H., L.M., and E.K.; methodology, L.H. and E.K.; software, L.H., and K.K.; validation, L.H., L.M., and K.K.; formal analysis, L.H., and K.K.; investigation, L.H., L.M., E.K., and K.K.; resources, L.H., and E.K.; data curation, L.H., and E.K.; writing—original draft preparation, L.H.; writing—review and editing, L.H.; visualization, L.H., and K.K.; supervision, E.K.; project administration, L.H., and E.K.; funding acquisition, E.K. All authors have read and agreed to the published version of the manuscript.

Funding: This research was funded by the Ministry of Agriculture of the Czech Republic, grant number RO 0418.

Informed Consent Statement: Informed consent was obtained from all subjects involved in the study.

Data Availability Statement: Data sharing not applicable.

Acknowledgments: We would like to thank the technicians who participated in the experiment and analyses.

Conflicts of Interest: The authors declare no conflict of interest.

References

1. Andrejčíková, M.; Macák, M.; Habán, M. Úrodový potenciál odrôd sladovníckeho jačmeňa (*Hordeum vulgare* L.) v pestovateľských podmienkach juhozápadného Slovenska. *J. Cent. Eur. Agric.* **2016**, *17*, 932–940. [CrossRef]
2. Götze, P.; Rücknagel, J.; Wensch-Dorendorf, M.; Märländer, B.; Christen, O. Crop rotation effects on yield, technological quality and yield stability of sugar beet after 45 trial years. *Eur. J. Agron.* **2017**, *82*, 50–59. [CrossRef]
3. Swędrzyńska, D.; Grześ, S. Impact of long-term tillage systems and different nitrogen fertilization on chemical and biological properties of soil and sugar beet yield. *Fragm. Agron.* **2017**, *34*, 92–106.
4. Koch, H.J.; Dieckmann, J.; Büchse, A.; Märländer, B. Yield decrease in sugar beet caused by reduced tillage and direct drilling. *Eur. J. Agron.* **2009**, *30*, 101–109. [CrossRef]
5. Laufer, D.; Koch, H.J. Growth and yield formation of sugar beet (*Beta vulgaris* L.) under strip tillage compared to full width tillage on silt loam soil in Central Europe. *Eur. J. Agron.* **2017**, *82*, 182–189. [CrossRef]
6. Afshar, R.K.; Nilahyane, A.; Chen, C.; He, H.; Bart Stevens, W.; Iversen, W.M. Impact of conservation tillage and nitrogen on sugarbeet yield and quality. *Soil Tillage Res.* **2019**, *191*, 216–223. [CrossRef]
7. Kenter, C.; Hoffmann, C.M.; Märländer, B. Effects of weather variables on sugar beet yield development (*Beta vulgaris* L.). *Eur. J. Agron.* **2006**, *24*, 62–69. [CrossRef]
8. Barłóg, P.; Grzebisz, W.; Szczepaniak, W.; Peplinski, K. Sugar beet response to balanced nitrogen fertilization with phosphorus and potassium. Part II. dynamics of beet quality. *Bulg. J. Agric. Sci.* **2014**, *20*, 1311–1318.
9. Hergert, G.W. Sugar Beet Fertilization. *Sugar. Tech.* **2010**, *12*, 256–266. [CrossRef]
10. Koch, H.J.; Laufer, D.; Nielsen, O.; Wilting, P. Nitrogen requirement of fodder and sugar beet (*Beta vulgaris* L.) cultivars under high-yielding conditions of northwestern Europe. *Arch. Agron. Soil Sci.* **2016**, *62*, 1222–1235. [CrossRef]
11. Zarski, J.; Kuśmierek-Tomaszewska, R.; Dudek, S. Impact of irrigation and fertigation on the yield and quality of sugar beet (*Beta vulgaris* L.) in a moderate climate. *Agronomy* **2020**, *10*, 166. [CrossRef]
12. Chatterjee, A.; Subedi, K.; Franzen, D.W.; Mickelson, H.; Cattanach, N. Nitrogen fertilizer optimization for sugarbeet in the Red River valley of north Dakota and Minnesota. *Agron. J.* **2018**, *110*, 1554–1560. [CrossRef]
13. DeBruyn, A.H.; O'Halloran, I.P.; Lauzon, J.D.; Van Eerd, L.L. Effect of sugarbeet density and harvest date on most profitable nitrogen rate. *Agron. J.* **2017**, *109*, 2343–2357. [CrossRef]
14. Islamgulov, D.; Alimgafarov, R.; Ismagilov, R.; Bakirova, A.; Muhametshin, A.; Enikiev, R.; Ahiyarov, B.; Ismagilov, K.; Kamilanov, A.; Nurligajnov, R. Productivity and technological features of sugar beet root crops when applying of different doses of nitrogen fertilizer under the conditions of the middle cis-ural region. *Bulg. J. Agric. Sci.* **2019**, *25*, 90–97.
15. Malnou, C.S.; Jaggard, K.W.; Sparkes, D.L. A canopy approach to nitrogen fertilizer recommendations for the sugar beet crop. *Eur. J. Agron.* **2006**, *25*, 254–263. [CrossRef]
16. Jaggard, K.W.; Qi, A.; Armstrong, M.J. A meta-analysis of sugarbeet yield responses to nitrogen fertilizer measured in England since 1980. *J. Agric. Sci.* **2009**, *147*, 287–301. [CrossRef]
17. George, H.; Hanlon, E.; Overman, A. Fertilizer Experimentation, Data Analyses, and Interpretation for Developing Fertilization Recommendations—Examples with Vegetable Crop Research 2017. Available online: https://edis.ifas.ufl.edu/pdffiles/SS/SS548 00.pdf (accessed on 12 December 2020).
18. Hlisnikovský, L.; Men ík, L.; Kunzová, E. The development of winter wheat yield and quality under different fertilizer regimes and soil-climatic conditions in the Czech Republic. *Agronomy* **2020**, *10*, 1160. [CrossRef]
19. Lim, W.; Sonn, Y.; Yoon, Y. The Selection of Yield Response Model of Sugar beet (*Beta vulgaris* var. Aaron) to Nitrogen Fertilizer and Pig Manure Compost in Reclaimed Tidal Land Soil. *Korean J. Soil Sci. Fertil.* **2010**, *43*, 174–179.
20. Eghball, B.; Wienhold, B.J.; Gilley, J.E.; Eigenberg, R.A. Mineralization of manure nutrients. *J. Soil Water Conserv.* **2002**, *57*, 470–473.
21. Maharjan, B.; Hergert, G.W. Composted cattle manure as a nitrogen source for sugar beet production. *Agron. J.* **2019**, *111*, 917–923. [CrossRef]
22. Davis, J.G.; Westfall, D.G. Fertilizing Sugar Beets. Available online: https://extension.colostate.edu/topic-areas/agriculture/fertilizing-sugar-beets-0-542/ (accessed on 8 October 2020).
23. Pişkin, A. The effect of sheep manure on yield and quality in production of organic sugar beet. *Isr. J. Plant Sci.* **2019**, *66*, 238–242. [CrossRef]
24. Du, Y.; Cui, B.; Zhang, Q.; Wang, Z.; Sun, J.; Niu, W. Effects of manure fertilizer on crop yield and soil properties in China: A meta-analysis. *Catena* **2020**, *193*. [CrossRef]
25. Li, H.; Feng, W.T.; He, X.H.; Zhu, P.; Gao, H.J.; Sun, N.; Xu, M. Gang Chemical fertilizers could be completely replaced by manure to maintain high maize yield and soil organic carbon (SOC) when SOC reaches a threshold in the Northeast China Plain. *J. Integr. Agric.* **2017**, *16*, 937–946. [CrossRef]

26. Menšík, L.; Hlisnikovský, L.; Kunzová, E. The State of the Soil Organic Matter and Nutrients in the Long-Term Field Experiments with Application of Organic and Mineral Fertilizers in Different Soil-Climate Conditions in the View of Expecting Climate Change. In *Organic Fertilizers—History, Production and Applications*; IntechOpen: London, UK, 2019.
27. Ren, F.; Sun, N.; Xu, M.; Zhang, X.; Wu, L.; Xu, M. Changes in soil microbial biomass with manure application in cropping systems: A meta-analysis. *Soil Tillage Res.* **2019**, *194*, 104291. [CrossRef]
28. Tong, X.; Xu, M.; Wang, X.; Bhattacharyya, R.; Zhang, W.; Cong, R. Long-term fertilization effects on organic carbon fractions in a red soil of China. *Catena* **2014**, *113*, 251–259. [CrossRef]
29. Zavattaro, L.; Bechini, L.; Grignani, C.; van Evert, F.K.; Mallast, J.; Spiegel, H.; Sandén, T.; Pecio, A.; Giráldez Cervera, J.V.; Guzmán, G.; et al. Agronomic effects of bovine manure: A review of long-term European field experiments. *Eur. J. Agron.* **2017**, *90*, 127–138. [CrossRef]
30. Kožnarová, V.; Klabzuba, J. Doporučení WMO pro popis meteorologických, resp. klimatologických podmínek definovaného období. *Rostl. Vyroba* **2002**, *48*, 190–192.
31. FAO. *World Reference Base for Soil Resources 2014. International Soil Classification System for Naming Soils and Creating Legends for Soil Maps*; FAO: Rome, Italy, 2014; ISBN 9789251083697.
32. Bradstreet, R.B. Kjeldahl Method for Organic Nitrogen. *Anal. Chem.* **1954**, *26*, 185–187. [CrossRef]
33. Mehlich, A. Mehlich 3 soil test extractant: A modification of Mehlich 2 extractant. *Commun. Soil Sci. Plant Anal.* **1984**, *15*, 1409–1416. [CrossRef]
34. Arnhold, E. *Easynls: Easy Nonlinear Model*, R Package Version 5.0; 2017. Available online: https://cran.r-project.org/web/packages/easynls/easynls.pdf (accessed on 11 January 2021).
35. Mangiafico, S. *Rcompanion: Functions to Support Extension Education Program Evaluation*, R Package Version 2.3.26; 2020. Available online: https://cran.r-project.org/web/packages/rcompanion/index.html (accessed on 11 January 2021).
36. Wickham, H.; Averick, M.; Bryan, J.; Chang, W.; McGowan, L.; François, R.; Grolemund, G.; Hayes, A.; Henry, L.; Hester, J.; et al. Welcome to the Tidyverse. *J. Open Source Softw.* **2019**, *4*, 1686. [CrossRef]
37. Olesen, J.E.; Trnka, M.; Kersebaum, K.C.; Skjelvåg, A.O.; Seguin, B.; Peltonen-Sainio, P.; Rossi, F.; Kozyra, J.; Micale, F. Impacts and adaptation of European crop production systems to climate change. *Eur. J. Agron.* **2011**, *34*, 96–112. [CrossRef]
38. Adekiya, A.O.; Ejue, W.S.; Olayanju, A.; Dunsin, O.; Aboyeji, C.M.; Aremu, C.; Adegbite, K.; Akinpelu, O. Different organic manure sources and NPK fertilizer on soil chemical properties, growth, yield and quality of okra. *Sci. Rep.* **2020**, *10*, 1–9. [CrossRef] [PubMed]
39. Zhang, W.; Xu, M.; Wang, B.; Wang, X. Soil organic carbon, total nitrogen and grain yields under long-term fertilizations in the upland red soil of southern China. *Nutr. Cycl. Agroecosyst.* **2009**, *84*, 59–69. [CrossRef]
40. Gai, X.; Liu, H.; Liu, J.; Zhai, L.; Yang, B.; Wu, S.; Ren, T.; Lei, Q.; Wang, H. Long-term benefits of combining chemical fertilizer and manure applications on crop yields and soil carbon and nitrogen stocks in North China Plain. *Agric. Water Manag.* **2018**, *208*, 384–392. [CrossRef]

Article

The Development of Winter Wheat Yield and Quality under Different Fertilizer Regimes and Soil-Climatic Conditions in the Czech Republic

Lukáš Hlisnikovský *, Ladislav Menšík and Eva Kunzová *

Crop Research Institute in Prague, Drnovská 507/73, Prague 6–Ruzyně, 161 06 Prague, Czech Republic; ladislav.mensik@vurv.cz

* Correspondence: l.hlisnik@vurv.cz (L.H.); kunzova@vurv.cz (E.K.); Tel.: +420-233-022-216 (L.H.)

Received: 20 July 2020; Accepted: 5 August 2020; Published: 7 August 2020

Abstract: Farmers must adapt to the changes brought about by the changing climate and market requirements. These adaptations are associated with fertilization—the availability of organic manures and mineral fertilizers and crop rotations. What is the effect of organic manures on wheat and soil? Is it necessary to apply mineral phosphorus P and potassium (K) fertilizers to the wheat? These questions are frequently asked in workshops in different growing areas. To provide a relevant answer on this issue, we evaluated how farmyard manure (FYM), mineral nitrogen (N) applied without phosphorus (P) and potassium (K) fertilizers, and application of NPK affected grain yield, grain quality, and soil properties under different soil-climate conditions (Ivanovice—Chernozem, Caslav—Phaeozem, Lukavec—Cambisol) between 2015 and 2018. The FYM significantly increased grain yield even after three years since being applied and incorporated into the soil in all localities, but its application didnot affect grain quality. In the soil, the FYM significantly increased total nitrogen Nt, P, and K content in all localities and oxidable carbon Cox content in two localities. Mineral nitrogen significantly affected grain yield and quality and increased concentrations of soil N and C, but decreased pH in Caslav. Application of mineral P and K wasnot connected with a positive effect on grain yield and quality, but increased the concentration of these elements in the soil, preventing depletion of these elements from the soil. Maximal yields were recorded when 70–98 kg N ha^{-1} was applied in Ivanovice, 55–72 kg N ha^{-1} in Caslav, and 155 kg N ha^{-1} in Lukavec.

Keywords: *Triticum aestivum* L.; farmyard manure; mineral fertilizers; crude protein content; soil properties, site-specific requirements

---

1. Introduction

Farmers have recently been exposed to many pressures and challenges that affect their decisions and activities. These pressures and challenges arise from currently changing climatic conditions, increased public awareness of the environment, the development on the market with meat and dairy products, new technologies, and financial possibilities of farmers themselves. As a reaction, we can see a huge shift from the traditional approaches of farmers, who are responsible for food and feedingstuff production and maintaining the function of the landscape. Such examples in the Czech Republic are a two-thirds decline in numbers of animals over the last two decades and changes in crop rotations connected with increased biogas production [1], a decline in production of organic manures, decline in doses of applied mineral phosphorus and potassium (Figure 1), and worsening of the soil organic matter quality in Chernozems [2].

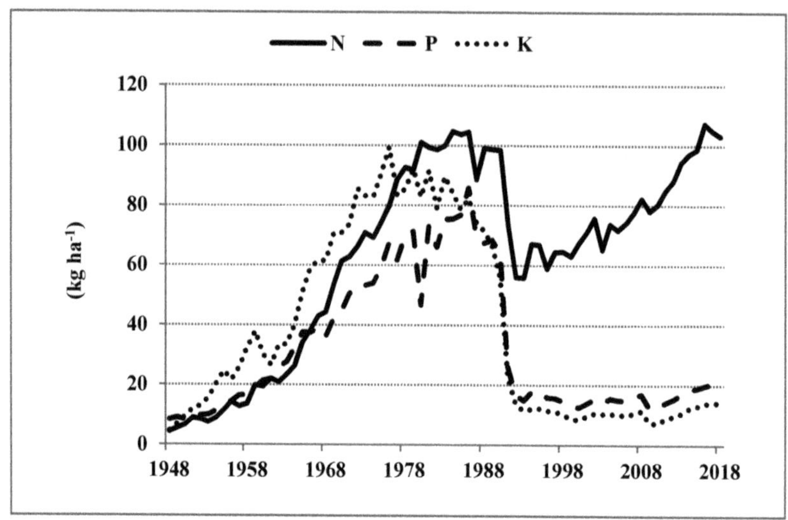

Figure 1. The mean doses of mineral N, P, and K applied on arable land in the Czech Republic.

The source of organic manures is significantly limited, because animal husbandry production decreased significantly since 1989 (the Velvet revolution, the transformation of the society from centrally planned economy into the market economy), and this trend continued when the Czech Republic joined the European Union (EU) [3]. Mineral P and K fertilizers are not applied in a sufficient amount owing to the high purchase prices, and crop rotations were adjusted to supply agricultural biogas stations. Moreover, with reduced animal husbandry, the need for clover and other grasses decreased, so their position in crop rotation was replaced.

Another challenge significantly affecting the behaviour and decision making of farmers representsupcoming weather change and weather instability. According to Olesen et al. [4], the farmers in Europe adapt to climate change by changing the timing of cultivation, and by selecting other crop species and cultivars.

In workshops with farmers, we often encounter questions about wheat fertilization and current weather changes. What are the benefits of organic manure? What will be the result of omitting P and K fertilizers? What dose of nitrogen should we choose? From the literature, we know that nitrogen is the most crucial element for wheat [5,6] and addition of mineral N was and is a standard way for securing high yields and grain quality of wheat. In the past, "the more fertilizer, the higher yield" strategy [7], which is characteristic by application of N doses greater than the crops need, led to the inefficiency and environmental problems and farmers and scientists focused on optimization of mineral fertilizer inputs. This optimizations depend on many factors, mainly on soil-climate conditions, so such optimization differs site by site [7,8]. Another issue represents the application of P and K fertilizers. Is it necessary? According to Duncan et al. [6], the co-application of P, K, and sulfur fertilizers increases the N efficiency and helps to achieve higher yields with higher protein content via "protection against protein dilution as yields increase". These questions are based on the above-mentioned issues. To answer some of these questions, we analysed suitable fertilizer treatments and the 4-year sequence (2015–2018) from the three long-term field experiments, established in the Czech Republic in 1955. In this paper, we evaluated the effect of seven different fertilizer treatments (control, farmyard manure—FYM, FYM+N1, FYM+N2, FYM+N1PK, FYM+N2PK, FYM+N3PK) and soil-climate conditions on the winter wheat grain yield and quality. The effect of different fertilization regimes on soil nutrient content and soil properties was also analysed in this study.

2. Materials and Methods

The long-term field fertilizer trials in Ivanovice, Caslav, and Lukavec were established in 1955. The trials aim to analyse the effect of different fertilizer treatments on yield and quality of arable crops and soil chemical properties under different soil-climate conditions.

2.1. Characteristics of Localities

2.1.1. Ivanovice

Ivanoviceis located in the South Moravia Region (49°19′ NL, 17°05′ EL), the altitude is 225 m a.s.l. The soil is naturally fertile, the soil type is degraded Chernozem, and the soil-forming substrate is loess. The topsoil is dark brown, with an average depth of 40 cm. The content of humus is approximately 4.39%. The climate is warm and dry, with mild winters. The average temperature is 9.14 °C (1969–2019), and the average annual sum of precipitation is 542 mm (Ivanovice Meteorological Station, measurements 1969–2019).

2.1.2. Caslav

Caslavis located in the Central Bohemian Region (49°85′ NL, 15°40′ EL), the altitude is 263 m a.s.l. The soil type is greyic Phaeozem. The topsoil average depth is 40–50 cm, deeper layers (80 cm) are sandy. The content of humus is approximately 2.98%. The climate is moderately warm and dry, with mild winters. The average annual temperature is 9.3 °C, and the average sum of precipitation is 572 mm (Caslav Meteorological Station, measurements 1974–2019).

2.1.3. Lukavec

Lukavecis located in the Vysočina Region (49°34′ NL, 14°59′ EL), the altitude is 620 m a.s.l. The soil type is sandy-loamy Cambisol. The topsoil average depth is 20 cm, with approximately 52% of sand, 20% of clay, and 28% of silt. The climate is moderately warm and humid. The average annual temperature is 7.7 °C, and the average annual sum of precipitation is 688 mm (Lukavec Meteorological Station, measurements 1969–2019).

2.1.4. Weather Course

The average monthly temperatures and the sum of precipitation at the time of the experiment are shown in Tables 1 and 2. Values below 30% of the long-term normal (warmer, less precipitation) are marked in yellow, while values 30% above the long-term normal (colder, more precipitation) are marked in green. From this point of view, we recorded that January, February, November, and December were significantly warmer in all localities compared with the long-term average. For precipitation, we recorded a significant lack of precipitation throughout the year in Ivanovice. Precipitations in Lukavec and Čáslav were also below average, especially from February to April 2015 and 2018 (Lukavec), and in May (Caslav).

Table 1. The mean monthly temperatures (°C). Yellow marked values are more than 30% below the long-term normal in the area, while green marked values are more than 30% higher than long-term normal.

	I.	II.	III.	IV.	V.	VI.	VII.	VIII.	IX.	X.	XI.	XII.	
							Ivanovice						
2015	1.1	0.9	4.8	9.5	13.8	17.7	21.6	22.7	15.4	8.7	5.7	2.4	10.4
2016	−1.6	4.5	4.7	8.9	14.7	19.0	20.3	18.9	17.1	8.5	3.9	−0.8	9.8
2017	−6.2	0.8	7.3	8.6	15.2	20.2	20.6	21.3	13.8	10.2	4.5	1.6	9.8
2018	2.0	−2.7	2.0	14.4	17.9	19.4	21.5	23.0	16.3	11.2	5.4	1.6	11.0
							Caslav						
2015	2.5	1.5	5.5	9.4	13.8	17.2	21.8	23.2	14.7	9.0	7.3	5.7	11.0
2016	0.1	4.5	4.6	9.1	14.9	18.8	20.3	19.1	18.0	8.8	3.7	0.8	10.2
2017	−4.5	2.3	7.5	8.5	15.5	19.6	20.1	20.4	13.0	11.0	5.1	2.5	10.1
2018	3.3	−2.0	2.5	14.2	17.5	19.1	22.0	23.1	16.4	14.6	5.3	2.6	11.6
							Lukavec						
2015	0.0	−1.2	3.2	6.9	11.5	15.3	19.7	20.5	12.2	6.9	5.2	3.8	8.7
2016	−1.5	2.1	2.3	6.9	12.2	16.3	17.9	16.6	15.3	6.2	1.8	−1.2	7.9
2017	−6.1	0.6	5.2	5.8	13.0	17.4	17.8	18.5	11.0	9.1	2.9	−0.2	7.9
2018	1.1	−4.9	−0.1	12.2	15.0	16.5	18.7	20.3	14.2	9.3	3.1	0.6	8.8

Table 2. The mean monthly precipitation (mm). Yellow marked values are more than 30% below the long-term normal in the area, while green marked values are more than 30% higher than long-term normal.

	I.	II.	III.	IV.	V.	VI.	VII.	VIII.	IX.	X.	XI.	XII.	
							Ivanovice						
2015	21.8	5.4	39.2	17.9	34.8	77.1	28.0	83.7	23.9	27.7	22.1	5.4	387.0
2016	15.9	61.1	17.6	43.1	36.0	27.6	108.8	42.7	14.3	43.1	30.4	7.3	447.0
2017	15.6	8.3	20.0	38.9	25.6	41.6	71.8	35.2	67.7	23.5	29.9	12.8	390.9
2018	38.0	22.6	36.2	20.2	27.6	52.2	43.6	22.0	65.4	24.6	17.2	18.0	387.6
							Caslav						
2015	38.5	6.3	35.0	17.2	62.3	59.6	23.3	51.4	18.6	42.5	68.7	19.0	442.4
2016	23.5	32.1	35.6	26.1	39.2	82.7	93.2	40.9	7.4	57.4	22.3	13.0	473.4
2017	28.1	15.2	33.6	95.9	40.6	84.0	93.2	65.9	59.7	80.0	58.3	40.2	694.7
2018	40.5	27.3	40.6	14.7	32.8	48.7	23.5	24.1	54.9	39.6	17.4	55.8	420.0
							Lukavec						
2015	54.9	7.9	24.2	23.5	55.6	62.6	20.8	94.3	24.4	75.4	113.6	18.9	576.1
2016	30.8	48.2	41.0	32.0	89.3	58.4	110.5	22.8	16.4	79.7	32.1	39.3	600.5
2017	30.1	33.3	68.4	102.1	33.0	66.0	135.4	67.1	35.0	110.1	51.6	44.8	776.9
2018	34.1	16.1	30.1	7.2	51.5	65.7	43.6	38.9	78.3	42.8	22.9	78.1	509.3

2.2. Field Trial Description

The design and methodology of all three trials (in three localities) are equal. The trial consists of four fields, where twelve fertilizer treatments are applied and analysed (twelve treatments per one field) in randomized complete block design. Each fertilizer treatment is replicated four times ($12 \times 4 = 48$ experimental plots per one field). The size of one experimental plot is 64 m^2 (8×8 m).

Out of twelve fertilizer treatments, seven treatments are evaluated in this paper: (1) the unfertilized control (control); (2) farmyard manure (FYM); (3–4) farmyard manure applied with mineral nitrogen (FYM+N1; FYM+N2); and (5–7) farmyard manure applied with mineral nitrogen, phosphorus, and potassium (FYM+N1PK; FYM+N2PK; FYM+N3PK). The doses of applied farmyard manure and mineral fertilizers are shown in Table 3. Mineral nitrogen was applied as lime ammonium nitrate, the mineral phosphorus as granulated superphosphate, and potassium as potassium chloride.

The FYM and mineral fertilizers were spread by hand. Applied FYM (cattle farmyard manure from local farmers) was incorporated into the soil within 24 h after application. The FYM was in the crop rotation applied to the maize, three years before the wheat was sown (2012—maize, 2013—spring barley, 2014—winter rapeseed, 2015—winter wheat), so winter wheat was the fourth crop following the FYM application. If the FYM and mineral fertilizers were applied together, mineral fertilizers were applied first (the case of maize in the crop rotation, no FYM was applied to spring barley, winter rapeseed, and winter wheat). The dose of FYM, applied to the maize, was 40 t ha^{-1} (approximately 200 kg N ha^{-1}, 56 kg P ha^{-1}, and 236 kg K ha^{-1}). Because winter wheat was sown three years after the FYM was applied, we estimate that the total amount of available nutrients from FYM to wheat was 5% (10 kg N ha^{-1}, 3 kg P ha^{-1}, and 12 kg K ha^{-1}) (for further information, see Section 4.4.).

Table 3. Doses of N, P, and K (kg ha^{-1}) applied in selected fertilizer treatments. FYM, farmyard manure; N, mineral nitrogen; P, phosphorus; K, potassium.

	N	P	K
Control	0	0	0
FYM	10	3	12
FYM+N1	(10)+40	(3)+0	(12)+0
FYM+N2	(10)+80	(3)+0	(12)+0
FYM+N1PK	(10)+40	(3)+60	(12)+60
FYM+N2PK	(10)+80	(3)+60	(12)+60
FYM+N3PK	(10)+120	(3)+60	(12)+60

Note: values in parentheses represent the amount of nutrients mineralised from manure.

Mineral nitrogen was applied in the autumn before the wheat was sown (40 kg N ha^{-1}, N1, N2, and N3 treatments), during the beginning of the spring for regeneration (40 kg N ha^{-1}, N2 and N3 treatments, late February, beginning of March), and during the May to support the grain production (40 kg N ha^{-1}, N3 treatment). The P and K fertilizers were applied during the autumn, before the wheat was sown.

Winter wheat cultivar, sown between 2014/2015 and 2017/2018, was Julie (Selgen a.s., Sibřina, the Czech Republic). It is a short straw cultivar offering high yields in all cropping areas, high frost resistance, offering class "E" (elite) grain quality, and a high volume of bakery products. Wheat was sown in October (according to the climate conditions), the depth of sowing ranged from three to four cm, the row spacing was 12.5 cm, and the sowing rate was 400 seeds per m^2. Pesticides were used during the experiments if necessary, while growth regulators havenot been introduced and applied. The harvest was done during the right BBCH stage, and was performed by the HEGE 180 (Hege Maschinen GmbH, Lichtenstein, Germany) in Ivanovice, Sampo Rosenlew 2010 (Sampo Rosenlew Ltd., Pori, Finland) in Caslav, and by Seedmaster Universal, Wintersteiger (Ried im Innkreis, Austria) in Lukavec.

2.3. Grain Quality Analyses

The crude protein content (CPC) was analysed according to the Kjeldahl method [9] (ČSN EN ISO 20483). The Zeleny's sedimentation test (ZST) was analysed according to the ČSN ISO 5529. Owing to the different issues, the quality parameters (CPC, ZST) from the Caslav were analysed only in grain from seasons 2015 and 2018.

2.4. Data Analyses

To compare the effect of fertilizer treatment and year on grain yield, grain quality, and soil chemical composition, we used analysis of variance (ANOVA) and multivariate analysis of variance MANOVA, followed by a post-hoc analysis performed by Tukey's HSD test (Statistica 13.3, TIBCO Software, Palo Alto, CA, USA). To analyse and visualize the relationships between fertilizer treatments, years, and soil chemical properties, we used principal component analysis (PCA), performed byAnalyse-it software (Analyse-it 5.30, Analyse-it Software, Ltd., Leeds, UK).

2.5. Soil Analyses

The soil chemical properties were analyzed at the beginning (2015) and in the end (2018) of the analyzed period. The soil samples were taken from the upper Ap horizons (0–30 cm) using the soil probe. Four soil samples were taken from each fertilizer treatment. The soil reaction was analyzed potentiometrically in 50 mL of 0.2 mol KCl(inoLab pH 730, Xylem Analytics-WTW, Weilheim, Germany). The soil organic carbon (SOC) content was measured according to [10,11]. The soil nitrogen content was analyzed using the sulfuric acid in the heating block (Tecator Inc., Hoganas, Sweden), followed by the Kjeldahl method [9]. The soil phosphorus, potassium, calcium, and magnesium were extracted by Mehlich III solution [12], followed by ICP–OES analysis (Thermo Scientific iCAP 7400 Duo, Thermo Fisher Scientific, Cambridge, UK).

3. Results

3.1. Effect of the FYM

Application of the FYM to the maize three years before wheat was sown (see the crop rotation description in Section 2.2.) significantly affected average wheat yields in all experimental sites (Table 4).

Table 4. Comparison of average grain yields (t ha^{-1}) as affected by locality, year, and fertilizer treatment.

	2015	2016	2017	2018	
			Ivanovice		
Control	3.79±0.20 Aa	4.48±0.10 Ab	5.14±0.19 Ac	4.37±0.07 Aab	4.44±0.14 A
FYM	4.32±0.14 Aa	5.00±0.10 Bb	5.74±0.19 Ac	4.93±0.12 Bb	5.00±0.14 B
			Caslav		
Control	5.87±0.14 Ad	5.31±0.06 Ac	4.48±0.13 Ab	3.83±0.09 Aa	4.87±0.21 A
FYM	6.60±0.21 Bc	5.91±0.08 Bb	5.10±0.15 Ba	4.98±0.06 Ba	5.65±0.18 B
			Lukavec		
Control	2.11±0.12 Aa	2.48±0.11 Aab	2.07±0.10 Aa	2.87±0.12 Ab	2.38±0.10 A
FYM	2.90±0.12 Bb	2.90±0.18 Ab	2.36±0.08 Aa	3.29±0.11 Bb	2.86±0.10 B

Mean values with the standard error of the mean SE, followed by the same letter (A vertically, a horizontally), are not significantly different ($\alpha < 0.05$).

The average grain yield increase between the control and FYM treatments ranged from 0.48 t ha^{-1} in Lukavec to 0.78 t ha^{-1} in Caslav. However, the highest percentage increase was recorded in Lukavec (20%), followed by Caslav (16%) and Ivanovice (13%). The differences in average grain yieldswere not achieved by the only application of the FYM, as the effect of the year also significantly affected the compared grain yields. While the overall effect of the fertilizer treatment and weather conditions on average yield was equal in Ivanovice (49% for fertilizer treatment; 51% for weather conditions) and Caslav (47% for fertilizer treatment; 52% for weather conditions), the higher effect of the fertilizer treatment was recorded in Lukavec (62% for fertilizer treatment; 35% for the weather conditions), showing a lower fluctuation of grain yields between the years.

3.2. Effect of the Mineral N Applied without P and K Mineral Fertilizers

In comparison with the control, application of mineral N fertilizers applied without mineral P and K fertilizers significantly increased the average grain yields in all experimental stations (Table 5).

Table 5. Comparison of average grain yields (t ha^{-1}) as affected by locality, year, and fertilizer treatment.

	2015	2016	2017	2018	
			Ivanovice		
Control	3.79±0.20 Aa	4.48±0.10 Ab	5.14±0.19 Ac	4.37±0.07 Aab	4.44±0.14 A
FYM+N1	7.30±0.10 Bb	7.79±0.17 Bc	7.53±0.09 Bbc	6.31±0.05 Ba	7.23±0.15 B
FYM+N2	8.87±0.29 Cc	9.08±0.12 Cc	7.28±0.14 Bb	6.00±0.14 Ba	7.81±0.33 C
			Caslav		
Control	5.87±0.14 Ad	5.31±0.06 Ac	4.48±0.13 Ab	3.83±0.09 Aa	4.87±0.21 A
FYM+N1	9.11±0.11 Cc	8.24±0.05 Bb	7.98±0.13 Bb	5.24±0.10 Ba	7.64±0.38 C
FYM+N2	7.88±0.23 Bb	8.50±0.06 Cc	7.75±0.11 Bb	4.87±0.11 Ba	7.25±0.37 B
			Lukavec		
Control	2.11±0.12 Aa	2.48±0.11 Aab	2.07±0.10 Aa	2.87±0.12 Ab	2.38±0.10 A
FYM+N1	4.76±0.12 Bbc	3.56±0.08 Ba	4.41±0.10 Bb	5.17±0.09 Bc	4.48±0.16 B
FYM+N2	6.94±0.26 Cb	5.20±0.11 Ca	5.38±0.36 Ca	5.63±0.20 Ba	5.79±0.21 C

Mean values with SE, followed by the same letter (A vertically, a horizontally), are not significantly different ($\alpha < 0.05$).

The average grain yield increase was 2.79 t ha^{-1} (+63%) for FYM+N1 treatment and 3.37 t ha^{-1} (+76%) for FYM+N2 treatment in Ivanovice. The effect of the fertilizer treatment on overall grain yields was dominant (86%) when compared with the effect of the season (9%). The difference between FYM+N1 and FYM+N2 treatments was significant. According to the quadratic model, the maximum grain yield was reached at FYM+70 kg N ha^{-1} (Figure 2a).

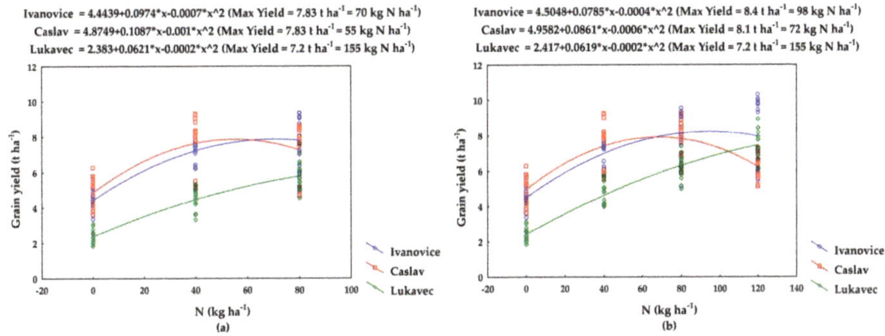

Figure 2. The average wheat grain yield as affected by fertilizer treatment ((**a**)—control, FYM+N1, FYM+N2; (**b**)—control, FYM+N1PK, FYM+N2PK, FYM+N3PK), and locality over the period 2015–2018.

3.3. Grain Quality

In Caslav, the average grain yield increase, following the application of mineral N, was lower than in Ivanovice. The average grain yield increase was 2.77 t ha^{-1} (+57%) for FYM+N1 treatment and 2.38 t ha^{-1} (+49%) for FYM+N2 treatment. The overall grain yields were affected by fertilizer treatment by 61%, while the effect of the season represents 37%. The difference between FYM+N1 and FYM+N2 treatments was significant. According to the quadratic model, the maximum grain yield was reached at FYM+55 kg N ha^{-1} (Figure 2a).

In Lukavec, the average grain yield increase, following the application of mineral N, was the highest out of all experimental stations. The average difference between the control and FYM+N1 was only 2.10 t ha^{-1}, but expressed as a percentage, it represents an 88% increase. Comparing the control and FYM+N2 treatments, the average grain yield increase was 3.41 t ha^{-1}, representing a 143% increase. The effect of the fertilizer treatment was, as in the case of Ivanovice, dominant (93%), while the effect of the season was marginal (4%). The difference between FYM+N1 and FYM+N2 treatments was significant. According to the quadratic model, the maximum grain yield was reached at FYM+155 kg N ha^{-1} (Figure 2a).

3.4. *Effect of the Mineral N Applied with P and K Mineral Fertilizers*

In comparison with the control, application of mineral NPK significantly increased average grain yields in all experimental stations (Table 6).

Table 6. Comparison of average grain yields (t ha^{-1}) as affected by locality, year, and fertilizer treatment.

	2015	2016	2017	2018	
			Ivanovice		
Control	3.79±0.20 Aa	4.48±0.10 Ab	5.14±0.19 Ac	4.37±0.07 Aab	4.44±0.14 A
FYM+N1PK	7.38±0.05 Bb	7.53±0.16 Bb	7.56±0.09 Bb	6.19±0.15 Ba	7.16±0.16 B
FYM+N2PK	9.26±0.11 Cc	9.03±0.14 Cc	7.50±0.21 Cb	5.99±0.29 Ba	7.95±0.35 C
FYM+N3PK	9.98±0.12 Dc	9.61±0.15 Cc	6.68±0.15 Cb	5.76±0.14 Ba	8.01±0.47 C
			Caslav		
Control	5.87±0.14 Ad	5.31±0.06 Ac	4.48±0.13 Ab	3.83±0.09 Aa	4.87±0.21 A
FYM+N1PK	8.79±0.36 Bb	8.00±0.04 Cb	7.96±0.12Cb	5.83±0.10 BCa	7.64±0.30 C
FYM+N2PK	8.71±0.25 Bc	8.27±0.11 Cc	7.28±0.21BCb	5.99±0.08 Ca	7.56±0.28 C
FYM+N3PK	6.21±0.46 Aab	6.72±0.16 Bb	6.70±0.19 Bb	5.56±0.13 Ba	6.29±0.17 B
			Lukavec		
Control	2.11±0.12 Aa	2.48±0.11 Aab	2.07±0.10 Aa	2.87±0.12 Ab	2.38±0.10 A
FYM+N1PK	4.76±0.33 Bab	4.06±0.04 Ba	4.79±0.21 Bab	5.31±0.17 Bb	4.73±0.15 B
FYM+N2PK	6.68±0.31 Ca	5.82±0.31 Ca	6.39±0.30 Ca	5.94±0.27 Ba	6.21±0.16 C
FYM+N3PK	8.33±0.25 Db	7.45±0.49 Dab	7.37±0.26 Dab	6.83±0.25 Ca	7.49±0.20 D

Mean values with SE, followed by the same letter (A vertically, a horizontally), are not significantly different ($\alpha < 0.05$).

In comparison with the control, the average grain yield increase varied from 2.72 t ha^{-1} (+61%, FYM+N1PK) to 3.57 t ha^{-1} (+80%, FYM+N3PK) in Ivanovice. The effect of fertilizer treatment on average grain yield was marginal (69%), while the season conditions affected average grain yields by 23%. No significant differences were recorded between FYM+N2PK and FYM+N3PK treatments. According to the quadratic model, the highest grain yield was achieved at FYM+98 kg N ha^{-1} (Figure 2b).

The lowest average grain yield increase, when compared with the control, was recorded in Caslav, ranging from 1.42 t ha^{-1} (+29%, FYM+N3PK) to 2.77 t ha^{-1} (+57%, FYM+N2PK). The effect of the fertilizer treatment on average grain yield was 62%, while the effect of the season was 36%. The highest grain yields were recorded in FYM+N1PK (7.64 t ha^{-1}) and FYM+N2PK (7.56 t ha^{-1}) treatments (without significant differences between these two treatments). According to the quadratic model, the highest grain yield was achieved at FYM+72 kg N ha^{-1} (Figure 2b).

The average grain yield increase in Lukavec ranged from 2.35 t ha^{-1} (+99%, FYM+N1PK) to 5.11 t ha^{-1} (+215%, FYM+N3PK). The factor "fertilizer treatment" dominantly affected the average grain yield (98%). According to the quadratic model, the highest grain yield ought to be achieved at FYM+155 kg N ha^{-1} (Figure 2b).

3.4.1. Ivanovice

The CPC was significantly affected by year ($p < 0.001$), fertilizer treatment ($p < 0.001$), and their interaction ($p < 0.001$). The most dominant factor influencing the CPC was fertilizer treatment (97%). The average CPC ranged from 8.1% (control) to 14.5% (FYM+N3PK) (Table 7). No differences between the control and FYM treatments were recorded. The year to year differences were significant in the FYM and FYM+N3PK treatments, especially in the FYM+N3PK treatment, showing a fluctuation based on the weather conditions of the season. However, the contribution of this factor was small when compared with the effect of fertilizer treatment (Table 7).

Table 7. The average crude protein content (CPC) (%) and the value of Zeleny's sedimentation test (ZST) (mL) of wheat grain as affected by locality, fertilizer treatment, and the year.

	2015	2016	2017	2018	
		Ivanovice			
		CPC (%)			
Control	7.9±0.1 Aa	8.3±0.1 Aa	8.1±0.1 Aa	8.1±0.2 Aa	8.1±0.1 A
FYM	8.1±0.1 Aa	8.7±0.1 Ab	8.0±0.2 Aa	8.3±0.2 Aab	8.3±0.1 A
FYM+N3PK	13.1±0.3 Ba	13.3±0.3 Ba	15.3±0.4 Bb	16.4±0.2 Bb	14.5±0.4 B
		ZST (mL)			
Control	15.5±0.6 Aa	20.0±0.7 Ac	16.8±0.6 Aab	18.3±0.5 Abc	17.6±0.5 A
FYM	15.8±0.6 Aa	20.0±0.4 Ab	16.5±0.3 Aa	19.3±0.5 Ab	17.9±0.5 A
FYM+N3PK	44.3±2.5 Ba	55.0±1.1 Bb	59.5±1.2 Bbc	67.0±2.0 Bc	56.4±2.3 B
		Caslav			
		CPC (%)			
Control	8.2±0.1 A	n.a.	n.a.	10.3±0.1 A	9.3±0.4 A
FYM	8.3±0.3 A	n.a.	n.a.	10.3±0.4 A	9.3±0.4 A
FYM+N3PK	14.0±0.2 B	n.a.	n.a.	12.9±0.8 B	13.4±0.4 B
		ZST (mL)			
Control	16.8±0.6 A	n.a.	n.a.	30.5±1.2 A	23.6±2.7 A
FYM	18.0±0.6 A	n.a.	n.a.	30.8±2.1 A	24.4±2.6 A
FYM+N3PK	48.5±1.2 B	n.a.	n.a.	52.3±8.0 B	50.4±3.8 B
		Lukavec			
		CPC (%)			
Control	9.4±0.2 Ab	8.6±0.2 Aa	8.1±0.1 Aa	8.6±0.2 Aa	8.7±0.1 A
FYM	8.9±0.1 Ab	8.9±0.1 Ab	8.1±0.1 Aa	9.2±0.2 Bb	8.8±0.1 A
FYM+N3PK	11.2±0.4 Bab	10.2±0.2 Ba	12.2±0.2 Bbc	13.0±0.1 Cc	11.6±0.3 B
		ZST (mL)			
Control	19.0±0.8 Ab	18.3±0.9 Ab	14.8±0.3 Aa	17.0 Aab	17.3±0.5 A
FYM	18.8±0.5 Ab	19.0±0.6 Ab	15.0±0.4 Aa	18.3±0.5 Ab	17.8±0.5 A
FYM+N3PK	34.3±2.5 Aab	30.3±1.5 Ba	37.0±3.1 Bab	43.0±1.2 Bb	36.1±1.6 B

Mean values with SE, followed by the same letter (A vertically, a horizontally), are not significantly different ($\alpha < 0.05$). n.a.—results not available.

The value of ZST developedsimilarly as the CPC. The average values ranged from 17.6 mL (control) to 56.4 mL (FYM+N3PK) (Table 7). The major factor affecting the ZST was fertilizer treatment (97%). No differences between the control and FYM treatments were recorded. The value of ZST fluctuated from year to year in the same treatment, but the generaleffect of this factor was negligible in comparison with the fertilizer treatment.

3.4.2. Caslav

The CPC was significantly affected by year ($p < 0.01$), fertilizer treatment ($p < 0.001$), and their interaction ($p < 0.01$). The major factor influencing the CPC was fertilizer treatment (80%), followed by the year (10%) and their interaction (10%). The average CPC ranged from 9.3% (control, FYM) to 13.4% (FYM+N3PK) (Table 7).

The value of ZST was significantly affected by year ($p < 0.01$), fertilizer treatment ($p < 0.001$), and their interaction ($p < 0.01$). The major factor influencing the ZST value was fertilizer treatment

(73%), followed by the year (24%) and their interaction (3%). The average ZST values ranged from 24 mL (control, FYM) to 50 mL (FYM+N3PK) (Table 7).

3.4.3. Lukavec

The CPC was significantly affected by the year, fertilizer treatment, and their interaction ($p < 0.001$). The dominant factor influencing the CPC in Lukavec was fertilizer treatment (93%). The average CPC ranged from 8.7% (control) to 11.6% (FYM+N3PK) (Table 7).

The ZST value was significantly affected by year, fertilizer treatment, and their interaction ($p < 0.001$), mainly by the fertilizer treatment (95%). The mean values ranged from 17.3 mL (control) to 36.1 mL (FYM+N3PK) (Table 7).

3.5. Soil Chemical Properties

The effects of fertilizer treatments on soil chemical properties are shown in Tables 8–10 and the relationship between the parameters, localities, and fertilizer treatments are shown in Figure 3. The value of pH was not affected by any treatment in Ivanovice (Chernozem) and Lukavec (Cambisol). On the other hand, the application of mineral fertilizers significantly decreased the soil pH value in Caslav (Phaeozem). Application of all fertilizers significantly increased the Nt (%) and Cox(%) in all localities. The same pattern can be seen in the case of P (mg kg^{-1}) and K (mg kg^{-1}). Only in Lukavec, we did not record any differences in P and K concentrations between the control, FYM+N1, and FYM+N2 treatments. The concentration of Ca (mg kg^{-1}) was not affected by any treatment, except for low, but significant differences in Caslav (Control and FYM+NPK treatments). On the other hand, the Mg (mg kg^{-1}) content varied significantly in Ivanovice and Caslav, while no fluctuation was recorded in any fertilizer treatment in Lukavec.

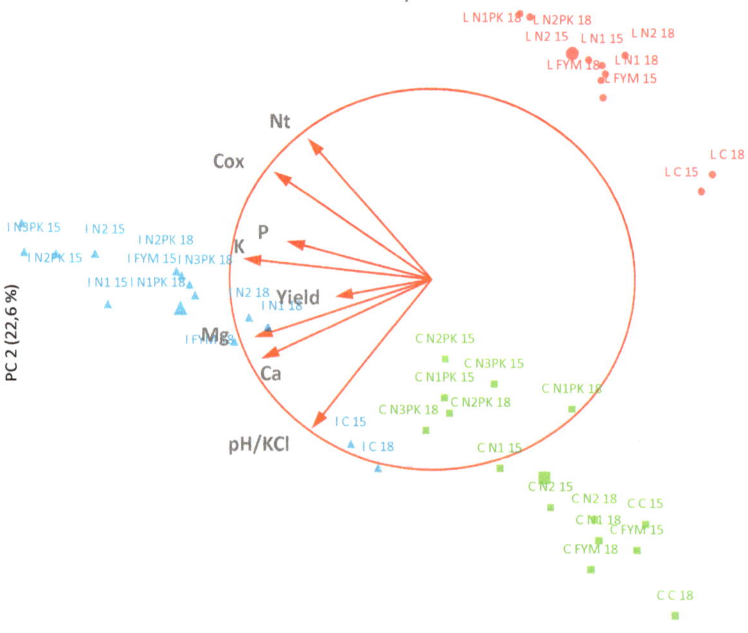

Figure 3. The relationships between soil chemical properties as affected by locality (I—Ivanovice, C—Caslav, L—Lukavec), fertilizer treatment, and year (15—2015, 18—2018).

Table 8. The pH;Nt; Cox; and content of P, K, Ca, and Mg, as affected by fertilizer treatment in Ivanovice.

Ivanovice							
	pH/KCl	Nt (%)	Cox (%)	P (mg kg^{-1})	K (mg kg^{-1})	Ca (mg kg^{-1})	Mg (mg kg^{-1})
Control	6.87±0.03A	0.19±0.01A	1.70±0.02A	97±6A	203±10A	4437±121A	198±3A
FYM	6.90±0.03A	0.23±0.01B	1.96±0.03B	159±12B	306±19B	4334±74A	217±6B
Control	6.87±0.03A	0.19±0.01A	1.70±0.02A	97±6A	203±10A	4437±121A	198±3A
FYM+N1	6.78±0.04A	0.22±0.01B	1.97±0.03B	153±17B	282±26B	4358±142A	235±9B
FYM+N2	6.83±0.03A	0.24±0.01C	2.03±0.04B	136±10AB	275±20B	4376±100A	232±7B
Control	6.87±0.03A	0.19±0.01A	1.70±0.02A	97±6A	203±10A	4437±121A	198±3A
FYM+N1PK	6.82±0.03A	0.23±0.01B	2.00±0.04B	205±13B	410±36B	4193±85A	226±5B
FYM+N2PK	6.78±0.02A	0.23±0.01B	1.99±0.03B	204±12B	379±30B	4201±97A	225±4B
FYM+N3PK	6.80±0.02A	0.24±0.01B	2.02±0.03B	187±12B	362±31B	4334±68A	242±8B

Mean values with SE, followed by the same letter (A vertically, a horizontally), are not significantly different ($\alpha < 0.05$). n.a.—results not available.

Table 9. The pH; Nt; Cox; and content of P, K, Ca, and Mg, as affected by fertilizer treatment in Caslav.

Caslav							
	pH/KCl	Nt (%)	Cox (%)	P (mg kg^{-1})	K (mg kg^{-1})	Ca (mg kg^{-1})	Mg (mg kg^{-1})
Control	6.72±0.02A	0.13±0.01A	1.16±0.04A	47±4A	107±3A	2876±23A	118±7A
FYM	6.69±0.05A	0.14±0.01B	1.19±0.02A	73±5B	129±5B	2917±30A	135±6A
Control	6.72±0.02B	0.13±0.01A	1.16±0.04A	47±4A	107±3A	2876±23A	118±7A
FYM+N1	6.65±0.01AB	0.16±0.01B	1.34±0.04B	73±10B	122±4B	2912±35A	161±3B
FYM+N2	6.68±0.03A	0.15±0.01B	1.29±0.01B	67±3AB	122±1B	2895±24A	154±6B
Control	6.72±0.02C	0.13±0.01A	1.16±0.04A	47±4A	107±3A	2876±23B	118±7A
FYM+N1PK	6.54±0.02A	0.16±0.00B	1.34±0.02B	171±5B	210±16B	2824±17AB	131±4AB
FYM+N2PK	6.51±0.01A	0.17±0.01B	1.38±0.02B	184±2B	237±5B	2746±39A	151±11AB
FYM+N3PK	6.61±0.01B	0.17±0.01B	1.42±0.03B	172±4B	213±14B	2810±18AB	163±11B

Mean values with SE, followed by the same letter (A vertically, a horizontally), are not significantly different ($\alpha < 0.05$). n.a.—results not available.

Table 10. The pH; Nt; Cox; and content of P, K, Ca, and Mg, as affected by fertilizer treatment in Lukavec.

Lukavec							
	pH/KCl	Nt (%)	Cox (%)	P (mg kg^{-1})	K (mg kg^{-1})	Ca (mg kg^{-1})	Mg (mg kg^{-1})
Control	5.77±0.04A	0.19±0.01A	1.44±0.05A	42±3A	125±9A	2055±31A	117±6A
FYM	5.86±0.06A	0.21±0.01B	1.67±0.05B	79±12B	169±5B	2079±42A	123±4A
Control	5.77±0.04A	0.19±0.01A	1.44±0.05A	42±3A	125±9A	2055±31A	117±6A
FYM+N1	5.79±0.05A	0.22±0.01B	1.70±0.04B	45±6A	147±7A	2098±50A	115±4A
FYM+N2	5.71±0.05A	0.22±0.01B	1.70±0.04B	37±2A	145±6A	2020±48A	108±5A
Control	5.77±0.04A	0.19±0.01A	1.44±0.05A	42±3A	125±9A	2055±31A	117±6A
FYM+N1PK	5.78±0.07A	0.22±0.01B	1.80±0.06B	193±16B	207±11B	2030±42A	111±4A
FYM+N2PK	5.79±0.05A	0.22±0.01B	1.79±0.06B	167±12B	183±10B	2056±37A	105±3A
FYM+N3PK	5.68±0.06A	0.22±0.01B	1.78±0.05B	173±9B	180±4B	2001±30A	114±5A

Mean values with SE, followed by the same letter (A vertically, a horizontally), are not significantly different ($\alpha < 0.05$). n.a.—results not available.

4. Discussion

4.1. *Effect of the FYM*

The effect of manure application on grain yield was significantly visible in all three localities with different soil-climatic conditions even three years after its incorporation into the soil. The differences in grain yield between the control and FYM treatments have not always been significant in particular years (Table 4). However, the average grain yields in the FYM treatment have always been higher in comparison with the control, whenthe wholetime period was evaluated (2015–2018). By comparing all three localities, we find that the greatest benefit of manure manifested itself in Lukavec (+20% mean grain yield increase in comparison with the control), followed by Caslav (+15%) and Ivanovice (+12%). In general, farmyard manure efficiency wasslightly higher in the locality with poorer soil type andmore humid climatic conditions and decreased on higher quality soils.The positive effect of the farmyard manure on the yield of arable crops three years after the manure applicationwas also recorded by [13], who analyzed the effect of mineral fertilizers and organic manures in the long-term fertilizer experiment in Switzerland. Application of the FYM affected not only yields (Table 4), but also the soil chemical properties (Tables 8–10), so the effect of the FYM was two-sided and both aspects synergized, resulting in higher yields than in the control even three years after FYM application. The same conclusions and explanations published [13] who recorded a higher SOC content in the soil treated with manure. A 6.40% higher SOC content in the soil treated with manure, together with higher microbial biomass, was also recorded by [14]. Positive influences on yields and soil properties of manure application have also been published in the meta-analysis from China, provided by [15], who analyzed more than 140 studies and more than 770 treatment comparisons.In our case, the soil fertilized with FYM showed a significantly higher content of Nt and Cox (significantly higher in Ivanovice and Lukavec), and a higher content of soil P and K (Tables 8–10). As the estimated amount of nutrients available directly from the FYM to the wheat was relatively low in our case (5%), we incline to the fact that the positive effect of manure consisted in the direct effect of released nutrients, which were mostly utilized by pre-crops and partially by wheat, and also in influencing the microbial part, nutrients turnover, and organic matter pool in soil. As we didnot perform the PLFA or other methods for evaluation of soil microbial activity, we cannot confirm these suggestions directly, but are in an agreement with the results of other researchers [16,17].

As mentioned above, the positive effect of the FYM on the wheat yield was significant in all localities. The opposite was true for qualitative parameters.On the basis of our results, we know that manure was able to provide nutrients and adjust soil conditions for higher yields of wheat grain (Table 4), but this was not enough to affect the CPC and ZST (Table 7). Except for the climate conditions, nitrogen is the most limiting factor for high CPC in the wheat grain [18,19]. From this point of view, the application of manure cannot replace mineral fertilizers for the production of bread-making quality grain, at least if wheat isnot fertilized with manure directly (which is not a common practise in the Czech Republic). On the other hand, the effect of the FYM can be taken as an advantage if wheat is produced for minor supply-chains (biscuit production) [19], or as the feedstuff.

4.2. The Effect of Mineral N

No wonder, application of mineral N directly to wheat significantly increased grain yield in all localities. Nitrogen is considered as the wheat's most important nutrient, and its addition positively influences the root length and density, water uptake, above-ground biomass production [20], phenology and leaf traits [21], and grain yield [22,23]. Focused on the site-specific nutrient management, Ivanovice and Lukavec provided the highest yields in the FYM+N2 treatment, with the maximum yield at 70 kg N ha^{-1} in Ivanovice and 155 kg N ha^{-1} in Lukavec. On the other hand, Caslav provided maximum yield in FYM+N1 treatment, and according to the quadratic model, the optimum dose represents 55 kg N ha^{-1}. This means that similar wheat yields can be harvested in all localities, but with extremely different nitrogen rates. It also means that nitrogen fertilization ought to be adjusted to the soil-climate conditions of the specific site—a single recommendation about the dose of mineral nitrogen cannot be applied nationwide. This knowledge can thus save the farmer's financial resources, the number of field operations, and the environment, as the excess of mineral nitrogen fertilizers applied to agricultural land is associated with negative impacts on soil nitrogen pool [24], leaching, and groundwater nitrate contamination [25,26]. Focused on the soil properties, application of mineral nitrogen significantly decreased the pH in Caslav, while no changes were recorded in Ivanovice and

Lukavec (Tables 8–10). The soil nitrogen concentration was higher in all localities, as well as the Cox content when compared with the control. Concentrations of P, K, and Mg were significantly higher in Ivanovice and Caslav, while no differences were recorded in Lukavec. From this point of view, the soil answer to N addition was similar in Ivanovice (Chernozem) and Caslav (Phaeozem)—these soils were more sensitive to the addition of mineral N. Both soil types are almost comparable, both represent the most fertile types in the Czech Republic, but Phaeozems are more prone to leaching during the wet seasons and donot contain so many carbonates in the topsoil layer [27], as can be seen in Tables 8 and 9. Another aspect is the soil buffering capacity against acidification, which is high in the case of Chernozems and lower in Phaozems [28]. Lukavec (Cambisol) offers similar values of Nt and Cox (even slightly higher)toCaslav, but the natural properties of the soil type in Lukavec are low pH value, the soil is lighter, and with lower sorption capacity [28]. Together with colder weather, the soil type creates a natural barrier that significantly affects wheat prosperity.

4.3. The Effect of Mineral N, P, and K

As with the effect of FYM+N, the application of mineral NPK fertilizers significantly increased wheat yields. In Ivanovice and Lukavec, the grain yield increased with increasing N dose (Control < FYM+N1PK < FYM+N2PK < FYM+N3PK). The difference between FYM+N2PK and FYM+N3PK was not significant in Ivanovice, and maximal yield was reached with 98 kg N ha^{-1}. This suggests that we have reached the maximum potential yield here, and increasing the nitrogen doses will not be connected with higher grain yield. From this point of view, we cannot influence the natural barriers of the locality and only new breakthroughs, such as the transformation of wheat from C3 to C4 pathway [29], could bring the new "green revolution", connected with significantly higher yields. Different behaviour was recorded in Caslav, where a significant decline in grain yields was connected with increasing doses of mineral N, and maximal yield was reached with 72 kg N ha^{-1}. Finally, the light soil in Lukavec responded to increasing nitrogen doses with significantly higher yields. According to the quadratic model, the maximum yield was reached with 155 kg N ha^{-1}. These results of the quadratic model specify the previous values obtained for the variants FYM+N1 and FYM+N2 (the variant FYM+N3 would be great for the comparison. However, it was not established when setting up the experiment). Another thing we can compare is the effect of added P and K fertilizers. As mentioned above, the average amount of applied P and K mineral fertilizers is very low in the Czech Republic. This creates a contrast between what is taught in schools and common agricultural practice. According to the results, no significant differences were recorded between FYM+N and FYM+NPK treatments in Ivanovice, where naturally fertile soil occurs and the pool of nutrients is high even in unfertilized Control treatment. Thus, the addition of 60 kg P and K ha^{-1} was not connected with any benefits if we speak about grain yields. In Čáslav, the high dose of nitrogen was counterproductive and the application of P and K again did not bring significant differences in yields, although the concentration of P and K were significantly lower in FYM+N treatments in comparison with FYM+NPK treatments. The high N dose counter-productivity theoretically stems from the fact that high doses of nitrogen supplied to the soil increase nitrogen losses from the soil, thereby reducing its usability by wheat [26], or from an increased risk of lodging and disease incidence and severity [30]. Finally, the third kind of reaction was recorded in Lukavec, where an increasing dose of mineral N increased grain yield, and yield reduction is expected at doses above 155 kg N ha^{-1}. Comparing the effect of P and K fertilizers in Lukavec, no differences between grain yields provided by FYM+N and FYM+NPK treatments were recorded, although the differences between the P and K concentrations in the soil were significant. According to the results, the application of P and K fertilizers significantly boosted the soil concentrations of P and K (Tables 8–10). This is a very important result showing that dependency on N, which is a current situation in the Czech Republic, is not long-term sustainable management and will slowly lead to soil depletion. Excellent review represents this paper [6]. According to [6], it was not clearly proven that application of P and K increases grain yield (some evaluated experiments proved that hypothesis, some did not), but it was stated that "improved soil P, K, or S availabilities has potential to increase grain yield and improve the efficiency of N fertiliser use". It is suggested that application of P and K fertilizers increases the efficiency of N, but we cannot confirm this conclusion, at least from the point of view of grain yields.

4.4. Grain Quality

Owing to the limited budget, we were able to analyze only three fertilization treatments in terms of quality (control, FYM, and FYM+N3PK). Our results are of limited value, yet they can provide important information. The first important finding is that, as the FYM significantly affected grain yield (in comparison with the control), it wasnot able to provide a sufficient amount of nutrients, or beneficially affect the soil environment to produce quality grain. At least in the Czech Republic, where the minimum CPC content for bread-making "elite class" quality is 12.6%. The lower threshold is, for example, in France, where the CPC limit required for organic breadmaking wheat increased during the time from 9 to 10.5% [31]. However, even this value wasnot reached in our experiment. Manure is a valuable source of organic matter and nutrients that are released over time from the organic form to the inorganic form via the mineralization process. The released nutrients thus become accessible to plants. Most of the nutrients are released into the soil environment in the first year after application and, in subsequent years, this amount gradually decreases. The yearly rates of mineralization, expressed as decay series, vary significantly between the authors [32–34], but generally range (for nitrogen) from 40 to 60% in the first year, 10 to 25% in the second year, and 5 to 10% in the third year. The expected very low amount of available nutrients released from the manure during the third year after application could not meet the needs of wheat to produce higher amounts of protein. On the other hand, application of NPK significantly increased the CPC and ZST in all localities. Application of mineral N represents an effective way to increase grain's CPC [6,35,36]. Suitable soil-climate conditions in Ivanovice providedthe highest CPC and ZST, followed by Caslav, which offers similar soil-climate conditions. Both sites provided grain meetingthe requirements for category "E" (min. CPC 12.6%, min. ZST value 49 mL). On the other hand, even high doses of mineral fertilizers cannot break naturally createdsoil-climate barriers, such as in Lukavec (Table 7), where these requirements werenot met in the currently evaluated period, as well as in the previous years (2011–2013) [37]. The CPC in the wheat grain is significantly affected mainly by two factors: (1) N supply and (2) weather conditions. In Lukavec, N supply was at the same level as in Ivanovice and Caslav, but the results were not satisfying. According toBarneix [35], the CPC cannot be increased, despite the ample N supply, because two main regulatory points are active at the same time during the grain-filling period. This means that wheat (*Triticum aestivum*) has a ceiling in terms of protein content that cannot be exceeded. This can be seen in Ivanovice and Caslav. Ancient wheat species, such as *Triticum monococcum, Triticum dicoccum*, and *Triticum. spelt*, have the ability to produce higher CPC in comparison with *T. aestivum* [38], so the CPC is influenced by wheat species, too. The weather conditions are the second factor affecting the grain's CPC. Both yield creation (accumulation of starch) and grain's CPC (accumulation of proteins) are independent processes. Dry and hot conditions usually lead to high CPC, while wet and colder conditions lead to lower CPC [39], because dry conditions decrease starch synthesis and deposition, while proteosynthesis is not inhibited as much as starch synthesis by these environmental factors [40]. From this point of view, it is unlikely that, in Lukavec and other localities with comparable soil–climatic conditions, we will achieve a harvest of high–quality grain, at least via the fertilization. However, this could change with upcoming weather changes, as is expected for example in the UK [41], or generally in Europe [42].The direction of such changes can be seen in the results from Lukavec from 2018. This season was dry in Ivanovice and Caslav, with relatively low yields and high quality. However, in Lukavec, the year 2018 was characterized by relatively good yields and, at the same time, high quality. This means that the drought has acted here as a positive factor, smoothing the negative effect of the local soil–climate barriers.

5. Conclusions

Application of farmyard manure three years before wheat was sown significantly affected wheat grain yields in all localities and positively influenced soil chemical properties, but has no further beneficial effect on grain quality.

Application of mineral nitrogen significantly increased grain yield and grain quality in all localities, and also positively affected Cox and Nt concentration. No effect on the pH value was recorded in Chernozem and Cambisol soil types, but Phaeozem reacted on N fertilizers with a decrease of the pH value.

Application of mineral nitrogen, phosphorus, and potassium was not connected with any significant increase in grain yields when compared with application of only mineral nitrogen. On the other hand, soil P and K concentrations were statistically significantly higher in treatments with P and K, and their application prevents depletion of these elements in the soil.

The highest yields were recorded between 70 and 98 kg N ha^{-1} in Ivanovice, 55 and 72 kg N ha^{-1} in Caslav, and with 155 kg N ha^{-1} in Lukavec.

Grain quality is mainly affected by nitrogen supply and limited by natural soil–climate barriers of the locality.

Author Contributions: Conceptualization, L.H., L.M., and E.K.; methodology, L.H., L.M., and E.K.; software, L.H.; validation, L.H. and L.M.; formal analysis, L.H.; investigation, L.H.; resources, E.K.; data curation, L.H.; writing—original draft preparation, L.H.; writing—review and editing, L.H.; visualization, L.H.; supervision, E.K.; project administration, E.K.; funding acquisition, E.K. All authors have read and agreed to the published version of the manuscript.

Funding: This research was funded by the Ministry of Agriculture of the Czech Republic via the project RO 0418.

Acknowledgments: We would like to thank the laboratory staff, technicians, and all their predecessors for their work on long-term fertilizer experiments.

Conflicts of Interest: The authors declare no conflict of interest.

References

1. Martinát, S.; Dvořák, P.; Frantál, B.; Klusáček, P.; Kunc, J.; Kulla, M.; Mintálová, T.; Navrátil, J.; Van Der Horst, D. Spatial consequences of biogas production and agricultural changes in the Czech Republic after EU accession: Mutual symbiosis, coexistence or parasitism? *AUPO Geogr.* **2013**, *44*, 75–92.
2. Horáček, J.; Novák, P.; Liebhard, P.; Strosser, E.; Babulicová, M. The long-term changes in soil organic matter contents and quality in chernozems. *Plant Soil Environ.* **2017**, *63*, 8–13.
3. Věžník, A.; Král, M.; Svobodová, H. Agriculture of the Czech Republic in the 21st century: From productivism to post-productivism. *Quaest. Geogr.* **2013**, *32*, 7–14. [CrossRef]
4. Olesen, J.E.; Trnka, M.; Kersebaum, K.C.; Skjelvåg, A.O.; Seguin, B.; Peltonen-Sainio, P.; Rossi, F.; Kozyra, J.; Micale, F. Impacts and adaptation of European crop production systems to climate change. *Eur. J. Agron.* **2011**, *34*, 96–112. [CrossRef]
5. De Oliveira Silva, A.; Ciampitti, I.A.; Slafer, G.A.; Lollato, R.P. Nitrogen utilization efficiency in wheat: A global perspective. *Eur. J. Agron.* **2020**, *114*. [CrossRef]
6. Duncan, E.G.; O'Sullivan, C.A.; Roper, M.M.; Biggs, J.S.; Peoples, M.B. Influence of co-application of nitrogen with phosphorus, potassium and sulphur on the apparent efficiency of nitrogen fertiliser use, grain yield and protein content of wheat: Review. *Field Crop. Res.* **2018**, *226*, 56–65. [CrossRef]
7. Liu, H.; Wang, Z.; Yu, R.; Li, F.; Li, K.; Cao, H.; Yang, N.; Li, M.; Dai, J.; Zan, Y.; et al. Optimal nitrogen input for higher efficiency and lower environmental impacts of winter wheat production in China. *Agric. Ecosyst. Environ.* **2016**, *224*, 1–11. [CrossRef]
8. Tedone, L.; Alhajj Ali, S.; Verdini, L.; De Mastro, G. Nitrogen management strategy for optimizing agronomic and environmental performance of rainfed durum wheat under Mediterranean climate. *J. Clean. Prod.* **2018**, *172*, 2058–2074. [CrossRef]
9. Kirk, P.L. Kjeldahl Method for Total Nitrogen. *Anal. Chem.* **1950**, *22*, 354–358. [CrossRef]

10. Sims, J.R.; Haby, V.A. Simplified colorimetric determination of soil organic matter. *Soil Sci.* **1971**, *112*, 137–141. [CrossRef]
11. Nelson, D.W.; Sommers, L.E. Total Carbon, Organic Carbon, and Organic Matter. In *Total Carbon, Organic Carbon, and Organic Matter: Methods of Soil Analysis Part 3—Chemical Methods*; Wiley: Hoboken, NJ, USA, 2018; pp. 961–1010.
12. Mehlich, A. Mehlich 3 soil test extractant: A modification of Mehlich 2 extractant. *Commun. Soil Sci. Plant Anal.* **1984**, 37–41. [CrossRef]
13. Maltas, A.; Kebli, H.; Oberholzer, H.R.; Weisskopf, P.; Sinaj, S. The effects of organic and mineral fertilizers on carbon sequestration, soil properties, and crop yields from a long-term field experiment under a Swiss conventional farming system. *Land Degrad. Dev.* **2018**, *29*, 926–938. [CrossRef]
14. Blanchet, G.; Gavazov, K.; Bragazza, L.; Sinaj, S. Responses of soil properties and crop yields to different inorganic and organic amendments in a Swiss conventional farming system. *Agric. Ecosyst. Environ.* **2016**, *230*, 116–126. [CrossRef]
15. Du, Y.; Cui, B.; Zhang, Q.; Wang, Z.; Sun, J.; Niu, W. Effects of manure fertilizer on crop yield and soil properties in China: A meta-analysis. *Catena* **2020**, *193*. [CrossRef]
16. Ma, Q.; Wen, Y.; Wang, D.; Sun, X.; Hill, P.W.; Macdonald, A.; Chadwick, D.R.; Wu, L.; Jones, D.L. Farmyard manure applications stimulate soil carbon and nitrogen cycling by boosting microbial biomass rather than changing its community composition. *Soil Biol. Biochem.* **2020**, *144*, 107760. [CrossRef]
17. Cai, A.; Xu, M.; Wang, B.; Zhang, W.; Liang, G.; Hou, E.; Luo, Y. Manure acts as a better fertilizer for increasing crop yields than synthetic fertilizer does by improving soil fertility. *Soil Tillage Res.* **2019**, *189*, 168–175. [CrossRef]
18. Vazquez, D.; Berger, A.; Prieto-Linde, M.L.; Johansson, E. Can nitrogen fertilization be used to modulate yield, protein content and bread-making quality in Uruguayan wheat? *J. Cereal Sci.* **2019**, *85*, 153–161. [CrossRef]
19. De Santis, M.A.; Giuliani, M.M.; Flagella, Z.; Reyneri, A.; Blandino, M. Impact of nitrogen fertilisation strategies on the protein content, gluten composition and rheological properties of wheat for biscuit production. *Field Crop. Res.* **2020**, *254*, 107829. [CrossRef]
20. Wang, L.; Palta, J.A.; Chen, W.; Chen, Y.; Deng, X. Nitrogen fertilization improved water-use efficiency of winter wheat through increasing water use during vegetative rather than grain filling. *Agric. Water Manag.* **2018**, *197*, 41–53. [CrossRef]
21. Fois, S.; Motzo, R.; Giunta, F. The effect of nitrogenous fertiliser application on leaf traits in durum wheat in relation to grain yield and development. *Field Crop. Res.* **2009**, *110*, 69–75. [CrossRef]
22. Szymańska, G.; Faligowska, A.; Panasiewicz, K.; Szukała, J.; Ratajczak, K.; Sulewska, H. The long-term effect of legumes as forecrops on the productivity of rotation winter triticale–winter rape with nitrogen fertilisation. *Acta Agric. Scand. Sect. B Soil Plant Sci.* **2020**, *70*, 128–134. [CrossRef]
23. Mourtzinis, S.; Marburger, D.; Gaska, J.; Diallo, T.; Lauer, J.G.; Conley, S. Corn, soybean, and wheat yield response to crop rotation, nitrogen rates, and foliar fungicide application. *Crop Sci.* **2017**, *57*, 983–992. [CrossRef]
24. Wang, L.; Zheng, X.; Tian, F.; Xin, J.; Nai, H. Soluble organic nitrogen cycling in soils after application of chemical/organic amendments and groundwater pollution implications. *J. Contam. Hydrol.* **2018**, *217*, 43–51. [CrossRef] [PubMed]
25. Ju, X.T.; Zhang, C. Nitrogen cycling and environmental impacts in upland agricultural soils in North China: A review. *J. Integr. Agric.* **2017**, *16*, 2848–2862. [CrossRef]
26. Wang, D.; Xu, Z.; Zhao, J.; Wang, Y.; Yu, Z. Excessive nitrogen application decreases grain yield and increases nitrogen loss in a wheat-soil system. *Acta Agric. Scand. Sect. B Soil Plant Sci.* **2011**, *61*, 681–692. [CrossRef]
27. Zorn, M.; Komac, B. Land degradation. *Encycl. Earth Sci. Ser.* **2013**, 580–583. [CrossRef]
28. Vašák, F.; Černý, J.; Buráňová, Š.; Kulhánek, M.; Balík, J. Soil pH changes in long-term field experiments with different fertilizing systems. *Soil Water Res.* **2015**, *10*, 19–23. [CrossRef]
29. Rangan, P.; Furtado, A.; Henry, R.J. New evidence for grain specific C4 photosynthesis in wheat. *Sci. Rep.* **2016**, *6*, 1–12.
30. Knott, C.A.; Van Sanford, D.A.; Ritchey, E.L.; Swiggart, E. Wheat yield response and plant structure following increased nitrogen rates and plant growth regulator applications in Kentucky. *Crop. Forage Turfgrass Manag.* **2016**, *2*. [CrossRef]

31. Casagrande, M.; David, C.; Valantin-Morison, M.; Makowski, D.; Jeuffroy, M.H. Factors limiting the grain protein content of organic winter wheat in south-eastern France: A mixed-model approach. *Agron. Sustain. Dev.* **2009**, *29*, 565–574. [CrossRef]
32. Pratt, P.F.; Broadbent, F.E.; Martin, J.P. Using organic wastes as nitrogen fertilizers. *Calif. Agric.* **1972**, *27*, 10–13.
33. Wilson, M. Manure Characteristics. Available online: https://extension.umn.edu/manure-land-application/manure-characteristics#phosphorus-and-potassium-817861 (accessed on 3 August 2020).
34. Eghball, B.; Wienhold, B.J.; Gilley, J.E.; Eigenberg, R.A. Biological Systems Engineering: Papers and Publications Mineralization of Manure Nutrients Mineralization of manure nutrients. *J. Soil Water Conserv.* **2002**, *57*, 470–473.
35. Barneix, A.J. Physiology and biochemistry of source-regulated protein accumulation in the wheat grain. *J. Plant Physiol.* **2007**, *164*, 581–590. [CrossRef] [PubMed]
36. Kindred, D.R.; Verhoeven, T.M.O.; Weightman, R.M.; Swanston, J.S.; Agu, R.C.; Brosnan, J.M.; Sylvester-Bradley, R. Effects of variety and fertiliser nitrogen on alcohol yield, grain yield, starch and protein content, and protein composition of winter wheat. *J. Cereal Sci.* **2008**, *48*, 46–57. [CrossRef]
37. Hlisnikovský, L.; Kunzová, E.; Hejcman, M.; Dvořáček, V. Effect of fertilizer application, soil type, and year on yield and technological parameters of winter wheat (*Triticum aestivum*) in the Czech Republic. *Arch. Agron. Soil Sci.* **2015**, *61*, 33–53. [CrossRef]
38. Hlisnikovský, L.; Hejcman, M.; Kunzová, E.; Menšík, L. The effect of soil-climate conditions on yielding parameters, chemical composition and baking quality of ancient wheat species *Triticum monococcum* L., *Triticum dicoccumSchrank* and *Triticum spelt* L. in comparison with modern *Triticum aestivum* L. *Arch. Agron. Soil Sci.* **2019**, *65*, 152–163.
39. Flagella, Z.; Giuliani, M.M.; Giuzio, L.; Volpi, C.; Masci, S. Influence of water deficit on durum wheat storage protein composition and technological quality. *Eur. J. Agron.* **2010**, *33*, 197–207. [CrossRef]
40. De Stefanis, E.; Sgrulletta, D.; De Vita, P.; Pucciarmati, S. Genetic variability to the effects of heat stress during grain filling on durum wheat quality. *Cereal Res. Commun.* **2002**, *30*, 117–124. [CrossRef]
41. Harkness, C.; Semenov, M.A.; Areal, F.; Senapati, N.; Trnka, M.; Balek, J.; Bishop, J. Adverse weather conditions for UK wheat production under climate change. *Agric. For. Meteorol.* **2020**, *282*, 107862. [CrossRef]
42. Olesen, J.E.; Bindi, M. Consequences of climate change for European agricultural productivity, land use and policy. *Eur. J. Agron.* **2002**, *16*, 239–262. [CrossRef]

Article

Economic Optimal Nitrogen Rate Variability of Maize in Response to Soil and Weather Conditions: Implications for Site-Specific Nitrogen Management

Xinbing Wang [1], Yuxin Miao [2,*], Rui Dong [1], Zhichao Chen [3], Krzysztof Kusnierek [4], Guohua Mi [1] and David J. Mulla [2]

1 College of Resources and Environmental Sciences, China Agricultural University, Beijing 100193, China; xbwang2020@cau.edu.cn (X.W.); bs20183030296@cau.edu.cn (R.D.); miguohua@cau.edu.cn (G.M.)
2 Precision Agriculture Center, Department of Soil, Water, and Climate, University of Minnesota, St. Paul, MN 55108, USA; mulla003@umn.edu
3 School of Surveying and Land Information Engineering, Henan Polytechnic University, Jiaozuo 454000, China; czc@hpu.edu.cn
4 Center for Precision Agriculture, Norwegian Institute of Bioeconomy Research (NIBIO), Nylinna 226, 2849 Kapp, Norway; krzysztof.kusnierek@nibio.no
* Correspondence: ymiao@umn.edu

Received: 1 August 2020; Accepted: 19 August 2020; Published: 21 August 2020

Abstract: The dynamic interactions between soil, weather and crop management have considerable influences on crop yield within a region, and should be considered in optimizing nitrogen (N) management. The objectives of this study were to determine the influence of soil type, weather conditions and planting density on economic optimal N rate (EONR), and to evaluate the potential benefits of site-specific N management strategies for maize production. The experiments were conducted in two soil types (black and aeolian sandy soils) from 2015 to 2017, involving different N rates (0 to 300 kg ha^{-1}) with three planting densities (55,000, 70,000, and 85,000 plant ha^{-1}) in Northeast China. The results showed that the average EONR was higher in black soil (265 kg ha^{-1}) than in aeolian sandy soil (186 kg ha^{-1}). Conversely, EONR showed higher variability in aeolian sandy soil (coefficient of variation (CV) = 30%) than in black soil (CV = 10%) across different weather conditions and planting densities. Compared with farmer N rate (FNR), applying soil-specific EONR (SS-EONR), soil- and year-specific EONR (SYS-EONR) and soil-, year-, and planting density-specific EONR (SYDS-EONR) would significantly reduce N rate by 25%, 30% and 38%, increase net return (NR) by 155 \$ ha^{-1}, 176 \$ ha^{-1}, and 163 \$ ha^{-1}, and improve N use efficiency (NUE) by 37–42%, 52%, and 67–71% across site-years, respectively. Compared with regional optimal N rate (RONR), applying SS-EONR, SYS-EONR and SYDS-EONR would significantly reduce N application rate by 6%, 12%, and 22%, while increasing NUE by 7–8%, 16–19% and 28–34% without significantly affecting yield or NR, respectively. It is concluded that soil-specific N management has the potential to improve maize NUE compared with both farmer practice and regional optimal N management in Northeast China, especially when each year's weather condition and planting density information is also considered. More studies are needed to develop practical in-season soil (site)-specific N management strategies using crop sensing and modeling technologies to better account for soil, weather and planting density variation under diverse on-farm conditions.

Keywords: yield; site-specific nitrogen management; regional optimal nitrogen management; net return; nitrogen use efficiency; spatial variability; temporal variability

1. Introduction

Improper nitrogen (N) management in current crop production systems has become a growing concern among governments, scientists and farmers around the world [1–3]. Optimizing N management in agriculture is crucially important for food security, environmental protection, and sustainable development [3–5]. This is particularly true for China, the world's largest producer, consumer and importer of chemical fertilizers [6,7]. Chinese scientists have been promoting a regional optimal N management (RONM) strategy to avoid significant over- or under-application problems [5,8]. If it were adopted for maize (*Zea mays* L.) production across China, more than 1.4 million tons N fertilizer and 18.6 million tons of greenhouse gas (GHG) emission would be reduced [9]. Such strategy can be easily adopted by farmers and won't increase their costs. However, due to the significant field-to-field and within-field variability of indigenous soil N supply and crop N demand, this fixed rate and timing strategy will unavoidably result in sub-optimal N management in different fields within a region [8,10]. There is a growing interest in China to develop alternative strategies to further improve N use efficiency (NUE) by better matching N supply with crop N requirement in both space and time [5,6]. Accordingly, it is necessary to determine key factors influencing maize response to N rate and evaluate the potential benefits of alternative N management strategies first.

The first and most important factor to consider is soil type differences, especially soil texture, which regulates many soil processes such as water retention and infiltration, soil organic matter mineralization and nutrient dynamics and, therefore, influences soil N availability and crop yield [11–14]. There are about 17 different soil types, according to the United States Department of Agriculture (USDA) Soil Taxonomy, in Lishu county, Jilin Province, Northeast China [15]. Recent research indicated that N requirements for maize varied spatially due to the spatial heterogeneity of soil texture [16]. The optimal N rate should be determined according to variability in these soil properties that influence soil N availability or crop response to available N [17]. Loamy clay and loamy sand are two representative soil textures in Northeast China. The loamy clay soil generally has a higher soil organic matter (SOM) content, higher water holding capacity, and stronger ability to fix NH_4-N than loamy sand [18]. Loamy sand soils, on the other hand, have generally lower SOM and water holding capacity, but due to greater aeration, they are usually characterized by a higher N mineralization rate than the loamy clay soils [19], causing higher risks of N leaching losses [20]. A recent study from Northeast China indicated that there was a weak parabolic relationship between N rate and maize root length in loamy clay and clay loam soils, but not in the loamy sand soil [21]. That study reported that root length and grain yield were both maximized at the optimal N rate (ONR) of 168–240 kg N ha^{-1} across soils and years. Results of Qiu et al. [22] indicated that ONR ranged between 140 and 210 kg ha^{-1} for maize in Northeast China across site-years. The results of studies conducted in North America indicated that the maize grown in fine-textured soils had significantly greater response to added N than the maize grown in medium-textured soils [23].

In addition to soil type and soil texture, weather conditions can also have a strong impact on crop growth, soil water and nutrient dynamics, and crop response to N fertilization. Precipitation and temperature have been found to significantly affect maize grain yield, soil mineral N, and maize response to N [24–27]. The interaction between soil properties and weather conditions controls the soil water and nutrient availability as well as crop yield potential during the growing season [28,29]. Due to the spatial and temporal variations in crop N demand and soil N supply and losses, crop responses to N fertilizer may vary both between and within soils under different weather conditions [30–32]. This can result in significant changes of ONR in space and time [33–35]. It has been found that maize yield response to N fertilization could be enhanced by abundant and well-distributed rainfall, and accumulated maize heat units [23]. Therefore, weather conditions should also be taken in account when determining the ONR for different soils.

Planting density is often considered one of the most important crop management practices to improve grain yield and NUE for maize production [36–38]. An optimal planting density is needed together with a matching optimal N rate to ensure appropriate aboveground and underground plant growth through different utilization of solar radiation and soil nutrients [38–40]. Hence, the maximum maize grain yield in a specific environment (related to soil and weather conditions) may be achieved [41].

So far, few studies have explored the effects of soil type (texture), weather condition, and planting density on the economic optimal N rate (EONR) for maize production, especially in Northeast China. Therefore, the objectives of this study were to (1) determine the EONR as affected by soil type, weather condition and planting density, and (2) evaluate the potential benefits of applying soil-specific (SS), soil- and year-specific (SYS), and soil-, year- and density-specific (SYDS) EONR for maize production in Northeast China.

2. Materials and Methods

2.1. Site Descriptions

The study was conducted in Lishu County (43°02′–43°46′ N, 123°45′–124°53′ E), Jilin Province in Northeast China. Two field locations within the study site with contrasting soil types were selected for this study: one field with a black soil (loamy clay) equivalent to typical Haploboroll and the other field with an aeolian sandy soil (loamy sand) equivalent to typical Cryopsamments according to the USDA Soil Taxonomy [42]. In Lishu County, about 54,700 hectares of black soil fields and about 13,900 hectares of aeolian sandy soil fields are used to grow spring maize [15]. The black soil field was fertile and fine-textured with higher field capacity (0.39 cm cm^{-3}), total N (1.35 g kg^{-1}), and SOM (26.2 g kg^{-1}) than the coarse-textured aeolian sandy soil field with lower field capacity (0.13 cm cm^{-3}), total N (0.65 g kg^{-1}), and SOM (9.7 g kg^{-1}) [43]. The daily precipitation (mm) and daily mean temperature (°C) during three maize growing seasons from 2015 to 2017 were reported in the previous study in Lishu County [43]. According to accumulated precipitation (APP) of maize growth season, year 2015, 2016, and 2017 were considered as dry (347.3 mm), wet (660.6 mm), and normal (509.9 mm) years, respectively.

2.2. Field Experiments and N Management Strategies

The same field experiment was conducted from 2015 to 2017 in the black soil and aeolian sandy soil fields. The experiment used a two-factor randomized complete block design with three replicates involving six N rates (from 0 to 300 kg N ha^{-1} for maize with an increment of 60 kg N ha^{-1}) with three planting densities (D1: 55,000 plants ha^{-1}, D2: 70,000 plants ha^{-1}, D3: 85,000 plants ha^{-1}) in each field. The plot size was 9 × 8 m^2 with wide-narrow row planting spacing of 0.40–0.80 m. For each N treatment, one-third of the N fertilizer in the form of urea and all the phosphorus in the form of calcium superphosphate (at rate of 90 kg P_2O_5 ha^{-1}) and potassium in the form of potassium sulphate (at rate of 90 kg K_2O ha^{-1}) fertilizers were blended into the top 20 cm soil as basal fertilizers before planting. The remaining two-thirds of the N fertilizer was side-dressed at the V8 growth stage.

To compare different N management strategies, we defined the treatments of 300 kg N ha^{-1} with 55,000 plants ha^{-1} and 240 kg N ha^{-1} with 70,000 plants ha^{-1} as the farmer N rate (FNR) and regional optimal N rate (RONR) management strategies, respectively. The treatment of 0 kg N ha^{-1} with 55,000 plants ha^{-1} was defined as check plot (CK). Three EONR management strategies were evaluated in this study: (1) soil-specific EONR (SS-EONR) adjusts N application rates according to different soil types; (2) soil- and year-specific EONR (SYS-EONR) adjusts N application rates according to different soil types and each year's weather conditions; and (3) soil-year-density-specific EONR (SYDS-EONR) adjusts N application rates according to different soil types, each year's weather conditions and different planting densities. The EONR was defined as the rate of N application where $1 of additional N fertilizer returned $1 in grain yield, and was based on the assumption that N fertilizer was the only variable cost and all other costs were fixed [44]. The optimal plant density was empirically determined at 70,000 and 55,000 plants ha^{-1} for the black and aeolian sandy soil fields, respectively. The SS-EONR, SYS-EONR and SYDS-EONR were determined based on the maize yield responses to the N application rate for specific soil, specific soil- and year, and specific soil-, year and density situations, respectively.

The local maize variety-Liangyu 66 was used in both fields. No irrigation was applied in the black soil field, while one-time irrigation of about 50 mm of water was applied before the anthesis growth stage in the aeolian sandy soil field each year. All plots were kept free of weeds, insects, and diseases with chemicals based on standard practices.

2.3. Sample Collection and Data Calculation

Before the start of the experimental series in 2015, soil samples were collected from each plot to determine the soil physical and chemical characteristics. At maize harvest stage (R6) for each growing season, three plant samples were randomly collected from each plot and split into stalks, leaves and grains. These three parts of plant samples were dried in the oven at 105 °C for one hour and then at 85 °C to a constant weight to determine dry aboveground biomass (AGB), which was the sum dry weight of talks, leaves and grains. Then they were ground into fine powder to determine plant N concentration (PNC) by the Kjeldahl digestion method [45], and the plant nitrogen uptake (PNU) was determined by multiplying PNC by AGB. Finally, the N nutrition index (NNI) for each plot was determined by the ratio of actual and critical PNC at harvest stage [46]. The critical PNC was calculated as following equation:

$$PNCc = 36.5 \times W^{-0.48} \quad (1)$$

where PNCc is the critical plant N concentration expressed as "g kg^{-1} dry matter (DM)" and W is the AGB expressed in "t DM ha^{-1}".

After sampling, grain yield was determined by harvesting the middle 20 m^2 area of each plot and standardized to 14% grain moisture content. Later, partial factor productivity (PFP), agronomic efficiency (AE), and recovery efficiency (RE) were calculated using the following equations:

$$PFP\ [kg\ kg^{-1}] = Y_N/N_F \quad (2)$$

$$AE\ [kg\ kg^{-1}] = (Y_N - Y_0)/N_F \quad (3)$$

$$RE\ [\%] = (PNU_N - PNU_0)/N_F \quad (4)$$

where Y_N and Y_0 are the yield in N fertilizer application plots and 0 kg N ha^{-1} plots, respectively, and PNU_N and PNU_0 are the plant N uptake (PNU) in N application plots and 0 kg N ha^{-1} plots, respectively, and N_F is the applied N fertilizer rate.

The economic income, defined as net return (NR, \$ ha^{-1}), was calculated according to Formula (4):

$$NR = GY \times GP - Cost \quad (5)$$

where GY is the grain yield (kg ha^{-1}), GP is the grain price (0.25 \$ kg^{-1}), and the Cost included field tillage (100 \$ ha^{-1}), sowing (127 \$ ha^{-1}), irrigation (423 \$ ha^{-1}), pesticide (100 \$ ha^{-1}), herbicide (100 \$ ha^{-1}), harvest (155 \$ ha^{-1}), N fertilizer (0.92 \$ N kg^{-1}), phosphorus fertilizer (0.52 \$ P_2O_5 kg^{-1}), potassium fertilizer (0.52 \$ K_2O kg^{-1}), and maize seeds (1.05 \$ 1000 $seeds^{-1}$).

2.4. Statistical Analysis

Analysis of variance (ANOVA) was conducted using the general linear model procedure in SPSS 25.0 software (SPSS Inc., Chicago, IL, USA). The main effects of soil, year, planting density, and N fertilizer rate on yield, AGB, PNC, and NNI were analyzed. Mean values of the aforementioned variables for each N treatment were compared using least significant difference test (LSD) at the $p < 0.05$ probability level. Three statistical models (quadratic, quadratic-plus-plateau and linear-plus-plateau) were selected to describe the crop yield response to N rate, AGB, PNC, and NNI. The PROC NLIN procedure in SAS software (Version 8.0, SAS, 2013), was used to build and analyze those models. The choice of the best model was based on the coefficients of determination (R^2) and root mean square error (RMSE). The quadratic model had the best fit to describe the crop yield response to AGB, and the linear-plus-plateau model had the best fit to describe the crop yield response to PNC and NNI at specific soil and specific soil-year respectively. The quadratic-plus-plateau model had the best fit and was therefore used to calculate the EONR and yield at EONR (EOY). The EONR (kg N ha^{-1}) was calculated as:

$$\text{EONR [kg ha}^{-1}] = (\text{CP} - \text{b})/2\text{c} \quad (6)$$

where CP was the ratio of the cost of N fertilizer to the price of maize grain, and b and c are the linear and quadratic coefficients from the quadratic-plus-plateau equation. The EOY was calculated by substituting the EONR value into the quadratic-plus-plateau equation [44].

Additionally, multiple linear regression was used to establish the relationships between EONR (obtained yield) and the soil total N (TN), planting density (D), growing degree days (GDD), and accumulated precipitation (APP) during maize growing season using the SPSS 25.0 software (SPSS Inc., Chicago, IL, USA). The GDD was calculated as follows:

$$\text{GDD} = \sum ((T_{max} - T_{min})/2 - T_{base}) \quad (7)$$

where T_{max}, T_{min}, T_{base} are the daily maximum, minimum, and base temperatures, respectively and T_{base} = 10 °C.

3. Results

3.1. The Description of Maize Agronomic Parameters

According to the results of ANOVA (Table 1), maize yield, AGB, PNC, and NNI, were all significantly affected by soil type, year with its weather pattern, N rate, and their interactions. However, the yield, PNC, and NNI were not directly affected by the planting density.

Table 1. Significance of mean squares in the analysis of variance (ANOVA) of yield, aboveground biomass (AGB), plant N concentration (PNC), and N nutrition index (NNI) across two soil types (S), three years (Y), three densities (D), and six N rates (N).

Source of Variation	Df	Significance of Mean Square			
		Yield	AGB	PNC	NNI
Soil (S)	1	***	***	***	***
Year (Y)	2	***	***	***	***
Density (D)	2	ns	***	ns	ns
Nitrogen (N)	5	***	***	***	***
S × Y	2	***	***	***	***
S × D	2	ns	ns	*	*
S × N	5	***	***	***	**
Y × D	4	*	**	ns	ns
Y × N	9	***	***	***	***
D × N	10	ns	ns	ns	ns
S × Y × D	4	ns	ns	ns	**
S × Y × N	9	ns	*	***	***
S × D × N	10	ns	ns	ns	ns
Y × D × N	18	ns	ns	ns	ns
S × Y × D × N	18	ns	ns	ns	ns

Note: *, **, and *** indicate significance at $p < 0.05$, $p < 0.01$, and $p < 0.001$ probability levels, respectively. ns was non-significant ($p > 0.05$).

The multiple comparisons of the analyzed agronomic parameters in data subsets aggregated by a given influencing factor are shown in Table 2. The maize yield, AGB, and NNI in the black soil field were significantly higher (by 3.43 t ha^{-1}, 5.91 t ha^{-1}, and 0.06) than in the aeolian sandy soil field. On the other hand, the PNC were significantly lower (by 1.31 kg kg^{-1}) in the black soil field than in the aeolian sandy soil field. The relatively wet season of 2016 brought the highest yield and AGB while NNI was the lowest among three years. In 2015, a relatively dry year, the yield and PNC were the lowest in the analyzed period. The yield, PNC, and NNI were not significantly affected by the three tested planting densities. The values of all of the parameters significantly improved with the increasing N rate, until the N4 treatment (240 kg ha^{-1}).

Table 2. The multiple comparisons of maize yield, net return (NR), aboveground biomass (AGB), plant N concentration (PNC), N nutrition index (NNI), and N surplus (NS) at two soil types (S), three years (Y), three densities (D), and six N rates (N) respectively.

Items	Treatments	Yield (t ha^{-1})	AGB (t ha^{-1})	PNC (kg kg^{-1})	NNI
Soil	B	10.22 ± 0.24 ^{a}	18.73 ± 0.45 ^{a}	8.85 ± 0.11 ^{b}	0.85 ± 0.01 ^{a}
	S	6.79 ± 0.17 ^{b}	12.82 ± 0.32 ^{b}	10.16 ± 0.15 ^{a}	0.79 ± 0.01 ^{b}
Year	2015	8.05 ± 0.30 ^{b}	16.25 ± 0.53 ^{a}	9.29 ± 0.18 $_{b}$	0.87 ± 0.01 ^{a}
	2016	9.13 ± 0.31 ^{a}	16.99 ± 0.59 ^{a}	9.37 ± 0.15 ^{ab}	0.78 ± 0.02 ^{b}
	2017	8.26 ± 0.29 ^{b}	14.16 ± 0.52 ^{b}	9.82 ± 0.18 ^{a}	0.81 ± 0.02 ^{b}
Density	D1	8.44 ± 0.29 ^{a}	14.87 ± 0.53 ^{b}	9.65 ± 0.18 ^{a}	0.80 ± 0.02 ^{a}
	D2	8.48 ± 0.32 ^{a}	15.68 ± 0.56 ^{ab}	9.45 ± 0.17 ^{a}	0.82 ± 0.02 ^{a}
	D3	8.59 ± 0.32 ^{a}	16.78 ± 0.58 ^{a}	9.41 ± 0.17 ^{a}	0.83 ± 0.02 ^{a}
Nitrogen	N0	4.57 ± 0.28 ^{e}	9.26 ± 0.50 ^{e}	7.15 ± 0.09 ^{e}	0.58 ± 0.02 ^{e}
	N1	7.12 ± 0.21 ^{d}	12.80 ± 0.42 ^{d}	8.50 ± 0.14 ^{d}	0.68 ± 0.01 ^{d}
	N2	8.95 ± 0.30 ^{c}	15.81 ± 0.58 ^{c}	9.68 ± 0.16 ^{c}	0.82 ± 0.01 ^{c}
	N3	9.88 ± 0.34 ^{b}	18.40 ± 0.59 ^{b}	10.32 ± 0.17 ^{b}	0.92 ± 0.01 ^{b}
	N4	10.28 ± 0.35 ^{ab}	19.58 ± 0.66 ^{ab}	10.81 ± 0.16 ^{a}	0.99 ± 0.01 ^{a}
	N5	11.08 ± 0.37 ^{a}	20.33 ± 0.74 ^{a}	11.08 ± 0.19 ^{a}	0.99 ± 0.01 ^{a}

Note: the notation for treatments within soil (B: black soil, S: aeolian sandy soil), year, density (D1: 55,000 plant ha^{-1}, D2: 70,000 plants ha^{-1}, and D3: 85,000 plants ha^{-1}), and nitrogen (N0: 0 kg ha^{-1}, N1: 60 kg ha^{-1}, N2: 120 kg ha^{-1}, N3: 180 kg ha^{-1}, N4: 240 kg ha^{-1}, and N5: 300 kg ha^{-1}). The number behind "±" is the standard error, and numbers for the same item followed by different letters indicate significant differences ($p < 0.05$).

An overview of the relationships between maize yield and agronomic parameters showed distinct crop response to growing conditions (Figures 1 and 2). Whether across the three years or in a specific year, for black soil and aeolian sandy soil fields the relationship between yield and AGB had a significant quadratic relationship. On the contrary, the relationships between yield and either PNC or NNI were modeled according to the linear-plus-plateau models. Across the three years, the yield was maximized when the PNC reached 9.6 kg kg^{-1} and 10.1 kg kg^{-1} in black soil and aeolian sandy soil, respectively. Correspondingly, in black soil field, the yield reached its maximum when the NNI was at 0.95, while in the aeolian sandy soil, the maximum yield was obtained at NNI of 0.81. Analyzed for a given year, in 2015, 2016, and 2017, the yield was maximized when the PNC reached 8.2, 9.2, and 9.9 kg kg^{-1} in black soil and 9.7, 9.5, and 10.1 kg kg^{-1} in aeolian sandy soil, respectively. Correspondingly, in the black soil field, the yield reached its maximum when the NNI was at 1.15, 0.84, and 0.90 in specific year of 2015, 2016, and 2017, while in the aeolian sandy soil, the yield was maximized at NNI of 0.74, 0.80, and 0.88, respectively.

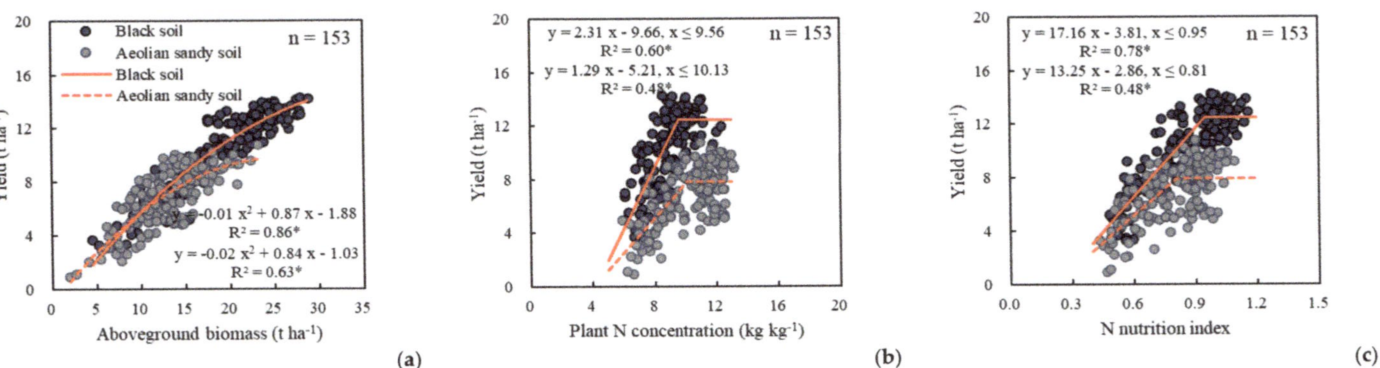

Figure 1. The relationships between crop yield and aboveground biomass (**a**), plant N concentration (**b**), or N nutrition index (**c**) for two soils across three years and three planting densities. (Note: the "n" is the number of samples for each soil type, the first equation is for black soil, the second equation is for aeolian sandy soil, * indicate significance at $p < 0.05$ probability level).

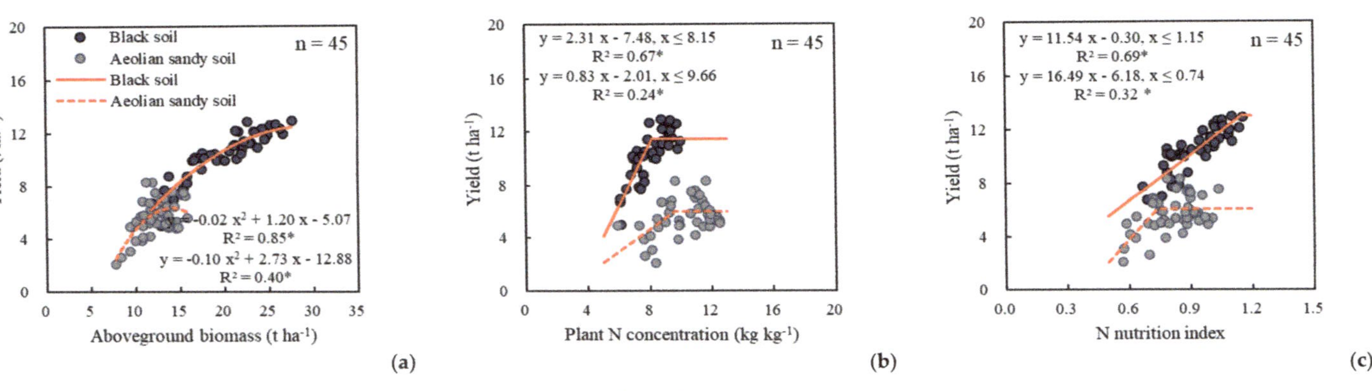

Figure 2. *Cont.*

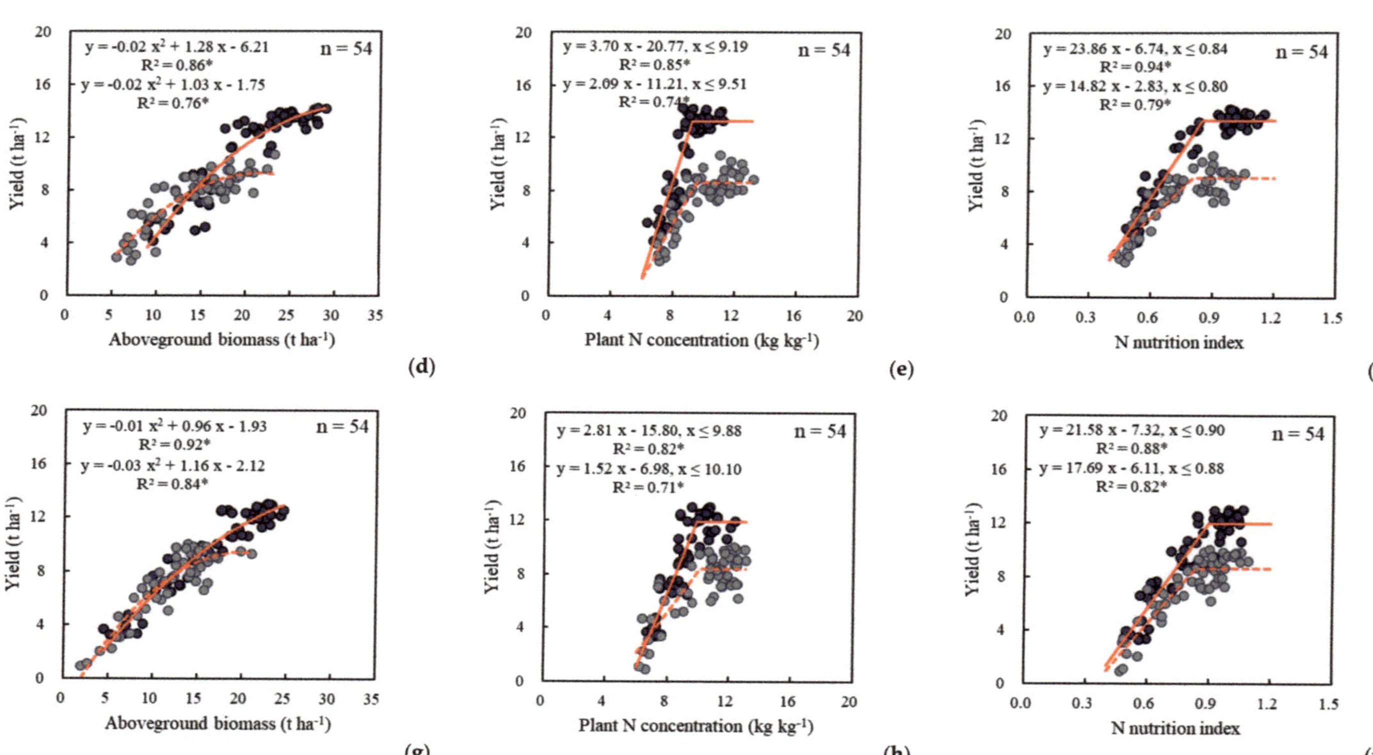

Figure 2. The relationships between crop yield and aboveground biomass (**a**,**d**,**g**), plant N concentration (**b**,**e**,**h**), or N nutrition index (**c**,**f**,**i**) in year of 2015 (**a**–**c**), 2016 (**d**–**f**), and 2017 (**g**–**i**) for two soils across three planting densities. (Note: the "n" is the number of samples for each soil type, the first equation is for black soil, the second equation is for aeolian sandy soil, * indicate significance at $p < 0.05$ probability level.

3.2. The Response of Maize Agronomic Parameters to N Application Rate

The maize yield was significantly higher in black soil than in aeolian sandy soil at each N application rate (Figure 3). According to the quadratic-plus-plateau model, the maize yield was maximized at the N rates of 285 and 201 kg ha^{-1} in black soil and aeolian sandy soils across three years, respectively. Furthermore, the lowest N rate for obtaining the maximum yield, or the agronomic optimal N rate (AONR), was not stable in either the black soil field or the aeolian sandy soil field and was influenced by the weather pattern in a given season. For specific year of 2015, 2016, and 2017, the soil-specific AONR of black soil and aeolian sandy soil fields were 300, 243, and 277 kg ha^{-1} and 112, 209, and 217 kg ha^{-1} in 2015, 2016, and 2017, respectively. The average EONRs were 265 kg ha^{-1} (276, 230, and 260 kg ha^{-1}) and 186 kg ha^{-1} (101, 193, and 203 kg ha^{-1}) in black soil and aeolian sandy soil fields, respectively, across three years (in specific year of 2015, 2016, and 2017).

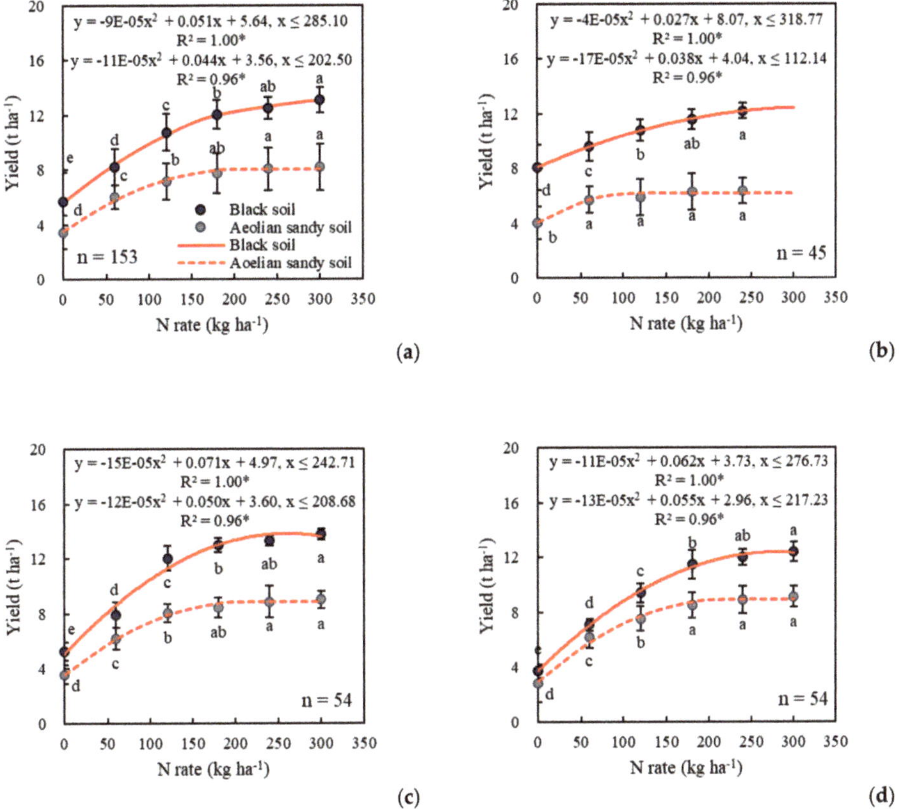

Figure 3. The responses of crop yield to N application rate across three years (**a**), and in specific year of 2015 (**b**), 2016 (**c**), and 2017 (**d**) in two soils across three planting densities. (Note: the "n" is the number of samples for each soil type, the first equation is for black soil, the second equation is for aeolian sandy soil, * indicate significance at $p < 0.05$ probability level, the lowercase letters in the table indicate the significant difference at 0.05 level).

Moreover, the soil-specific EONR was also influenced by the year and planting density interaction (Figure 4). In the black soil field, the soil-specific EONR had a coefficient of variation (CV) of 10% and reached 210, 225, and 240 kg ha^{-1}, 234, 214, and 252 kg ha^{-1}, and 266, 250, and 266 kg ha^{-1} at the planting density of 55,000, 70,000, and 85,000 plants ha^{-1} in year of 2015, 2016, and 2017, respectively. The NUE analysis using PFP and AE showed that the highest values were obtained at the planting density of 70,000 plants ha^{-1} (64 and 42 kg kg^{-1}) in all three years compared with 50,000 plants ha^{-1} (54 and 20 kg kg^{-1}) and 85,000 plants ha^{-1} (50 and 37 kg kg^{-1}). In the aeolian sandy soil field, the soil-specific EONR varied with the CV of 30% and reached 88, 150, and 96 kg ha^{-1}, 158, 209, and 215 kg ha^{-1}, and 168, 177, and 208 kg ha^{-1} at the planting density of 55,000, 70,000, and 85,000 plants ha^{-1} in year of 2015, 2016, and 2017, respectively. Interestingly, the NUE analysis (PFP and AE), showed that the highest values were obtained at the planting density of 55,000 plants ha^{-1} (75 and 23 kg kg^{-1}) in all three years compared with 70,000 plants ha^{-1} (55 and 34 kg kg^{-1}) and 85,000 (50 and 36 kg kg^{-1}) (Figure 4). Meanwhile, according to the multiple linear regression (Figure 5), soil-specific EONR (R^2 = 0.77) and obtained yield (R^2 = 0.95) showed significant relationships with the soil total N, growing degree days, accumulated precipitation, and planting density.

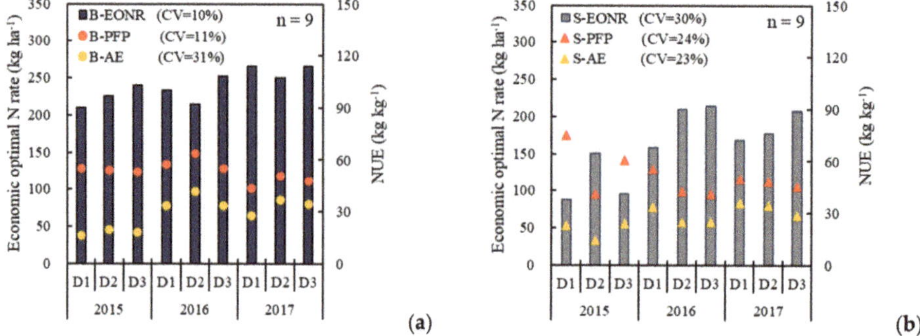

Figure 4. The variation of economic optimal N rate (EONR) and N use efficiency (NUE) (partial factor productivity (PFP) and agronomic efficiency (AE)) in a specific soil (B: black soil, (**a**); S: aeolian sandy soil, (**b**), year (2015, 2016, and 2017), and planting density (D1: 55,000 plant ha^{-1}, D2: 70,000 plant ha^{-1}, and D3: 85,000 plant ha^{-1}). (Note: the "n" is the number of samples for EONR, PFP, and AE respectively).

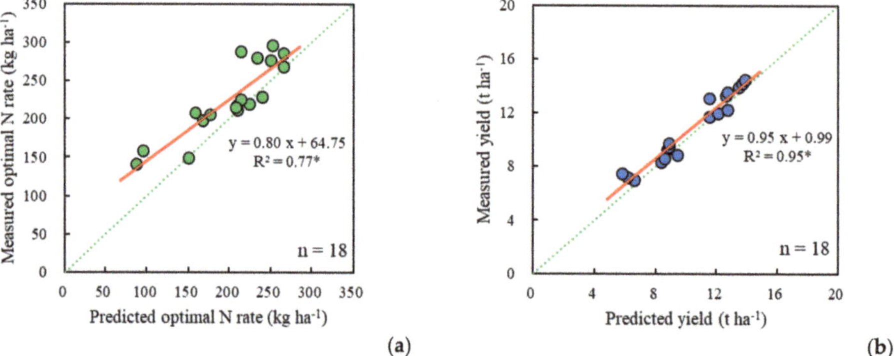

Figure 5. The relationships for measured and predicted soil-year-density specific economic optimal N rates (**a**) or yield (**b**) using soil, weather, and planting information (Note: the "n" is the number of samples, EONR = 101.24 × TN + 100.68 × GDD − 10.12 × APP + 0.57 × D − 164796.76, Y = 6.74 × TN + 0.91 × GDD − 0.09 × APP + 0.02 × D − 1486.83, where EONR is economic optimal N rate, Y is grain yield, TN is soil total N, GDD is growing degree days, APP is accumulated precipitation, D is planting density).

3.3. The Potential Benefits of Site-Specific N Management Strategies

Based on the EONR specific to different soil types, years and planting densities, as described above (Figure 4), three site-specific N management strategies were proposed. The results of the SS-EONR, SYS-EONR, and SYDS-EONR strategies with their explicit N-rates and optimal planting densities at 70,000 and 55,000 plants ha^{-1} for the black and aeolian sandy soil fields, respectively, were averaged across the soils and years (Table 3). This facilitated the comparison with FNR at 300 kg N ha^{-1} and 55,000 plants ha^{-1} and RONR at 240 kg N ha^{-1} and 70,000 plants ha^{-1}. The variation between the different strategies at the two soil types are given in Figure 6.

Table 3. The comparison of the N rate, yield, net return, partial factor productivity (PFP), agronomic efficiency (AE), and recovery efficiency (RE) from different N management strategies across soils and years.

Management	N Rate (kg ha^{-1})	Yield (t ha^{-1})	Net Return ($ ha^{-1})	PFP (kg kg^{-1})	AE (kg kg^{-1})	RE (%)
CK	0 ± 0 ^{d}	4.77 ± 0.80 ^{b}	459 ± 100 ^{c}			
FNR	300 ± 0 ^{a}	10.07 ± 1.02 ^{a}	1508 ± 155 ^{b}	33.57 ± 3.40 ^{c}	17.67 ± 3.12 ^{b}	53.79 ± 7.71 ^{a}
RONR	240 ± 0 ^{b}	10.33 ± 1.16 ^{a}	1612 ± 190 ^{a}	43.03 ± 4.83 ^{bc}	23.15 ± 4.28 ^{ab}	57.48 ± 7.22 ^{a}
SS-EONR	225 ± 18 ^{bc}	10.45 ± 1.14 ^{a}	1664 ± 167 ^{a}	45.93 ± 2.18 ^{b}	25.05 ± 3.81 ^{ab}	56.18 ± 7.85 ^{a}
SYS-EONR	211 ± 25 ^{bc}	10.48 ± 1.14 ^{a}	1685 ± 161 ^{a}	51.15 ± 3.71 ^{ab}	26.81 ± 3.15 ^{ab}	56.50 ± 7.99 ^{a}
SYDS-EONR	187 ± 25 ^{c}	10.34 ± 1.13 ^{a}	1671 ± 157 ^{a}	57.46 ± 4.14 ^{a}	29.52 ± 3.11 ^{a}	58.23 ± 8.31 ^{a}

Note: CK: check, zero N rate; FNR: farmer N rate; RONR: regional optimal N rate, SS-EONR: soil-specific economic optimal N rate; SYS-EONR: soil-, and year-specific economic optimal N rate; SYDS-EONR: soil-, year- and density-specific economic optimal N rate. The number behind "±" is the standard error. Different lowercase letters in the same column indicate significant difference at 0.05 level ($p < 0.05$).

In comparison with FNR across the two soil types and three years (Table 3), the SS-EONR, SYS-EONR, and SYDS-EONR strategies significantly reduced N rate by 25%, 30%, and 38%, increased NR by 155, 176, and 163 $ ha^{-1}, and improved NUE parameters (PFP and AE) by 37–42%, 52%, and 67–71%, respectively. Meanwhile, these three strategies showed no significant effects on maize yield. When compared with RONR, the SS-EONR, SYS-EONR, and SYDS-EONR strategies significantly reduced N rate by 6%, 12%, and 22%, and improved NUE parameters (PFP and AE) by 7–8%, 16–19%, and 28–34%, respectively, without significantly affecting maize yield and NR.

Analyzed for each soil type separately, the SS-EONR, SYS-EONR, and SYDS-EONR strategies performed differently when compared with FNR and RONR across the three years (Figure 6). In the black soil field, in comparison with FNR, the SS-EONR, SYS-EONR, and SYDS-EONR strategies significantly reduced N rate by 12%, 15%, and 22%, increased NR by 201, 212, and 193 $ ha^{-1}, and improved PFP by 20%, 26%, and 35%, respectively, without significantly affecting maize yield, AE, and RE. However, when compared with RONR, the SS-EONR, SYS-EONR, and SYDS-EONR strategies did not significantly affected N rate, maize yield, NR, and NUE parameters (PFP, AE, and AE).

In the aeolian sandy soil field, in comparison with FNR, the SS-EONR, SYS-EONR, and SYDS-EONR strategies reduced N rate by 38%, 45%, and 54%, increased NR by 109, 140, and 133 $ ha^{-1}, and improved NUE parameters (PFP and AE) by 62%, 76–93%, and 105–126%, respectively. When compared with RONR, the SS-EONR, SYS-EONR, and SYDS-EONR strategies reduced N rate by 23%, 31%, and 42%, increased NR by 95, 126, and 119 $ ha^{-1}, and improved NUE parameters (PFP and AE) by 31–33%, 44–56%, and 68–83%, respectively. It is worth noting that the SS-EONR, SYS-EONR, and SYDS-EONR strategies showed no significant difference in maize yield whether comparing with FNR or RONR in the aeolian sandy soil field.

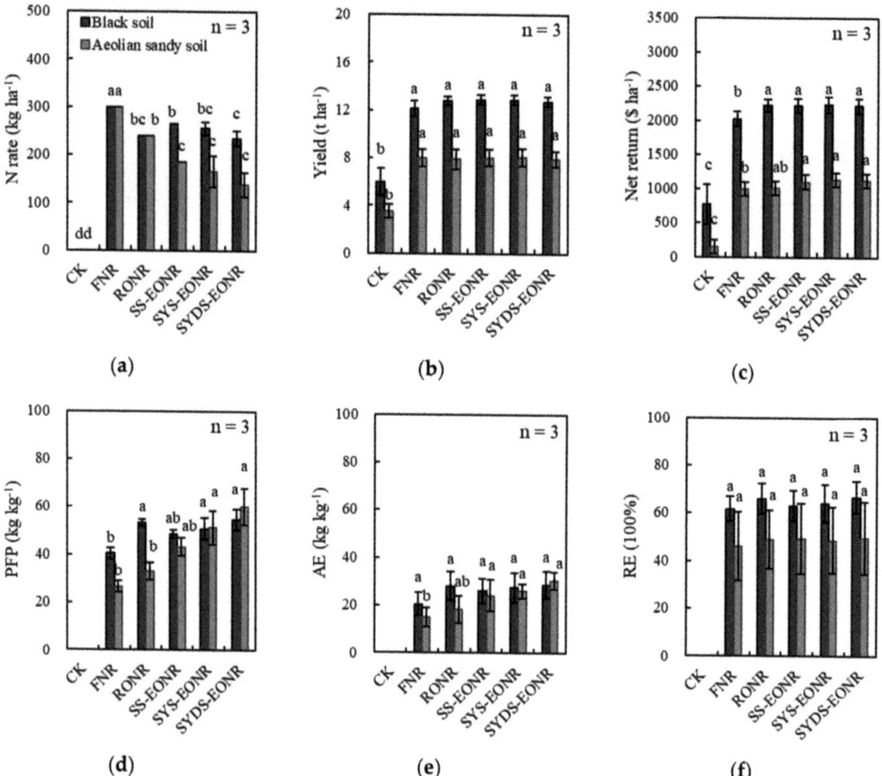

Figure 6. The comparison of the N rates (**a**), yield (**b**), net return (**c**), partial factor productivity (PFP, **d**), agronomic efficiency (AE, **e**), and recovery efficiency (RE, **f**) from different N management strategies at specific soil type across three years. (Note: CK: zero N rate, FNR: farmer application N rate, RONR: regional optimal N rate, SS-EONR: soil-specific economic optimal N rate, SYS-EONR: soil-year-specific economic optimal N rate, SYDS-EONR: soil-year- density-specific economic optimal N rate. The "n" is the number of samples for each N management strategies at specific soil type respectively. The lowercase letters in the table indicate the significant difference at 0.05 level ($p < 0.05$)).

4. Discussion

4.1. The Soil-Specific Economic Optimal N Rate

In this study, the black soil field was characterized by a higher water holding capacity and soil fertility than the aeolian sandy soil field (Table 1). This led to a more efficient nutrient supply to the maize crop during the growing season and resulted in larger AGB and NNI in the black soil field (Table 2). These findings were in agreement with previous studies conducted in this study region [47,48]. According to the relationship between grain yield and AGB or NNI (Figure 1), a higher yield was recorded in the black soil field than in aeolian sandy soil field (Table 2). That is despite the fact that the PNC was higher in the aeolian sandy soil field than in the black soil field (Table 2). The NNI is generally used during the growing season to diagnose crop N status (deficient, optimal or surplus) for guiding in-season N application [49], however, the concept can also be extended to the maturity stage to guide adjustment of N management in the following season [50].

It is usually assumed that the ONRs are higher in coarse-textured soil fields than in fine-textured soil fields, due to the disability in coarse-textured soil fields to retain moisture leading to higher N leaching potential [51]. As a result, most farmers apply more N fertilizer in the aeolian sandy soil field than in the black soil field [52]. Furthermore, in this study location, maize production is rain-fed and water deficit frequently occurs during the maize growing season, hence the drought has been the main limiting factor of crop growth in the aeolian sandy soil field [53,54]. In order to avoid the overuse of N fertilizers, many researchers tend to use the linear plus plateau model to determine the ONR in China [22,55]. In this study, according to the R^2 and RMSE, the quadratic-plus-plateau model had the best fit and was, therefore, used to calculate the EONR. The EONR across three years and three planting densities was considerably higher in fine-textured black soil field (265 kg ha^{-1}) than in coarse-textured aeolian sandy soil field (186 kg ha^{-1}) (Figure 3a). According to the relationships between yield and PNC or NNI (Figure 1), plants with a given level of PNC and NNI could produce much more yield in black soil than in aeolian sandy soil. The minimum NNI to obtain the maximum yield in the aeolian sandy soil field (0.81) was significantly lower than in the black soil field (0.95). In other words, adding more N fertilizer would not lead to substantial increase of maize yield in aeolian sandy soil field. Therefore, N fertilizer was not considered the main limiting factor there. This result was in agreement with the previous studies stating that the ONR was lower in coarse-textured soil fields than in fine-texture soil fields and showing great soil-specific variability [50,56].

4.2. The Influence of Weather Conditions and Planting Density on Soil-Specific Economic Optimal N Rate

The interaction between soil properties and weather conditions had the greatest influence on the response of crop yield to N fertilizer [23,24,57]. According to the previous research [58–60], the relationship between soil properties and yield was mainly affected by the spatial and temporal variability in soil water holding capacity and precipitation. Therefore, ONR should be adjusted based on the interaction between soil properties and weather conditions. Precipitation was significantly different among three years covered by this study (Figure 1), and had a significant effect on yield, AGB, PNC, and NNI (Tables 1 and 2). Meanwhile, the minimum PNC and NNI to obtain the maximum maize yield also showed inter-annual variation in both fields (Figure 2). This resulted in the year-to-year variability of soil-specific EONR (Figure 3b–d). For the year of 2016, in black soil field with high soil buffering capacity and fertility (total N and SOM), the relatively high GDD with well-distributed precipitation would lead to a higher AGB and grain yield potential than in 2015 and 2017, a phenomenon noted also in several other studies [40,61,62]. Furthermore, the synchronization of high GDD and well-distributed precipitation in 2016 would lead to a higher soil nitrification rate [19,63] and would provide relatively more soil N for the maize growth than in 2015 and 2017. As a consequence, in 2016 the minimum NNI to obtained the maximum maize yield was the lowest among the three years. Therefore, the SS-EONR for the black soil was lower in 2016 than in 2015 and 2017. On the other hand, in the year of 2015, in aeolian sandy soil with low soil buffering capacity and fertility (total N and SOM), the severe drought restricted the crop growth and yield formation, a phenomenon well described in another study [64]. Due to the low AGB and yield potential, the minimum NNI to obtain the maximum maize yield in the aeolian sandy soil field was the lowest in 2015 among the three tested growing seasons. Therefore, the SS-EONR for the aeolian soil was lower in the dry year (2015) than in 2016 and 2017.

Another question faced by scientists and the farmers is how planting density should be adjusted for different soil types and weather conditions. Although, in this study, the planting density did not have any significant effect on the yield, PNC, and NNI (Table 1), the soil-specific EONR still varied among three weather conditions and planting densities, along with PFP and AE (Figure 4). Also, the variability of the parameters was higher in the aeolian sandy soil field than in the black soil field. The buffering capacity mainly comes from the texture and organic carbon. Therefore, in the fertile black soil field with a higher buffering capacity, the production would in theory be less affected by the varying conditions than in the barren aeolian sandy soil field. The barren aeolian sandy soil field had a low yield potential and high variation in soil conditions, leading to high variation in AGB and yield, which translated to high variation in EONR. Due to the relatively higher N uptake and AGB accumulation at the relatively higher planting densities [37,65], the highest soil-specific AONRs were defined in this study under the high (85,000 plants ha^{-1}) planting density in the fertile black soil field and under the middle (70,000 plants ha^{-1}) and high (85,000 plants ha^{-1}) planting density in the aeolian sandy soil field. Therefore, according to the PFP and AE with the highest values among three planting densities, the middle (70,000 plants ha^{-1}) and low (55,000 plants ha^{-1}) planting densities with their corresponding SYS-EONR would be the optimal N management strategy for maize production in the black soil and aeolian sandy soil fields, respectively. The SS-EONR could be adjusted based on the information about the soil properties, weather conditions, and planting density [66]. Through the multiple linear regression analysis (Figure 5) performed in this study, the SYDS-EONR and the obtained grain yield could be determined preliminarily using soil N, GDD, APP, and planting density.

4.3. The Potential Benefits of Applying Soil-Specific Economic Optimal N Rate

Currently, the RONR strategy recommended about 240 kg N ha^{-1} with 70,000 plant ha^{-1} for this study region [8,9]. In this strategy, the N fertilizer is applied at a fixed rate and timing without accounting for spatial and temporal variability in soil N supply and crop N demand. According to the results of this study, the EONR changed dramatically from the black soil field to the sandy soil field and from year to year, which confirmed the findings of the previous studies showing that an ONR varied significantly in space and time [33–35]. The previous research demonstrated that soil-specific N management could adjust the N fertilizer application to match crop requirement by identifying the gap between soil N supply and crop N demand according to their spatial and temporal variation in a particular growing season for a specific soil type [67,68]. Therefore, it is of great interest to learn how much we can further improve N management using alternative strategies that are more complex and accurate than the simple FNR and RONR strategies.

Across the two typical soils in this study region, with distinctly different soil properties, compared with FNR, the soil-specific EONR strategies would decrease the N application rates with no negative effect on maize yield, while increasing NR and NUE (Table 3 and Figure 6). When compared with RONR, the soil-specific EONR strategies still showed the potential to decrease the N application rates and increase NUE but with no negative effect on maize yield and NR. Meanwhile, because the EONR showed higher variability in aeolian sandy soil than in black soil across different weather conditions and planting densities (Figure 4), the soil-specific EONR strategies showed greater potential in decreasing N application rates and increasing NUE in aeolian sandy soil than in black soil. Therefore, the soil-specific EONR strategies have a great potential to be implemented to achieve high-yield and high-efficiency maize production in China. Furthermore, because of the variation in weather conditions, especially precipitation, EONR varied among different years (Figure 3). The SYS-EONR strategy would perform better in increasing NUE than the SS-EONR strategy. Although the planting density had no significant effects on grain yield and NR in this study (Table 1), the EONR was significantly affected by it and the interaction among soil type, weather conditions, and planting density (Figure 4). Therefore, the SYDS-EONR strategy would result in the highest potential benefits in reducing the N application rate, and increasing NUE than the SS-EONR and SYS-EONR strategies (Table 3 and Figure 6).

These results indicated that soil-specific N management had the potential to increase N management and improve NUE. The best improvement may be achieved in the coarse-textured aeolian sandy soils and implementing the soil-, year- and planting density-specific EONR strategy. However, it is a great challenge to determine soil- and year-specific planting densities and corresponding optimal N rates across different farmers' fields. Future studies are needed to use crop-sensing technologies and crop growth modeling methods to guide in-season soil-specific N management under on-farm conditions [5,43,69,70].

5. Conclusions

The future direction of world agriculture is towards high yield, profitability and resource use efficiency. Due to the variation in soil properties, weather conditions, and planting densities, the optimal N rate should be adjusted according to specific soil, weather, and planting density combinations. The results of this study indicated that the average EONR in a fertile black soil field (265 kg ha^{-1})

was higher than in an aeolian sandy soil field (186 kg ha^{-1}). The variation in weather conditions and planting density had significant effect on EONR, resulting in CV of 10% and 30% in black and aeolian sandy soil fields, respectively. The optimal planting density was defined at 70,000 and 55,000 plants ha^{-1} for the black soil and aeolian sandy soil fields, respectively. The soil-specific EONR management strategy performed better than RONR in reducing N application rate and improving NUE. The best improvement was achieved using the SYDS-EONR strategy which considered the soil, weather, and planting density combinations. More studies are needed to develop practical in-season soil (site)-specific N management strategies to better account for soil, weather and planting density variation under diverse on-farm conditions.

Author Contributions: Y.M. and D.J.M. designed the experiment. X.W. conducted the experiment, performed the analysis, and wrote the original paper, R.D. and Z.C. assisted in the experiment, plant and soil sampling, and sample processing. Y.M., K.K., and G.M. reviewed and revised the manuscript. All authors have read and agreed to the published version of the manuscript.

Funding: The research was financially supported by National Key Research and Development Program of China (2016 YFD0200600, 2016YFD0200602), Norwegian Ministry of Foreign Affairs (SINOGRAIN II, CHN-17/0019), and the UK Biotechnology and Biological Sciences Research Council (BB/P004555/1).

Acknowledgments: We would like to thank Yanjie Guan, Xuezhi Yue, Zheng Fang, Weidong Lou, and Guiyin Jiang for their help with the field experiments.

Conflicts of Interest: The authors declare no conflict of interest.

References

1. Dhital, S.; Raun, W.R. Variability in optimum nitrogen rates for maize. *Agron. J.* **2015**, *108*, 2165–2173. [CrossRef]
2. Davidson, E.A.; Suddick, E.C.; Rice, C.W.; Prokopy, L.S. More food, low pollution (Mo Fo Lo Po): A grand challenge for the 21st century. *J. Environ. Qual.* **2015**, *44*, 305–311. [CrossRef]
3. Zhang, X.; Davidson, E.A.; Mauzerall, D.L.; Searchinger, T.D.; Dumas, P.; Shen, Y. Managing nitrogen for sustainable development. *Nature* **2015**, *528*, 51–59. [CrossRef]
4. Rockström, J.; Steffen, W.; Noone, K.; Persson, Å.; Chapin, F.S.; Lambin, E.F.; Lenton, T.M.; Scheffer, M.; Folke, C.; Schellnhuber, H.J.; et al. A safe operating space for humanity. *Nature* **2009**, *461*, 472–475. [CrossRef]
5. Cao, Q.; Miao, Y.; Feng, G.; Gao, X.; Liu, B.; Liu, Y.; Li, F.; Khosla, R.; Mulla, D.J.; Zhang, F. Improving nitrogen use efficiency with minimal environmental risks using an active canopy sensor in a wheat-maize cropping system. *Field Crop. Res.* **2017**, *214*, 365–372. [CrossRef]
6. Miao, Y.X.; Stewart, B.A.; Zhang, F.S. Long-term experiments for sustainable nutrient management in China: A review. *Agron. Sustain. Dev.* **2011**, *31*, 397–414. [CrossRef]
7. Zhang, W.F.; Dou, Z.X.; He, P.; Ju, X.T.; Powlson, D.; Chadwick, D.; Norse, D.; Lu, Y.L.; Zhang, Y.; Wu, L.; et al. New technologies reduce greenhouse gas emissions from nitrogenous fertilizer in China. *Proc. Natl. Acad. Sci. USA* **2013**, *110*, 8375–8380. [CrossRef]
8. Cui, Z.; Yue, S.; Wang, G.; Meng, Q.; Wu, L.; Yang, Z.; Zhang, Q.; Li, S.; Zhang, F.; Chen, X. Closing the yield gap could reduce projected greenhouse gas emissions: A case study of maize production in China. *Glob. Chang. Biol.* **2013**, *19*, 2467–2477. [CrossRef]
9. Wu, L.; Chen, X.P.; Cui, Z.L.; Zhang, W.F.; Zhang, F.S. Establishing a regional nitrogen management approach to mitigate greenhouse gas emission intensity from intensive smallholder maize production. *PLoS ONE* **2014**, *9*, e98481. [CrossRef]
10. Cao, Q.; Cui, Z.L.; Chen, X.P.; Khosla, R.; Dao, T.H.; Miao, Y.X. Quantifying spatial variability of indigenous nitrogen supply for precision nitrogen management in small scale farming. *Precis. Agric.* **2012**, *13*, 45–61. [CrossRef]
11. Rabot, E.; Wiesmeier, M.; Schlüter, S.; Vogel, H.J. Soil structure as an indicator of soil functions: A review. *Geoderma* **2018**, *314*, 122–137. [CrossRef]
12. Qian, P.; Schoenau, J.J. Assessing nitrogen mineralization from soil organic matter using anion exchange membranes. *Fertil. Res.* **1994**, *40*, 143–148. [CrossRef]

13. Sogbedji, J.M.; Es, H.M.V.; Klausner, S.D.; Bouldin, D.R.; Cox, W.J. Spatial and temporal processes affecting nitrogen availability at the landscape scale. *Soil Tillage Res.* **2001**, *58*, 233–244. [CrossRef]
14. Dharmakeerthi, R.S.; Kay, B.D.; Beauchamp, E.G. Spatial variability of in-season nitrogen uptake by corn across a variable landscape as affected by management. *Agron. J.* **2006**, *98*, 255–264. [CrossRef]
15. Lishu County Bureau of Agriculture in Jilin Province, China. *Soil J. Lishu Cty.* **1985**, *4*, 33–37.
16. Zhu, Q.; Schmidt, J.P.; Bryant, R.B. Maize (Zea mays, L.) yield response to nitrogen as influenced by spatio-temporal variations of soil–water-topography dynamics. *Soil Tillage Res.* **2015**, *146*, 174–183. [CrossRef]
17. Schmidt, J.P.; Sripada, R.P.; Beegle, D.B.; Rotz, C.A.; Hong, N. Within-field variability in optimum nitrogen rate for corn linked to soil moisture availability. *Soil Sci. Soc. Am.* **2011**, *75*, 306–316. [CrossRef]
18. Chantigny, M.H.; Rochette, P.; Angers, D.A.; Massé, D.; Côté, D. Ammonia volatilization and selected soil characteristics following application of anaerobically digested pig slurry. *Soil Sci. Soc. Am.* **2004**, *68*, 306–312. [CrossRef]
19. Sahrawat, K.L. Factors affecting nitrification in soils. *Commun. Soil Sci. Plant Anal.* **2008**, *39*, 1436–1446. [CrossRef]
20. St Luce, M.; Whalen, J.K.; Ziadi, N.; Zebarth, B.J. Chapter two—Nitrogen dynamics and indices to predict soil nitrogen supply in humid temperate soils. In *Advances in Agronomy*; Academic Press: Cambridge, MA, USA, 2011; Volume 112, pp. 55–102.
21. Feng, G.Z.; Zhang, Y.J.; Chen, Y.L.; Li, Q.; Chen, F.J.; Gao, Q.; Mi, G.H. Effects of nitrogen application on root length and grain yield of rain-fed maize under different soil types. *Agron. J.* **2006**, *108*, 1656–1665. [CrossRef]
22. Qiu, S.J.; He, P.; Zhao, S.C.; Li, W.J.; Xie, J.G.; Hou, Y.P. Impact of nitrogen rate on maize yield and nitrogen use efficiencies in northeast china. *Agron. J.* **2014**, *107*, 305–313. [CrossRef]
23. Tremblay, N.; Bouroubi, Y.M.; Bélec, C.; Mullen, R.W.; Kitchen, N.R.; Thomason, W.E.; Ebelhar, S.; Mengel, D.B.; Raun, W.R.; Francis, D.D.; et al. Corn response to nitrogen is influenced by soil texture and weather. *Agron. J.* **2012**, *104*, 1658–1671. [CrossRef]
24. Tremblay, N.; Pandalai, S.G. Determining nitrogen requirements from crops characteristics: Benefits and challenges. *Recent Res. Dev. Agron. Hortic.* **2004**, *1*, 157–182.
25. Bélec, C.; Tremblay, N. Adapting nitrogen fertilization to unpredictable seasonal conditions with the least impact on the environment. *Horttechnology* **2006**, *16*, 408–412.
26. Shanahan, J.F.; Kitchen, N.R.; Raun, W.R.; Schepers, J.S. Responsive in-season nitrogen management for cereals. *Comput. Electron. Agric.* **2008**, *61*, 51–62. [CrossRef]
27. Kyveryga, P.M.; Blackmer, A.M.; Morris, T.F. Alternative benchmarks for economically optimal rates of nitrogen fertilization for corn. *Agron. J.* **2007**, *99*, 1057–1065. [CrossRef]
28. Schröder, J.J.; Neeteson, J.J.; Oenema, O.; Stuik, P.C. Does the crop or soil indicate how to save nitrogen in maize production? Reviewing the state of the art. *Field Crop. Res.* **2000**, *66*, 151–164. [CrossRef]
29. Kay, B.D.; Mahboubi, A.A.; Beauchamp, E.G.; Dharmakeerthi, R.S. Integrating soil and weather data to describe variability in plant available nitrogen. *Soil Sci. Soc. Am.* **2006**, *70*, 1210–1221. [CrossRef]
30. Fiez, T.E.; Pan, W.L.; Miller, B.C. Nitrogen use efficiency of winter wheat among landscape positions. *Soil Sci. Soc. Am.* **1995**, *59*, 1666–1671. [CrossRef]
31. Hergert, G.W.; Ferguson, R.B.; Shapiro, C.A.; Penas, E.J.; Anderson, F.B. Classical statistical and geostatistical analysis of soil nitrate-N spatial variability. In *Site-Specific Management for Agricultural Systems*; American Society of Agronomy; Crop Science Society of America; Soil Science Society of America: Madison, WI, USA, 1995; Volume 677, pp. 175–186.
32. Mamo, M.; Malzer, G.L.; Mulla, D.J.; Huggins, D.R.; Strock, J. Spatial and temporal variation in economically optimum nitrogen rate for corn. *Agron. J.* **2003**, *95*, 958–964. [CrossRef]
33. Scharf, P.C.; Kitchen, N.R.; Sudduth, K.A.; Davis, J.G.; Hubbard, V.C.; Lory, J.A. Field-scale variability in optimal nitrogen fertilizer rate for corn. *Agron. J.* **2005**, *97*, 452–461. [CrossRef]
34. Miao, Y.; Mulla, D.J.; Hernandez, J.A.; Wiebers, M.; Robert, P.C. Potential impact of precision nitrogen management on corn yield, protein content, and test weight. *Soil Sci. Soc. Am. J.* **2007**, *71*, 1490–1499. [CrossRef]
35. Miao, Y.X.; Mulla, J.D.; Batchelor, D.W.; Paz, O.J.; Robert, C.P.; Wiebers, M. Evaluating management zone optimal nitrogen rates with a crop growth model. *Agron. J.* **2006**, *98*, 545–553. [CrossRef]
36. Tollenaar, M.; Lee, E.A. Yield potential, yield stability and stress tolerance in maize. *Field Crop. Res.* **2020**, *75*, 161–169. [CrossRef]

37. Lee, E.A.; Tollenaar, M. Physiological basis of successful breeding strategies for maize grain yield. *Crop Sci.* **2007**, *47*, 202–215. [CrossRef]
38. Yan, P.; Pan, J.; Zhang, W.; Shi, J.; Chen, X.; Cui, Z. A high plant density reduces the ability of maize to use soil nitrogen. *PLoS ONE* **2017**, *12*, e0172717. [CrossRef]
39. Mahdi, A.H.; Ismail, S.K. Maize productivity as affected by plant density and nitrogen fertilizer. *Int. J. Curr. Microbiol. Appl. Sci.* **2015**, *4*, 870–877.
40. Ciampitti, I.A.; Vyn, T.J. A comprehensive study of plant density consequences on nitrogen uptake dynamics of maize plants from vegetative to reproductive stages. *Field Crop. Res.* **2011**, *121*, 2–18. [CrossRef]
41. Tokatlidis, I.S.; Koutroubas, S.D. A review of maize hybrids' dependence on high plant populations and its implications for crop yield stability. *Field Crop. Res.* **2004**, *88*, 103–114. [CrossRef]
42. Staff, S. *Keys to Soil Taxonomy*; United States Department of Agriculture, Natural Resources Conservation Services: Washington, DC, USA, 1998; p. 328.
43. Wang, X.B.; Miao, Y.X.; Dong, R.; Chen, Z.C.; Guan, Y.J.; Yue, X.Z.; Fang, Z.; Mulla, D. Developing active canopy sensor-based precision nitrogen management strategies for maize in Northeast China. *Sustainability* **2019**, *11*, 706. [CrossRef]
44. Colwell, J.D. *Estimating Fertilizer Requirements: A Quantitative Approach*; Centre for Agriculture and Bioscience International: Wallingford, UK, 1994.
45. Nelson, D.W.; Sommers, L.E. Determination of total nitrogen in plant material. *Agron. J.* **1962**, *65*, 423–425. [CrossRef]
46. Li, W.; He, P.; Jin, J. Critical nitrogen curve and nitrogen nutrition index for spring maize in North-East China. *J. Plant Nutr.* **2012**, *35*, 1747–1761. [CrossRef]
47. Sun, Z.; Li, Z.; Lu, X.; Bu, Q.; Ma, X.; Wang, Y. Modeling soil type effects to improve rainfed corn yields in Northeast China. *Agron. J.* **2016**, *108*, 498–508. [CrossRef]
48. Wu, D.; Xu, X.; Chen, Y.; Shao, H.; Sokolowski, E.; Mi, G. Effect of different drip fertigation methods on maize yield, nutrient and water productivity in two-soils in Northeast China. *Agric. Water Manag.* **2019**, *213*, 200–211. [CrossRef]
49. Xia, T.T.; Miao, Y.X.; Wu, D.L.; Shao, H.; Khosla, R.; Mi, G.H. Active optical sensing of spring maize for in-season diagnosis of nitrogen status based on nitrogen nutrition index. *Remote Sens.* **2016**, *8*, 605. [CrossRef]
50. Herrmann, A.; Taube, F. The range of the critical nitrogen dilution curve for maize (Zea mays L.) can be extended until silage maturity. *Agron. J.* **2004**, *96*, 1131–1138. [CrossRef]
51. Alotaibi, K.D.; Cambouris, A.N.; St Luce, M.; Ziadi, N.; Tremblay, N. Economic optimum nitrogen fertilizer rate and residual soil nitrate as influenced by soil texture in corn production. *Agron. J.* **2018**, *110*, 2233–2242. [CrossRef]
52. Zhao, Y.J. Limiting Factors Identification and Production System Design of Spring Maize for High Yield and High Nitrogen Use Efficiency in Smallholder Farmers' Fields in the Northeast China—A Case Study in Lishu County. Ph.D. Thesis, China Agricultural University, Beijing, China, 2019.
53. Dong, Q.; Li, M.; Liu, J.; Wang, C. Spatio-temporal evolution characteristics of drought of spring maize in northeast China in recent 50 years. *Int. J. Nat. Disasters Health Secur.* **2011**, *20*, 52–59.
54. Lu, X.; Li, Z.; Bu, Q.; Cheng, D.; Duan, W.; Sun, Z. Effects of rainfall harvesting and mulching on corn yield and water use in the corn belt of Northeast China. *Agron. J.* **2014**, *106*, 2175–2184. [CrossRef]
55. Chen, Y.; Xiao, C.; Wu, D.; Xia, T.; Chen, Q.; Chen, F.; Mi, G. Effects of nitrogen application rate on grain yield and grain nitrogen concentration in two maize hybrids with contrasting nitrogen remobilization efficiency. *Eur. J. Agron.* **2015**, *62*, 79–89. [CrossRef]
56. Ziadi, N.; Cambouris, A.N.; Nyiraneza, J.; Nolin, M.C. Across a landscape, soil texture controls the optimum rate of N fertilizer for maize production. *Field Crop. Res.* **2013**, *148*, 78–85. [CrossRef]
57. Power, J.F.; Wiese, R.; Flowerday, D. Managing farming systems for nitrate control: A research review from management systems evaluation areas. *J. Environ. Qual.* **2001**, *30*, 1866–1880. [CrossRef] [PubMed]
58. Taylor, J.C.; Wood, G.A.; Earl, R.; Godwin, R.J. Soil factors and their influence on within-field crop variability: II. Spatial analysis and determination of management zones. *Biosyst. Eng.* **2003**, *84*, 441–453. [CrossRef]
59. Armstrong, R.D.; Fitzpatrick, J.; Rab, M.A.; Abuzar, M.; Fisher, P.D.; O'Leary, G.J. Advances in precision agriculture in south-eastern Australia: III. Interactions between soil properties and water use help explain spatial variability of crop production in the Victorian Mallee. *Crop Pasture Sci.* **2009**, *60*, 870–884. [CrossRef]

60. Shahandeh, H.; Wright, A.L.; Hons, F.M. Use of soil nitrogen parameters and texture for spatially-variable nitrogen fertilization. *Precis. Agric.* **2011**, *12*, 146–163. [CrossRef]
61. Ciampitti, I.A.; Roger, W.E.; Joe, L. Corn Growth and Development. Kansas State University Agricultural Experiment Station and Cooperative Extension Service. MF3305. 2016. Available online: https://bookstore.ksre.ksu.edu/pubs/MF3305.pdf (accessed on 18 August 2020).
62. Meng, Q.; Cui, Z.; Yang, H.; Zhang, F.; Chen, X. Establishing high-yielding maize system for sustainable intensification in China. *Adv. Agron.* **2018**, *145*, 85–109.
63. Grundmann, G.L.; Renault, P.; Rosso, L.; Bardin, R. Differential effects of soil water content and temperature on nitrification and aeration. *Soil Sci. Soc. Am.* **1995**, *59*, 1342. [CrossRef]
64. Hao, W.P. Influence of Water Stress and Re-Watering on Maize WUE and Compensation Effects. Ph.D. Thesis, Chinese Academy of Agricultural Sciences, Beijing, China, 2013.
65. Xu, C.; Huang, S.; Tian, B.; Ren, J.; Meng, Q.; Wang, P. Manipulating planting density and nitrogen fertilizer application to improve yield and reduce environmental impact in Chinese maize production. *Front. Plant Sci.* **2017**, *8*, 1234. [CrossRef]
66. Bean, G.M.; Kitchen, N.R.; Camberato, J.J.; Ferguson, R.B.; Fernandez, F.G.; Franzen, D.W.; Laboski, C.A.M.; Nafziger, E.D.; Sawyer, J.E.; Scharf, P.C.; et al. Improving an active-optical reflectance sensor algorithm using soil and weather information. *Agron. J.* **2018**, *110*, 2541–2551. [CrossRef]
67. Pasuquin, J.M.; Pampolino, M.F.; Witt, C.; Dobermann, A.; Oberthür, T.; Fisher, M.J.; Inubushi, K. Closing yield gaps in maize production in southeast ASIA through site-specific nutrient management. *Field Crop. Res.* **2014**, *156*, 219–230. [CrossRef]
68. Muschietti-Piana, M.D.P.; Cipriotti, P.A.; Urricariet, S.; Peralta, N.R.; Niborski, M. Using site-specific nitrogen management in rainfed corn to reduce the risk of nitrate leaching. *Agric. Water Manag.* **2018**, *199*, 61–70. [CrossRef]
69. Thompson, L.J.; Ferguson, R.B.; Kitchen, N.; Franzen, D.W.; Mamo, M.; Yang, H.; Schepers, J.S. Model and sensor-based recommendation approaches for in-season nitrogen management in corn. *Agron. J.* **2015**, *107*, 2020–2030. [CrossRef]
70. Sela, S.; Woodbury, P.B.; van Es, H.M. Dynamic model-based N management reduces surplus nitrogen and improves the environmental performance of corn production. *Environ. Res. Lett.* **2018**, *13*, 054010. [CrossRef]

Article

Effect of Site Specific Nitrogen Management on Seed Nitrogen—A Driving Factor of Winter Oilseed Rape (*Brassica napus* L.) Yield

Remigiusz Łukowiak * **and Witold Grzebisz**

Department of Agricultural Chemistry and Environmental Biogeochemistry, Poznan University of Life Sciences, Wojska Polskiego 28, 60-637 Poznan, Poland; witold.grzebisz@up.poznan.pl

* Correspondence: remigiusz.lukowiak@up.poznan.pl; Tel.: +48-61-846-6287

Received: 31 July 2020; Accepted: 7 September 2020; Published: 10 September 2020

Abstract: It has been assumed that the management of both soil and fertilizer N in winter oilseed rape (WOSR) is crucial for N accumulation in seeds (N_{se}) and yield. This hypothesis was evaluated based on field experiments conducted in 2008/09, 2009/10, 2010/11 seasons, each year at two sites, differing in soil fertility, including indigenous N (N_i) supply. The experimental factors consisted of two N fertilizers: N and NS, and four N_f rates: 0, 80, 120, 160 kg ha^{-1}. Yield, as governed by site × N_f rate interaction, responded linearly to N_{se} at harvest. The maximum N_{se} (N_{semax}), as evaluated by N input ($N_{in} = N_i + N_f$) to WOSR at spring regrowth, varied from 95 to 153 kg ha^{-1}, and determined 80% of yield variability. The basic reason of site diversity in N_{semax} was N_i efficiency, ranging from 46% to 70%, respectively. The second cause of N_{se} variability was a shortage of N supply from + 9.5 soil to −8.8 kg ha^{-1} to the growing seeds during the seed filling period (SFP). This N pool supports the N concentration in seeds, resulting in both seed density and a seed weight increase, finally leading to a yield increase.

Keywords: seed density; N uptake; indices of N productivity; mineral N; indigenous N_{min} at spring; post-harvest N_{min}; N balance; N efficiency

1. Introduction

Over the last four decades, oilseed rape (*Brassica napus* L., OSR) has become one of the most important global oil crops. The main reason for the rise in OSR production was an intensive breeding progress, resulting in new double 00 varieties which deliver plant oil of high consumption value [1,2]. Between 2009–2018, the world OSR harvested area increased from 31 million (mln) to 37 mln ha. The world average yield for this period increased to about 2.0 t ha^{-1}, being only slightly lower than that recorded recently in Canada (2.2 t ha^{-1}) [3]. The leading producers of OSR are Canada, China, and the European Union (EU). Canada, which delivers about 25% of the world rapeseed production, increased the sown area of this crop from 6.5 mln ha in 2009 to 9.1 mln ha in 2018 [3,4]. In the EU, the leading producers of winter oilseed rape (WOSR) are Germany, France, and Poland. Seed yields in these countries, in spite of high breeding progress, stagnated in the period extending from 2009 to 2018 [3,5].

The main constraints in WOSR production in the EU are weather conditions during the growing season and soil fertility level [6]. The resistance of WOSR to frost does not depend only on temperatures during winter but also on plants' physiological status just before winter, which significantly affects plant density [7,8]. Two basic yield components, i.e., seed density (SD, number of seed per m^2) and seed weight (thousand seed weight, TSW, g) are responsible for the final yield of WOSR. The first, dominant component is SD, which is indirectly defined by plant density, number of pods (pod density, PD, pods per m^2) [9]. The critical period of yield formation, referring to the development of primary yield components, such as inflorescences and succeeding pods, extends from the budding stage up to the pod full size [10]. One of the most specific characteristics of the yield development of WOSR is a strong compensation mechanism, occurring between yield components during the period extending from the onset of flowering towards the end of seed growth, i.e., maturity [11]. As reported by Berry and Spink [12], the amount of water required by WOSR during this period to exploit its yielding potential is 300 mm. Any unfavorable weather conditions during the spring vegetation lead to disturbance in the development of yield components (PD and SD). Weather disturbance during pod and seed growth negatively affects TSW [13,14].

The key nutrient responsible for yield formation by WOSR is nitrogen (N) [15,16]. This nutrient affects the number of inflorescences, flowers, and finally pods and seeds. A balanced structure of yield components depends on synchronization of a crop N requirement with its supply from both soil and applied fertilizers [17,18]. In practice, N supply to a given crop is, in general, oriented on the amount applied in fertilizers without considering soil resources. As a result of this fertilization strategy, N fertilizer productivity is highly variable, being both in shortage or in excess with respect to WOSR requirements during the critical stages of yield formation, leading in both cases to yield reduction [16,19]. In some EU countries like Germany and the Netherlands, the N fertilization strategy of crop plants is based on the measurement of the content of mineral N (N_{min}) before the spring WOSR regrowth [20]. This strategy, as has been documented recently, clearly shows that N use efficiency (NUE) depends not only on the N_{min} content in the root zone, but also on the content of 16 other nutrients, like P, K, and Mg, being responsible for both N uptake, and its utilization by plants [21–24]. Any shortage of this set of nutrients at the onset of flowering and during the seed filling period (SFP) leads to yield depression [11,25].

In spite of the extensive study on N supply to WOSR, and its impact on the development of yield components, the yield prognosis based on N_{min} content is good, but not satisfactory. The *black box* in the effective management of N during the growing season of WOSR is a lack of knowledge on N release from its soil resources during the spring vegetation [26]. The main reason for the necessity of the N_f rate optimization is a huge variability in soil potential for release of N_{min}, both in the period preceding the spring vegetation and during the full season of WOSR growth [23]. The key question remains, however, to what extent does the development of basic yield components depend on the indigenous resources of N during WOSR spring vegetation? Is the soil N supply at the onset of flowering to WOSR plants sufficient to meet the requirements of the growing pods and seeds? The scientific problem focuses on the impact of three N sources, i.e., (i) N mineral soil resources, indigenous N (N_i) at the beginning of spring WOSR vegetation; (ii) the optimum N_f rate; and (iii) the quantity of the soil N_{min} released during the spring vegetation on the degree of WOSR yield components expression, i.e., on the sink capacity development as a prerequisite of high seed yield achievement.

The objective of the study was to define the impact of site-specific variability in in-season N management during the WOSR growing season based on the amount of N accumulated in seeds, and its relationship with the final yield.

2. Materials and Methods

2.1. Experiments Set Up

Three series of field experiments with winter oilseed rape (*Brassica napus* L.) were conducted during the 2008/09, 2009/10, and 2010/11 seasons. Each year, two different fields (sites) were investigated. Field experiments were conducted on soils with texture ranging from sand/sandy loam to sandy clay loam, classified as Albic Luvisol. The content of the available nutrients (measured each year just after a fore-crop harvesting from two soil depths of 0.0–0.3, and 0.3–0.6 m) ranged, depending on the nutrient, from low to very high, and it was, in general, sufficiently high for covering the nutrient requirements of the high-yielding WOSR (Table 1).

Table 1. Soil agrochemical properties before WOSR sowing in consecutive study years.

Year	Site/Location (Acronym)	Variety	Soil Textural Class/ Agronomy Class	pH (1 M KCl)	Organic Matter [1], g kg^{-1} Soil	P_2O_5 [2]	K_2O [2]	Mg [3]	$S\text{-}SO_4$ [4]	N_{min} [5]
						mg kg^{-1} Soil				kg ha^{-1}
2009	Gostyń (Go) 51°52′06′′ N 16°51′55′′ E	Californium	Loamy sand Light [7]	6.0/5.7 [6]	11/7	149/124 M/M [8]	249/123 VH/M	46/60 M/H	5.6/7.5 L/M	58.4/50.8
	Kołaczkowo (Ko) 52° 13′00′′ N 17°37′33′′ E	Nelson	Sand/ loamy sand/ Very light/light	6.2/6.4	9/6	127/68 M/L	154/175 H/H	86/80 VH/VH	8.1/7.7 M/M	32.4/27.3
2010	Buszewo (Bu) 51°32′41′′ N 16°22′11′′ E	Californium	Sand clay loam Medium	6.6/6.4	13/5	264/262 VH/VH	138/86 M/L	95/83 VH/H	9.5/8.6 M/M	40.3/25.7
	Wieszczyczyn (Wi) 51°02′03′′ N 17°05′38′′ E	Nemax	Loamy sand/ sandy loam Light	5.8/6.0	12/7	265/227 VH/VH	138/65 M/L	57/42 H/M	8.5/7.6 M/M	34.8/14.2
2011	Wenecja (Ve) 51°48′45′′ N 17°45′51′′ E	Californium	Sandy loam/ loam Medium	5.6/5.6	16/12	121/103 M/M	116/82 L/L	85/74 H/H	12.3/12.2 H/H	64.8/40.2
	Donatowo (Do) 51°04′51′′ N 16°51′37′′ E	ES Mercure F1	Loamy sand Light	6.3/6.2	12/6	172/149 H/M	204/150 VH/M	53/44 H/M	7.2/6.4 M/M	35.3/19.7

Methods of available nutrients determination: [1] loss of ignation; [2] Egner–Riehm method; [3] Schachtschabel method; [4,1] nephelometric method of Bardsley [27] [5] 0.01 M $CaCl_2$ solution (soil/solution ratio 1:5; m/v); [6] Soil layers: 0.0–0.3/0.3–0.6 m; [7] Soil agronomy class; [8] classes of available nutrient content: L—low, M—medium, H—high, VH—very high.

The local climate, classified as intermediate between Atlantic and Continental, is seasonally variable (Table 2). Precipitation during the period extending from January to July was for most of the sites slightly higher than the long-term average, which amounted to 347.5 mm. Each year, the highest amount of precipitation was recorded in July. In 2010, a severe shortage of precipitation was recorded in June, a critical month with respect to pods and seed growth [14], but it was preceded by high precipitation in May. Air temperatures were, with the exception of Wieszczyczyn in 2010, around the long-term average.

2.2. Experimental Design

The field experiment was a balanced 2 × 4 factorial trial conducted on two different sites each year. The first factor was the type of N fertilizer: (i) ammonium nitrate, 34-0-0, (ii) ammonium saltpeter with ammonium sulfur, 32-0-0-5. The second factor was composed of three rates of nitrogen/sulfur plus absolute N/S control. The arrangement of the 2nd factor was as follows:

(1) Control—PK + 0 N + 0.00 kg S ha;
(2) PK + 80 kg N ha^{-1} + 12.50 kg S ha^{-1};
(3) PK + 120 kg N ha^{-1} + 18.75 kg S ha^{-1};
(4) PK + 160 kg N ha^{-1} + 25.00 kg S ha^{-1}.

The first N or NS rate of 80 kg N ha^{-1}, together with the respective amount of S, was applied before WOSR spring regrowth. The second N rate of 40 or 80 kg ha^{-1} was applied at the end of March or the beginning of April. Amounts of applied phosphorus (P) and potassium (K) fertilizers were adjusted to the actual status of P and K soil fertility level, and applied before WOSR sowing. P was used as di-ammonium phosphate (18% N, 46% P_2O_5), and K as muriate of potash (60% K_2O). The size of an individual experimental plot was 400 m^2 (9 × 44.4 m) The fore-crop for WOSR at all sites was winter wheat. Standard tillage technology was applied for soil preparation for WOSR. Immediately after the winter wheat harvest, phosphorus and potassium fertilizers were broadcast on the entire field and shallow stubble ploughing (10–12 cm) + harrowing was done. Two or three weeks later, depending on the planned sowing date, a standard ploughing to a depth of 25 cm was done with simultaneous soil compaction with a Cambell roller. Seedbed preparation and seeding was conducted immediately after ploughing. The amount of seeds, based on 1000 seed weight, was adjusted to reach a plant density at emergence of 40–50 plants m^{-2}. The row spacing was 12.5 or 25 cm, depending on the farm equipment. A robust program to control all weeds, pests, and diseases was employed in accordance with standard farm practice during the growing season, following integrated pest management (IPM) principles. In order to achieve a homogenous stage of seed maturation (seed moisture of 8%), desiccants were applied. Harvest was performed the earliest seven days later by a combine harvester from an area of 300 m^2.

Table 2. Main characteristics of meteorological conditions during the study on the background of the long-term averages.

Year	Site	Meteorological Characteristics	August	September	October	November	December	January	February	March	April	May	June	July	Total [1]/ Average
2009	**Go**	[2] P, mm	95.5	18.0	64.2	24.2	30.0	24.5	45.3	48.5	7.7	70.0	91.0	81.0	**368.0**
		[3] T. °C	18.3	13.2	9.1	5.6	1.4	−3.1	0.1	4.2	11.9	13.1	15.7	19.0	**8.7**
	Ko	P. mm	60.4	20.9	67.5	22.8	16.2	20.6	25.7	50.5	2.1	85.4	101.5	105.3	**391.1**
		T. °C	18.9	13.3	9.0	5.3	1.1	−3.4	−0.1	3.4	11.0	13.4	15.7	19.6	**8.5**
2010	**Bu**	P. mm	109.6	52.4	76.6	43.2	41.1	43.6	20.0	57.6	39.8	92.3	18.1	109.6	**381.0**
		T. °C	18.2	12.4	6.2	4.4	−5.6	−6.3	−0.4	4.4	10.0	12.5	18.7	21.6	**8.6**
	Wi	P. mm	31.9	58.2	62.8	40.3	49.1	19.1	15.3	38.5	33.4	83.6	21.2	121.6	**332.7**
		T. °C	19.1	16.4	7.2	6.5	−1.4	−7.3	−1.2	4.2	9.1	11.4	16.9	22.3	**7.9**
2011	**Ve**	P. mm	96.8	25.5	57.3	19.4	22.6	19.0	25.7	57.1	1.2	66.2	70.6	116.5	**356.3**
		T. °C	18.1	12.6	8.6	4.7	0.6	−3.8	−0.6	3.0	10.6	12.8	15.1	18.8	**8.0**
	Do	P. mm	109.8	88.8	9.1	86.8	36.8	27.4	18.5	22.6	10.1	29.8	67.8	109.2	**285.4**
		T. °C	17.4	13.1	6.3	6.1	3.4	1.1	−3.5	2.4	7.9	14.8	18.8	17.6	**8.4**
Long-term [4] 1961	2009	P. mm	66.7	48.8	42.0	45.3	48.4	40.1	32.6	40.1	38.1	56.7	62.7	77.2	**347.5**
		T. °C	17.5	13.3	8.6	3.6	0	−1.6	−0.5	2.9	7.9	13.2	16.4	18.1	**8.1**

[1] January–July; [2] Precipitation; [3] Temperature; [4] Poznan-Ławica Meteorological Station.

2.3. Source of Primary Materials

Plant materials for the determination of dry matter and measurement of N concentration were collected from an area of 1.0 m^2 at BBCH 89 (maturity-nearly all pods are ripe, with dark, hard seeds). Plant samples were taken from the same sowing rows across a particular experimental block. The harvested plant sample was partitioned into a sub-sample of seeds, and post-harvest residues (stem, fallen + remaining leaves, and threshed pods). At the stage of BBCH 89, three yield components were analyzed: (i) the number of pods per m^2 (pod density, PD), (ii) the number of seeds per pod, Se/PD; (iii) the thousand-seed weight (TSW, g). The number of seeds per m^2 (seed density, SD) was calculated based on the first two components. Nitrogen concentration (N_c) in the plant samples was determined using a standard macro-Kjeldahl procedure.

The soil content of NH_4-N and NO_3-N was determined in field-fresh (not air dried) soil samples within 24 h after sampling. Twenty-gram soil samples were shaken for 1 h with 100 mL of 0.01 M $CaCl_2$ solution (soil/solution ratio 1:5 m/v). Concentrations of NH_4-N and NO_3-N were determined by the colorimetric method using flow injection analyses (FIAstar5000, FOSS). The method of NO_3-N concentration analysis consists of two basic steps: reduction from nitrate to nitrite by using a cadmium column, followed by colorimetric determination of nitrite, based on the Griess–Ilosvay reaction with N-(1-Naphtyl) ethylene-diamine dichloride as a diazotizing agent. The measurement was performed at a wavelength of 540 nm. To determine NH_4-N, a special FOSS ammonia indicator (mixture of cresol red, bromcresol purple, and bromothymol blue) was used. The measurement was made at a wavelength of 590 nm. The total N_{min} was calculated as the sum of NH_4-N and NO_3-N, and expressed in kg ha^{-1}. The N_{min} content was calculated for a given soil layer, using indices which were constructed based on the soil textural class and soil bulk density [28].

2.4. Calculated Indices/Parameters

Based on the amount of N in the respective parts of WOSR plants at physiological maturity (BBCH 89) and in the soil before spring regrowth and after harvest, the following indices have been calculated:

A Plant and nutrient indices:

1. Harvest index: $HI = Y/TB \cdot 100\%$, (1)
2. Nitrogen Harvest Index: $NHI = (N_{se}/TN) \cdot 100\%$, (2)
3. Partial factor productivity of N_f: $PFP_{Nf} = Y/N_f$ (kg seeds kg^{-1} $N_{f,}$ (3)
4. Agronomic N efficiency: $AE_N = (Y_i - Y_0)/N_f$ (kg seeds kg^{-1} N_f), (4)
5. Physiological N_f efficiency: $PE_N = (Y_i - Y_0)/(TN_i - TN_0)$ (kg seeds kg^{-1} N_T), (5)
6. Apparent N_f recovery: $R_N = (TN_i - TN_0)/N_f \cdot 100\%$, (6)
7. Unit Nitrogen Accumulation: $UNA = N_{se}/Y$ (kg N t^{-1} seeds), (7)
8. Unit Nitrogen productivity: $UNP = Y/N_{se}$ (kg seeds kg N_{se}), (8)

where: Y, Y_0, Y_i—seed yield, seed yield on the N control plot and N fertilized plots, t ha^{-1} or kg ha^{-1}; TB—total biomass—t ha^{-1} or kg ha^{-1}; TN—total N uptake, kg ha^{-1}; N_{se}—amount of N in seeds, kg ha^{-1}; N_f – N fertilizer rate, kg ha^{-1}; TN, TN_0, TN_i—total amount of N in WOSR at harvest for the N control and N fertilized plots.

B Soil nitrogen parameters:

1. N input: $N_{in} = N_{mins} + N_f$ (kg ha^{-1}), (9)
2. Mineral N balance: $N_b = N_{in} - TN$ (kg ha^{-1}), (10)
3. Net N gain: $N_{gain} = N_{minr} - N_b$ (kg ha^{-1}), (11)
4. Total N input: $N_{int} = N_{in} + N_{gain}$ (kg ha^{-1}), (12)
5. Nitrogen input efficiency: $NE_{in} = N_{se}/N_{in} \cdot 100\%$, (13)
6. Total N input efficiency: $NE_{int} = N_{se}/N_{int} \cdot 100\%$, (14)

where: N_{mins}—the amount of mineral N at the WOSR spring regrowth, kg ha^{-1}; N_{minr}—the amount of mineral N after WOSR harvest, kg ha^{-1}.

2.5. Statistical Analyses

The data were subjected to conventional analysis of variance using STATISTICA® 10 (StatSoft, Krakow, Poland). The distribution of the data (normality) was checked using the Shapiro–Wilk test. The homogeneity of variance was checked by Levene's test. The differences between treatments were evaluated with Tukey's test. In the second step principal component analysis (PCA) was used to illustrate the dependence between amounts of N in a particular WOSR part, or in soil and yield and its components. In the third step of the diagnostic procedure, stepwise regression was applied to define an optimal set of variables for a given N characteristic. In the computational procedure, a consecutive variable was removed from the multiple linear regressions in a step-by-step manner. The best regression model was chosen based on the highest F-value for the model and the significance of all independent variables.

3. Results

3.1. Yield and Its Components

Yield of WOSR was significantly determined by the interaction of Y × S × N or by the interaction of Y × F × N_f (Table 3). It is necessary to stress that weather conditions during the spring vegetation were favorable for WOSR growth. In each year of the study, the total amount of precipitation during the period extending from the onset of flowering to seed maturity (FL-SM) was below 300 mm, as reported by Berry and Spink [12] for British conditions, but it was above the long-term average for this region in Poland (197 mm). The highest yield of 4.41 t ha^{-1} was obtained at Venetia with 299 mm of rainfall during FL-SM, directly supporting the prognosis of Berry and Spink [12]. The effect of site on yield was revealed in 2009 and 2011, and that of N_f in 2010 and 2011. Based on a simple regression analysis, two distinct patterns of yield response to the increasing N_f rate were observed (Table A1). The quadratic regression model, which prevailed in four sites, indicates a saturation status of N_f supply with respect to the achieved yield. This conclusion is supported by the well-defined optimum N fertilizer rate (N_{fop}) for the maximum achievable yield (Y_{max}) for a particular site. The calculated N_{fop} of 103.1, 142.7, 104.3, and 128.0 kg ha^{-1}, resulted in a Y_{max} of 3.142, 3.629, 3.589, and 4.012 t ha^{-1}. The pattern of yield response to the N_f rate in the other two sites, representing Wi in 2010, and Ve in 2011, fitted the linear regression model the best. This type of yield pattern clearly indicates that the applied N_f rate was too low to reach the highest yield.

Total biomass of WOSR (TB) and harvest index (HI) were significantly affected by all studied factors. In 2009 and 2010, the effect of weather on TB and HI was the most pronounced. Any increase in TB resulted in a simultaneous, and at the same time, a significant drop in HI. In 2011, TB was on average much lower compared to both previous years, but HI was significantly higher (29.8% vs. 22.6% in 2009, and vs. 23.1% in 2010). In spite of the non-significant impact of HI on yield, a much higher yield was recorded in 2011 (Table 3).

The key yield component, i.e., SD responded significantly, as in the case of yield, to the interaction of Y × S × N, and also to Y × F × N. In 2009, the difference between sites reached 23.4%, whereas the positive impact of NS fertilizer as compared to N alone was much lower, i.e., 5.2%. In 2010, neither factors affected SD. In 2011, a pronounced effect on SD was exerted by the type of N_f fertilizer (+8% for NS versus N alone). The impact of site on SD, and in consequence on yield was year-dependent. A detailed analysis of SD response to the applied N_f rate clearly showed the occurrence of two regression models, i.e., quadratic and linear (Table A1), which covered the same set of sites as described previously for yield. This fact was documented by the strongest value of the correlation coefficient between yield and SD (r = 0.84, Table A2).

Table 3. Yield and yield components of winter oilseed rape.

Year (Y)	Treatments	Treatment Level	Y	TB	HI	PD	SD	Se/Po	TSW
			t ha^{-1}		%	No. m^{-2}		No. pod^{-1}	g
2009	Site	Go	2.935a[1]	12.3a	25.7b	7878b	60.062a	8.1a	4.88b
	(S)	Ko	3.329b	17.2b	19.4a	7187a	74.128b	10.5b	4.53a
	F-value		***	***	***	***	***	***	***
	Fertilizer (F)	N	3.068	14.1a	23.0b	7250a	65.403a	9.3	4.73
		NS	3.195	15.4b	22.0a	7815b	68.787b	9.2	4.68
	F-value		n.s.	***	*	***	*	n.s.	n.s.
	N rate (N)	0	2.506a	10.8a	25.9c	5259a	52.328	10.0b	4.78
	kg ha^{-1}	80	3.360b	15.4b	22.6b	7569b	72.697	10.0b	4.66
		120	3.347b	16.8c	20.1a	9374d	70.929	7.8a	4.73
		160	3.314b	15.9b	21.3ab	7929c	72.426	9.3b	4.65
	F-value		***	***	***	***	***	***	n.s.
2010	Site	Bu	2.823	11.4a	24.8b	6217b	48.511	8.3a	5.76a
	(S)	Wi	2.921	15.1b	21.4a	5054a	46.635	9.4b	6.25b
	F-value		n.s.	***	***	***	n.s.	**	***
	Fertilizer (F)	N	2.729a	13.6b	21.3a	5985b	46.028	8.3a	5.92a
		NS	3.016b	12.9a	24.9b	5286a	49.119	9.4b	6.09b
	F-value		**	**	***	***	n.s.	**	**
	N rate	0	1.806a	6.82a	26.2b	2618a	31.610a	10.3b	5.65a
	kg ha^{-1}	80	3.117b	15.97c	21.2a	7096c	51.258b	7.2a	6.11b
		120	3.211b	15.75c	21.9a	7004c	53.108b	8.2a	6.08b
		180	3.355b	14.49b	23.1a	5823b	54.318b	9.6b	6.19b
	F-value		***	***	***	***	***	***	***
2011	Site	Ve	3.366a	12.2	28.4a	7494b	60.478	8.9a	5.51a
	(S)	Do	3.685b	12.0	31.1b	5658a	62.289	11.4b	5.93b
	F-value		23.9 ***	0.94	13.7 ***	271 ***	1.84	49.4 ***	54.0 ***
	Fertilizer (F)	N	3.415a	12.1	28.9a	6585	59.015a	9.4a	5.77
		NS	3.636b	12.0	30.6b	6567	63.753b	10.8b	5.67
	F-value		**	n.s.	*	n.s.	**	***	n.s.
	N rate	0	2.447a	9.5a	27.5a	4319a	44.552	10.9b	5.45a
	kg ha^{-1}	80	3.609b	12.3b	29.5ab	7130b	62.724	9.3a	5.77b
		120	3.858c	12.8b	30.5b	6971b	65.904	10.1ab	5.86b
		160	4.189d	13.7c	31.5b	7884c	72.355	10.2ab	5.80b
	F-value		***	***	**	***	***	*	**
F-values for selected interactions									
Year × Site × Fertilizer			n.s.	***	***	***	n.s.	***	n.s.
Year × Site × N rate			***	***	***	***	***	***	n.s.
Year × Fertilizer × N rate			***	***	***	***	***	***	n.s.
Site × Fertilizer × N rate			n.s.	*	**	***	n.s.	*	n.s.
Year × Site × Fertilizer × N rate			n.s.	***	***	***	n.s.	***	n.s.

***, **, * significant at $p < 0.001$; $p < 0.01$; $p < 0.05$, respectively; n.s.—non significant; a[1] within a year, means within a column followed by the same letter indicate a lack of significant difference between the treatments. Y—yield; TB—total biomass; HI—harvest index; PD—pod density; SD—seed density; Se/Po—number of seeds per pod; TSW—thousand seed weight.

The second basic yield component, i.e., TSW responded mainly to site in particular years of study. In 2009, a slightly lower TSW was the attribute of Ko, whose yield was significantly lower when compared to Go. In 2010, a significantly higher TSW was recorded for Wi, which yielded at the same level as Bu. In 2011, a significantly higher TSW for Do was in accordance with a higher yield for this site. In general, TSW showed a negative, but not significant relationship with SD, which indirectly indicates the presence of a yield compensation mechanism, which is revealed during the SFP [9].

The other yield components, i.e., pod density (PD), and the number of seeds per pod (Se/Po) exerted a much weaker impact on yield when compared to SD (Table A2). In 2009, the difference between sites with respect to the degree of PD and TSW expression were more pronounced for TSW than for PD (29.6% vs. 9.6%). In 2010, the opposite pattern was observed (13.3% vs. 23%). In 2011, the yield compensation mechanisms were the highest, reaching 28% for TSW, and 38% for PD.

3.2. *Characteristics of Nitrogen Accumulation Patterns at Harvest*

The concentration of N in seeds (N_c) in all the study years responded mainly to site, and in 2010 and 2011, to the rate of applied N_f (Table 4). In 2009, a significantly higher N_c was determined in seeds for Ko than for Go. A significant, but a slightly lower difference between sites was observed in 2010. In 2011, the average seed N_c was the highest and the difference between fields was also very high. In 2010 and 2011, the increasing N_f rate resulted in a progressive N_c increase. In spite of the significant impact of site and the N_f rate, the interactional impact of all experimental factors on the N_c in seeds was negligible. The impact of this seed characteristic on yield was positive, but not decisive (Table A2).

The strongest impact on yield was exerted by the amount of N accumulated in seeds at harvest (N_{se}) (Table A2). This WOSR characteristic responded to the interaction of Y × S × N, and Y × F × N. The first interaction was significantly stronger ($p \leq 0.001$ vs. $p \leq 0.01$). A significant difference between sites were recorded in 2009 and 2011. In 2009, the recorded difference reached 35.1%, but in 2011, only 20%. However, both N_{se} and yield were higher in 2011. The effect of the N_f rate was recorded in all years. As in the case of yield, N_{se} patterns followed the same type of regression model (Table A1). The quadratic regression model, presenting a saturation level of N_{se} accumulation, was recorded in three sites, i.e., at Go in 2009, at Bu in 2010, and at Do in 2011. The optimum N_{fop} rate for the N_{semax} was 92.3 for Go, 110.2 for Bu, and 157.2 kg ha^{-1} for Do. The respective N_{semax} were as 96.2, 124.1, and 154.1 kg ha^{-1}. The linear regression model, representing the unsaturation pattern of N_{se} was recorded in the other three sites.

The amount of N in WOSR residues (N_{res}) and the nitrogen harvest index (NHI) were highly variable between sites within a particular year. The greatest difference between studied fields was recorded in 2009, and resulted in the lowest yields. An opposite trend was recorded in 2011, when the difference between fields was much lower, reaching only 30%, but WOSR yielded the highest. In 2009 and 2010, a significantly higher N_{res} was recorded for the plot fertilized with NS fertilizer. An opposite trend was recorded in 2011. In this particular year, the applied NS fertilizer resulted in a significant decrease of N_{res}, which corresponded to the higher seed yield. The NHI indicates the relative share of N_{se} in the total WOSR biomass at harvest. A significant difference between sites was recorded in 2009 and 2010, but not in 2011, the year with the highest NHI. It is necessary to stress that NHI did not show any significant relationship with the N_c in seeds and N_{se}, but was strongly, and negatively correlated with N_{res}. Finally, NHI did not impact seed yield (Table A3).

Total N uptake (TN) was significantly driven by all the studied factors. The effect of N fertilizers on TN was only significant in 2009, and 2010. The strongest response of TN to the increasing N_f rate was observed in 2009, in which WOSR yielded the lowest, in spite of a very high value on the N control plot. TN affected yield significantly, but at a much lower level as observed for N_{se} (Table A2).

Table 4. Indices of nitrogen management by winter oilseed rape.

Year	Treatments	Level of Treatment	N_c	N_{se}	N_{res}	TN	NHI	UNA	UNP
			%	kg ha^{-1}	kg ha^{-1}		%	kg N t^{-1}	kg Seeds kg^{-1} N
2009	Site	Go	2.91a[1]	85.5a	55.1a	140.7a	64.3b	47.4a	22.1b
	(S)	Ko	3.47b	115.5b	113.1b	228.6b	50.6a	69.2b	14.6a
	F-value		***	***	***	***	***	***	***
	Fertilizer (F)	N	3.18	97.9a	80.4a	178.3a	57.6	57.4	18.5
		NS	3.21	103.2b	87.9b	191.0b	57.3	59.1	18.2
	F-value		n.s.	*	*	***	n.s.	n.s.	n.s.
	N rate (N)	0	3.22	81.2a	55.7a	136.8a	66.9b	53.9a	21.5b
	kg ha^{-1}	80	3.18	107.5b	86.9b	194.4b	56.4a	57.4ab	17.9a
		120	3.18	106.4b	93.5bc	199.9bc	54.4a	59.8b	17.3a
		160	3.19	107.1b	100.5c	207.5c	52.1a	62.1b	16.7a
	F-value		n.s.	***	***	***	***	**	***
2010	Site	Bu	3.49b	99.1	58.2a	157.3a	63.0b	56.2a	18.2b
	(S)	Wi	3.37a	99.1	100.6b	199.7b	50.9a	68.0b	15.2a
	F-value		***	n.s.	***	***	***	***	***
	Fertilizer (F)	N	3.43	94.6a	77.1a	171.8a	56.0a	62.4	16.4a
		NS	3.42	103.6b	81.7b	185.3b	57.9b	61.7	16.9b
	F-value		n.s.	*	**	***	*	n.s.	*
	N rate (N)	0	3.25a	58.5a	39.1a	97.6a	60.4c	53.9a	18.6c
	kg ha^{-1}	80	3.39b	105.8b	105.5d	211.3b	51.4a	70.2c	15.2a
		120	3.40b	109.4b	90.3c	199.8b	55.3b	63.4b	16.3b
		160	3.67c	122.6c	82.8b	205.4b	60.8c	60.9b	16.6b
	F-value		***	***	***	***	***	***	***
2011	Site	Ve	3.29a	114.7a	58.1a	172.8a	65.8	49.7a	20.5b
	(S)	Do	3.75b	137.8b	75.8b	213.6b	64.5	58.1b	17.4a
	F-value		***	***	***	***	n.s.	***	***
	Fertilizer (F)	N	3.56	124.1	70.3b	194.4	64.3	55.4b	18.5
		NS	3.48	128.4	63.6a	192.0	65.9	52.4a	19.5
	F-value		n.s.	n.s.	*	n.s.	n.s.	*	n.s.
	N rate (N)	0	3.24a	84.3a	49.4a	133.7a	64.2	51.1	20.1
	kg ha^{-1}	80	3.52ab	128.0b	70.0b	197.9b	64.5	54.9	18.4
		120	3.53ab	134.4b	71.3b	205.7b	64.2	53.2	19.4
		160	3.79b	158.3c	77.1b	235.4c	67.7	56.5	18.0
	F-value		*	***	***	***	n.s.	n.s.	n.s.
F-value for selected interactions									
Year × Site × Fertilizer			n.s.	n.s.	***	***	***	***	**
Year × Site × N rate			n.s.	***	***	***	***	***	***
Year × Fertilizer × N rate			n.s.	**	**	n.s.	***	**	n.s.
Site × Fertilizer × N rate			n.s.	n.s.	***	***	**	***	***
Year × Site × Fertilizer × N rate			n.s.	n.s.	***	***	***	***	***

***, **, * significant at $p < 0.001$; $p < 0.01$; $p < 0.05$, respectively; n.s.—non significant; a[1] within a year, means within a column followed by the same letter indicate a lack of significant difference between the treatments. N—N content in seeds; N_{se}, N_{res}—the amount of N in seeds, harvest residues, respectively; TN—total amount of N in WOSR at harvest; NHI—nitrogen harvest index; UNA—unit N accumulation, UNP—unit N productivity.

In agronomic practice, two indices are used which are reciprocal to each other, i.e., unit N accumulation (UNA) and unit N productivity (UNP). In spite of this, the relationship between both indices was not linear, and was expressed by the power function the best:

$$UNP = 993.7UNA^{-0.996} \text{ for } n = 24, R^2 = 0.99 \text{ and } p < 0.001. \quad (15)$$

This type of relationship clearly shows that the unit productivity of N accumulated by WOSR at harvest was the highest when its accumulation was low, decreasing exponentially with its increase. Both indices responded to the interaction of all studied factors. The greatest difference for UNA, reaching 46%, between sites was recorded in 2009. This difference was slightly higher, with respect to UNP (51%). The smallest difference was recorded in 2011 (17%, 18%, for UNA and UNP, respectively). The effect of N_f type was, in general, low, being in most cases not significant. The effect of the increasing N_f rate was site specific. The most conspicuous impact of UNA was recorded for TN and HI. The first WOSR characteristic responded positively, but the second negatively to UNA increase (Table A2).

The applied principal component analysis (PCA) clearly revealed the distinct impact of yield components and crop N indices on the WOSR final yield. PCs with eigenvalues greater than 1.0 were used as a primary criterion to determine the number of PCs. The first four PCs accounted for 92.27% of the total variance in the data (Table A3). However, only the variables with scores on PCs over 0.70 ($R^2 > 0.50$) were taken into consideration. PC1 and PC2 contributed to 49.43% and 22.12% of the total variance, respectively. Four of the nine variables had high loadings on PC1, fulfilling the chosen criteria. The highest, and at the same time positive loadings, were recorded in descending order (*r*) for Y = N_{se} = TN > SD. The required criteria for PC2 were reached for PD, and for PC3 for Se/Po (negative). PC4 had a positive loading for UNA. The studied variable weight was evaluated by the Eigen vector, which varies between −1 to +1. The Eigen vectors for the examined variables were broadly scattered on the two first PCA axes (Figure 1a). The closest to an absolute of 1 were TN, N_{se}, and Y. The distance of N_{se} to Y, as seen from Figure 1b, was the closest among the studied N indices and yield components. The N_c in seeds exerted the same loading on both PC1 and PC2 axes (Table A3).

3.3. Indices of Fertilizer N Management

The partial factor productivity of fertilizer N (PFP_N) is the basic index, describing the productivity of applied N_f [29]. In the studied case, PFP_N was significantly driven by Y × S × N, and Y × F × N interactions (Table 5). The PFP_N indices in 2011 were the highest, and the difference between sites was small (5.3%). The lowest PFP_N indices, but at the same time, the highest difference between sites, were recorded in 2009 and 2010 (13%). The PFP_N decreased in accordance with the progressively increasing N_f rate, irrespective of other factors.

The next index, i.e., agronomic efficiency of N_f (AE_N) was driven by the same set of factors as PFP_N. A significant, and at the same time, a very strong difference between fields was recorded in 2010 and 2011, reaching 137% and 81%, respectively. The effect of the type of N fertilizer was significant in all of the studied years, clearly indicating a significantly higher N net productivity in the presence of sulfur. The highest AE_N increase in response to the NS fertilizer was recorded in 2011, a year with the highest yields. The lowest drop in AE_N in response to the increasing N rates was also recorded in 2011. It was twice as low as recorded in 2010, when it was the highest.

Table 5. Indices of nitrogen management by WOSR.

Year	Treatments	Level of Treatment	PFP_N	AE_N kg kg^{-1}	PE_N	R_N %
2009	Site	Go	28.3a[1]	7.3	6.3b	72.6b
	(S)	Ko	32.1b	8.3	3.5a	39.8a
	F-value		***	n.s.	***	***
	Fertilizer	N	29.4a	7.0a	3.2a	46.4a
	(F)	NS	31.0b	8.6b	6.5b	66.0b
	F-value		*	*	**	***
	N rate	80	42.0c	11.0b	8.5b	71.9b
	(N)	120	27.9ab	7.2a	3.9a	52.5a
	kg ha^{-1}	160	20.7	5.2a	2.3a	44.2a
	F-value		***	***	***	***
2010	Site	Bu	30.8b	18.0b	21.8b	103.9b
	(S)	Wi	27.5a	7.6a	7.5a	94.2a
	F-value		**	***	***	*
	Fertilizer	N	27.6a	11.3a	13.2	90.9a
	(F)	NS	30.6b	14.3b	16.2	107.2b
	F-value		**	**	n.s.	***
	N rate	80	39.7c	17.1b	25.6b	144.6c
	(N)	120	26.8b	11.7a	11.7a	85.1b
	kg ha^{-1}	160	21.0a	9.7a	6.7a	67.4a
	F-value		***	***	***	***
2011	Site	Ve	33.6a	15.9b	13.1b	79.2b
	(S)	Do	35.4b	8.8a	5.2a	50.0a
	F-value		*	***	***	***
	Fertilizer	N	32.7a	10.6a	7.2a	61.9
	(F)	NS	36.2b	14.2b	11.2b	67.3
	F-value		***	***	***	n.s.
	N rate	80	45.1c	14.5b	12.7b	75.7
	(N)	120	32.2b	11.8a	7.2a	56.9
	kg ha^{-1}	160	26.2a	10.9a	7.7a	61.2
	F-value		***	***	***	n.s.
F-value for selected interactions Year × Site × Fertilizer			n.s.	n.s.	n.s.	***
Year × Site × N rate			***	***	***	***
Year × Fertilizer × N rate			***	****	****	n.s.
Site × Fertilizer × N rate			n.s	n.s	n.s	**
Year × Site × Fertilizer × N rate			n.s	n.s	n.s	**

***, **, * significant at $p < 0.001$; $p < 0.01$; $p < 0.05$, respectively; n.s.—non significant; a[1] within a year, means within a column followed by the same letter indicate a lack of significant difference between the treatments. PFP_N—partial factor productivity of fertilizer N; AE_N agronomic efficiency of fertilizer N; PE_N—physiological efficiency of fertilizer N; recovery of fertilizer N.

The physiological N use efficiency index (PE_N), describing the utilization efficiency of N taken up by a plant during the growing season, significantly responded to the same set of factors as described for AE_N. Each year, the differences between sites were extremely high. In 2009, a net productivity of 1.0 kg of N taken up by WOSR plants was 80% higher in the field located at Go, as compared to Ko. In 2010, the observed difference between sites was twice as high. In 2011, the difference between sites was much smaller, but also high, amounting to 152%. The effect of NS fertilizer was as a rule positive, and a significant response to S was observed in 2009 and 2011. The effect of the increasing N_f rate on PE_N was clearly demonstrated in 2009 and 2010, decreasing in accordance with the increased N_f rate. In 2011, index stagnation in plots fertilized with N at the rate of 120 and 160 kg ha^{-1} was recorded.

Nitrogen recovery (R_N) reveals the contribution of N fertilizer in the total N taken by a crop during the growing season. R_N indices responded significantly to all the studied factors, including years. The highest R_N values were recorded in 2010, exceeding 100% in the field located at Bu (105%). In this particular year, the difference between sites was low (9.7%). The lowest R_N of 39.8% was recorded in 2009 at Ko, being, however, twice as high at Go. The same trend was recorded in 2010, but the difference between sites was 29.2%. The effect of NS fertilizer on R_N was as a rule positive, but not significant in 2009. A positive impact of NS on R_N was noted in 2009 at Ko; in 2010 at Wi, and in 2011 at Ve. The effect of the increasing N_f rate on R_N showed the same pattern, as presented for the other N indices. In 2010, it exceeded 100% on plots fertilized with N_f at the rate of 80 kg ha^{-1}.

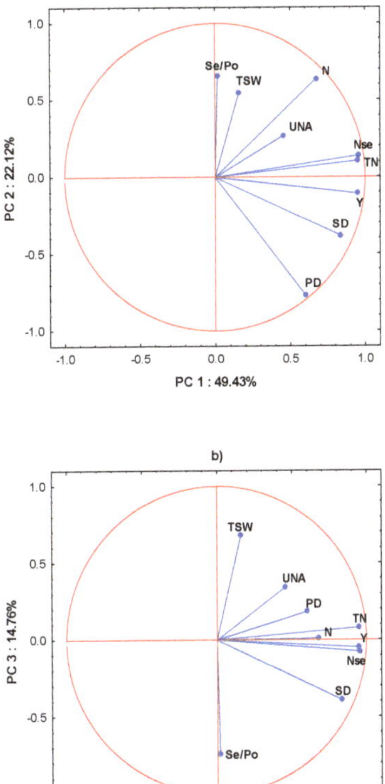

Figure 1. Score plot of WOSR components and nitrogen indices in PC1 and PC2 axes (**a**) and PC1 and PC3 axes (**b**). Y—yield, SD—seed density, PD—pod density, TSW—thousand seed weight, TN—total nitrogen, N_c—N concentration in seeds, N_{se}—N accumulated in seeds, UNA—unit N accumulation.

3.4. Indices of Soil N Management

Nitrogen input (N_{in}) is composed of two N mineral sources. The first, internal (indigenous) source, is soil mineral N measured in spring (N_{mins}), which is recorded in the soil at the onset of WOSR spring regrowth. The second one is the amount of N_f applied to the growing crop during spring vegetation [23]. In the studied case, the N_f rate was 80, 120, and 160 kg ha^{-1} (Table 6). Due to a fixed amount of applied N_f, it was not possible to evaluate N_{in} statistically. In spite of this, the difference between sites was the most pronounced in 2011 (90%), followed by 2009 (33%), and the least pronounced in 2010 (12%).

The residual N (N_{minr}) is the amount of N_{min} measured just after WOSR harvest. This N characteristic showed a significant impact of all studied factors. As in the case of yield, N_{minr} was evaluated based on the Y × S × N interaction. The difference between sites was evaluated based on two indices, referring to the N_{minr} content on the N control plot and to its maximum value, as an indicator of the applied rate of N_f:

1. The Nminr content on the N control plot:

   a. Go (52.0) < Do (83.5 < Bu (98.3) ≤ Ko (102.5) < Wi (117.2) < Ve (146.1 kg ha^{-1}).

2. The maximum N_{minr} content:

   b. Go (85.6) < Ko (102.5) < Bu (111.8) ≤ Wi (119.5) ≤ Do (121.1) < Ve (146.1 kg ha^{-1}).

3. The net N_{minr} increase with respect to the N control:

   c. Do (+37.6) > Go (+33.5) > Ko (+18.6) > Bu (+13.5) > Wi (+2.3) > Ve (+0.0 kg ha^{-1})

The highest amount of residual N_{min} was recorded, irrespective of the applied N_f rate, at Ve. The lack of differences between the control N plot and the N_f highest rate indicates that N was taken up by WOSR plants with the same efficiency. For other fields, the net effect of N_f was highly diversified between years and sites. The difference between sites in 2011 was very strong (47%). In the case of Do, the N_{minr} content increased progressively with the increased N_f rate. The lowest amount of N_{minr} was recorded in 2009 and the difference between sites was significant (27%). In the case of Go, the amount of N_{minr} increased linearly with the applied N_f rate.

Nitrogen balance (N_b) was significantly driven by all factors in the study. A positive N_b value indicates the presence of a sufficiently high pool of N_{min} present in spring. A negative N_b value indicates N_{min} net release from soil resources during the growing season. The differences between sites in particular years were extremely high. A positive N_b was registered in two sites, i.e., in 2009 for Go, and in 2011 for Ve. In the other four sites, a high release of N_{min} from soil resources during the growing season was recorded. In the first two years, N_b was negative, irrespective of the type of applied N fertilizer. A quite opposite trend was revealed in 2011. The effect of increasing N_f rates was year specific. In 2009, negative N_b values were recorded on plots with low N rates (up to 80 kg ha^{-1}). In 2010, the observed trend was opposite. In 2011, N_b decreased with the N_f rate increase.

The next characteristic of N management during the growing season is N_{gain}, i.e., the net amount of N taken up by a crop from soil N resources during the growing season. Its variability was most affected by the interactional effect of Y × S × N. The impact of years was tremendous. In 2009, N_{gain} was positive, and the difference between fields was 10-fold. In 2010, N_{gain} was extremely high, and the difference between sites was high, but much smaller when compared to 2009. A 4-fold difference between sites was observed in 2011. The effect of N fertilizer type appeared only in 2009 when the application of NS fertilizer increased the net amount of N_{min} by 19%. The effect of increasing N rates on N_{gain} followed two patterns (Table A4), linear (Ko, Ve, Do), and quadratic (Go, Bu, Wi).

Table 6. Indices of nitrogen management in the soil/plant system.

Year	Treatments	Level of Treatments	N_{in}	N_{minr}	N_b kg ha^{-1}	N_{gain}	N_{int}
2009	Site	Go	199.2	74.1a[1]	56.6b	17.5a	216.7a
	(S)	Ko	149.7	94.4b	−78.9a	173.3b	323.0b
	F-value		n.a.	***	***	***	***
	Fertilizer (F)	N	174.5	82.3	−4.8b	87.1a	261.5a
		NS	174.5	86.2	−17.5a	103.7b	278.2b
	F-value		n.a.	n.s.	***	***	***
	N rate (N)	0	84.5	77.3a	−56.1a	133.4d	217.9a
		80	164.5	80.0a	−29.9b	109.9c	274.3b
		120	204.5	87.8ab	4.6c	83.3b	287.7bc
		180	244.5	92.0b	36.9d	55.1a	299.5c
	F-value		n.a.	**	***	***	***
2010	Site	Bu	156.0	104.9a	−1.3b	106.2a	262.2a
	(S)	Wi	139.0	115.6b	−60.7a	176.3b	315.3b
	F-value		n.a.	**	***	***	***
	Fertilizer (F)	N	147.5	116.8b	−24.3	141.1	288.6
		NS	147.5	103.6a	−37.8	141.3	288.8
	F-value		n.a.	***	n.s.	n.s.	n.s.
	N rate (N)	0	147.5	104.0a	49.9b	54.1a	201.6a
		80	147.5	113.4ab	−63.8a	177.2b	324.7b
		120	147.5	117.1b	−52.3a	169.4b	316.9ab
		180	147.5	106.4ab	−57.9a	164.2b	311.7ab
	F-value		n.a.	*	***	***	***
2011	Site	Ve	275.0	147.2b	102.2b	45.0a	320.0
	(S)	Do	145.0	100.4a	−68.6a	169.0b	314.0
	F-value		n.a.	***	***	***	n.s.
	Fertilizer (F)	N	210.0	122.2	15.6	106.6	316.6
		NS	210.0	125.4	18.0	107.4	317.4
	F-value		n.a.	n.s.	n.s.	n.s.	n.s.
	N rate (N)	0	210.0	114.8ab	76.3b	38.5a	248.5a
		80	210.0	113.1a	12.1ab	101.0ab	311.0b
		120	210.0	133.2ab	4.3a	128.9b	338.9c
		180	210.0	134.1b	−25.4a	159.5b	369.5d
	F-value		n.a.	**	**	**	***
F-values for selected interactions							
Year × Site × Fertilizer			n.a.	***	***	n.s.	n.s.
Year × Site × N rate			n.a.	***	***	***	***
Year × Fertilizer × N rate			n.a.	**	n.e.	**	**
Site × Fertilizer × N rate			n.a.	**	***	**	**
Year × Site × Fertilizer × N rate			n.a.	n.s.	***	*	*

***, **, * significant at $p < 0.001$; $p < 0.01$; $p < 0.05$, respectively; n.s.—non significant; n.a.—non analyzed. a[1] within a year, means within a column followed by the same letter indicate a lack of significant difference between the treatments. N_{in}—indigenous N (N_i) + fertilizer N (N_f); N_{minr}—residual N_{min} (post-harvest N_{min}); N_b—N balance; N_{gain}—N mineralized during the growing season and incorporated into WOSR biomass; N_{int}—total N input into the soil/plant system during the growing season.

Total N input (N_{int}) responded almost to the same set of factors and their interactions as recorded for N_{gain}. A significant impact of site was recorded in 2009 and 2010 but not in 2011, during which WOSR yielded the highest. The effect of N fertilizer type was the same as observed for N_{gain}. The effect of N rates was year specific. The effect of increasing N rates on N_{int} followed two patterns linear (Ko, Ve, Do), and quadratic (Go, Bu, Wi), i.e., showing a high resemblance to N_{gain} patterns (Table A4).

4. Discussion

The yield of any crop, including WOSR, is highly variable, being affected by three key factors: (i) the course of weather during the growing season, (ii) soil fertility level, and (iii) N supply to plants during the critical periods of formation of yield components [22,30]. Four criteria were chosen for the evaluation of year-site differences in yield:

1. Seed density (SD);
2. Amount of N in seeds (N_{se});
3. N sources: indigenous N in spring, fertilizer N, in-season N (N_{in}, N_{int});
4. N productivity ($PFPN_{in}$, $PFPN_{int}$).

4.1. Seed Density and Yield

The basis of a soil productivity evaluation in a field experiment is the yield obtained on a plot without application of fertilizer nitrogen (N_f), i.e., N control plot. A crop response to the applied N_f indicates its potential productivity, assuming an optimization of N supply [15,30]. The study clearly showed that the natural, indigenous productivity of the studied sites, with respect to its content in spring (N_{mins}), differed significantly. The order of sites, assuming a 10% difference between sites [$(S_n-S_{n-1})/S_n < 0.1$], as the discrimination criterion was as follows:

$$\text{Do } (2.93) \geq \text{Ko } (2.65) > \text{Go } (2.28) \geq \text{Wi } (2.10) \geq \text{Ve } (2.01) > \text{Bu } (1.41 \text{ t ha}^{-1}).$$

The order obtained clearly indicates that the fields located at Do and Ko were, irrespective of year, the most productive. On both these sites, the indigenous N content in spring was low compared to the other sites, but the content of available K and Mg was high, consequently creating favorable conditions for enhancing N productivity, as corroborated by high values of the PFP_N index (Tables 2 and 5). It has been recently documented that a shortage of these nutrients disturbs the development of yield components during both the onset of flowering and SFP [11,25,31,32]. The second group of studied sites, showing a relatively high N_i productivity, comprised three sites, i.e., Go (2009), Wi (2010), and Ve (2011). The main reason for the high yield was the high content of N_{mins} at Go and Ve, which exceeded 100 kg ha^{-1} (Table 1). The fertility level of soil at Wi, including N_{mins} content, was only moderate. The third group, with the lowest yield, comprised only one site, i.e., Bu (2010).

A different order of sites was obtained based on the N_f optimum (N_{fop}) or N_{fmax}, taking into account the mode of yield response to the applied N_f, i.e., quadratic and linear, respectively:

1. Ve (4.46) > Do (4.01) ≥ Wi (3.79) ≥ Ko (3.63) = Bu (3.59) > Go (3.14 t ha^{-1});
2. Yield increase: Ve (+2.45) > Bu (+2.18) > Wi (+1.69) > Go (+1.38) > Do (+1.08) ≥ Ko (+0.98).

The highest yield was harvested at Ve, where it increased linearly in relation to the applied N_f. The same type of response was found for Wi, but at a much lower level. This type of WOSR response to N_f indicates that the N_f rate of 160 kg ha^{-1} was too low to maximize crop productivity in these two sites. A limited supply of N to WOSR plants during pre-flowering growth results in a significant decrease in PD and SD, i.e., yield components determining the sink strength [11,33]. A slightly lower yield response to N_f, as compared to Ve, was recorded for Bu. In this site, a N_f of 104.3 kg ha^{-1} was sufficient to reach the maximum yield. The same type of response was the attribute of the other three sites, located at Go, Do, and Ko. The quadratic response model of N_f impact on WOSR yield indicates the occurrence of factors constraining N_f productivity. In this case, the most probable reason for the limited N_f productivity was the shortage of nutrients other than N during SFP, which are responsible for the supply of assimilates to the growing pods and seeds [34,35].

Yield of WOSR was significantly driven by SD. The course of weather in a particular growing season was revealed as a decisive factor, impacting the course of the obtained trends. As shown in Figure 2, WOSR yield in 2009 fitted the quadratic regression model the best, indicating a saturation status of SD. In this exact year, an SD of 90.2 m^{-2} resulted in the Y_{max} of 3.661 t ha^{-1}. In the other two years, irrespective of the site, yield increased linearly with the increased SD. The impact of the N pools on SD was analyzed based on the effect of:

(1) Indigenous N (N_{mins}, control N plot):

Ko (57707) > Do (49646) ≥ Go (49562) > Ve (39935) > Wi (34135) > Bu (27745 seeds m^{-2}).

(2) N input (N_{in}) for N_{fop} or N_{fmax} for respective treatments:

Ko (83612) ≥ Ve (76463) >Do (67810) ≥ Go (66312) ≥ Bu (59755) ≥ Wi (56359 seeds m^{-2}).

The net SD increase in response to N_{fop} or N_{fmax} was as follows:

Ve (36528) > Bu (32010) > Ko (25905) > Wi (22224) > Do (18164) ≥ Go (16750 seeds m^{-2}).

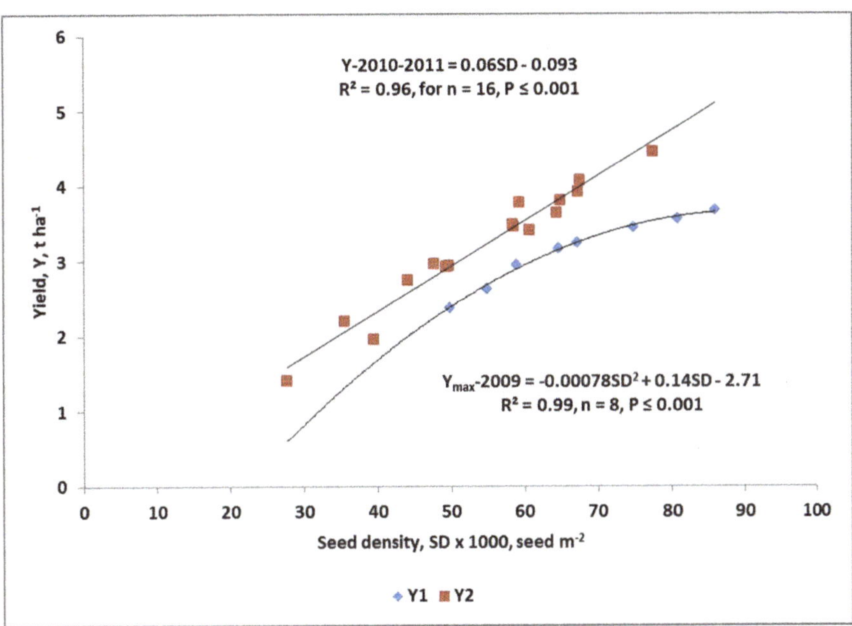

Figure 2. Regression models of seed yield response to seed density in particular years.

The obtained order of sites clearly stresses the impact of N_f on SD, which was the highest in the most productive sites with respect to the agronomy class, i.e., at Ve, and Bu. However, SD depends not only on the supply of N but also on the supply of other nutrients such as K and Mg during SFP [25,31,36]. The final yield of WOSR does not depend only on SD but is significantly corrected by the supply of assimilate to the growing seeds during the SFP [34]. The lower yield, as recorded in 2009, irrespective of the site, was due to lower TSW, in spite of a reasonably high SD (Table 3). The observed phenomenon, known as the yield compensation mechanism, can be explained by a natural decrease in radiation use efficiency in WOSR during SFP [37]. In fact, the lower TSW was due to a significantly higher SD, resulting in a dilution of dry matter in the growing seeds.

4.2. Seed Nitrogen as Yield Driver

As shown in Figure 1a,b, the amount of N accumulated in seeds at WOSR harvest, i.e., seed nitrogen (N_{se}), showed the strongest impact on yield. A close relationship was observed between N_{se}, treated as a single predictor, and yield. The relationship obtained followed the linear regression model:

$$Y = 0.023N_{se} + 0.\ 685 \text{ for n} = 24, R^2 = 0.92, \text{and } p \leq 0.001. \quad (16)$$

The regression model developed clearly shows that any increase in the N_{se} at harvest resulted in the higher yield of seeds. The relationship obtained is corroborated by the simultaneous increase in both N_{se} and yield in response to the progressive N_f rate (Tables 3 and 4, and Table A3). An N_{se} increase is also recorded after application of other nutrients, resulting in an increase of N content in WOSR seeds [35,38]. As reported by Fordoński et al. [19], the significant relationship between N_c in seeds and yield was only found when wheat was a preceding crop for WOSR, but not when this crop followed peas or faba beans. This type of dependency indicates a lower supply of N to WOSR when cereals preceded WOSR. The finding obtained as shown by Equation (16), is indirectly supported by the study by Hoffmann et al. [13], who presented data on N_{se}, but without an analysis of its relationship with yield. The regression model developed based on Hoffmann's data gave an even better estimation of WOSR yield dependence on N_{se} than our own:

$$Y = 0.038N_{se} + 0.948 \text{ for n} = 16, R^2 = 0.95, \text{and } p \leq 0.001. \quad (17)$$

The high reliability of these two presented regression models indicates the significant response of N_{se} to the applied N_f rates. There remains, however, a question concerning the yield forming importance of N_{se}, which summarizes the effect of two components, i.e., SD and the N_c in seeds at harvest. As shown in Tables 3 and A1, SD responded significantly to the applied N_f rate, but the type of response was site-specific. In four of six sites, the effect of N_f followed the quadratic regression model, indicating a saturation SD status with respect to the rate of applied N_f. In the other two cases, the N_f rate was too low to maximize SD, subsequently indicating a shortage of N supply to the growing pods and seeds during the SFP. This model also indicates a lack of synchronization between the rate of seed growth and the rate of N remobilization from vegetative WOSR organs during SFP [11]. The N_c in seeds, with the exception of 2009, increased progressively with the applied N_f rate. As a result, WOSR seeds in these two years, in spite of the same SD, accumulated significantly more N (Figure 3). The direction coefficient for the 2010/2011 regression model was by 50% higher as compared to that developed for 2009. The difference obtained indicates, irrespective of weather, a better supply of N to the growing seeds during SFP, consequently resulting in higher yield (Equation (16). These two models clearly support the hypothesis of the sink strength dominance over the source strength with respect to seed yield [39]. The result obtained clearly demonstrates that WOSR plants well-supplied with N during the SFP have a potential to minimize the competition for assimilates between growing seeds and their weight (TSW), consequently leading to a higher seed yield [11]. In this study, yield showed, in spite of the important impact of weather in consecutive years, a positive and significant response to all yield components:

$$Y = -4.26^{***} + 0.0002PD^{**} + 0.11Se/PD^{**} + 0.0003SD^{***} + 0.61TSW^{***} \text{ for n} = 24, R^2 = 0.99. \quad (18)$$

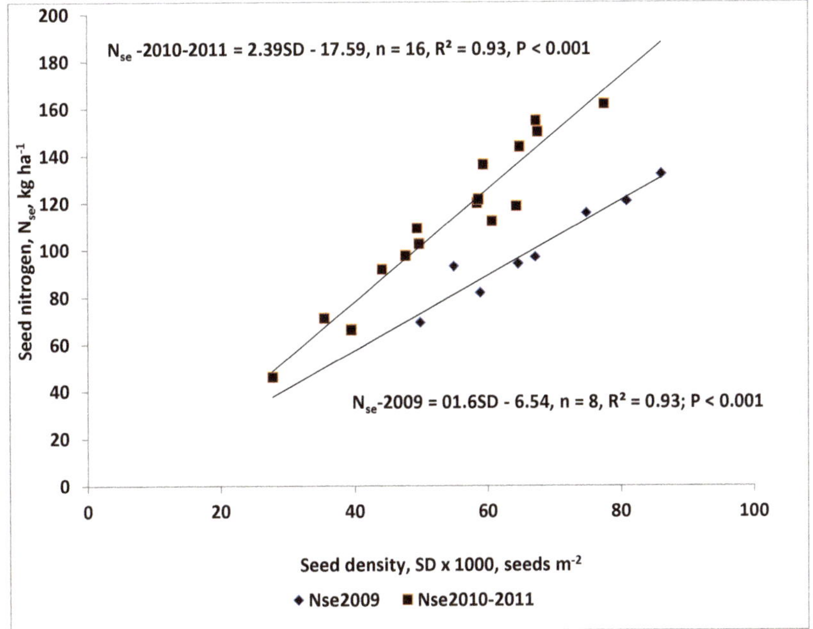

Figure 3. Regression models of seed density impact on the seed nitrogen.

4.3. Impact of the In-Season N Supply on N_{se} and Yield

In this study, it has been assumed that N_{se} is significantly affected by the N supply to WOSR plants during the growing season. The total pool of N, termed as total N input (N_{int}) is composed of three sub-pools. The first N source, defined in this study as the N input (N_{in}), comprised two sub-pools of N. The first one, the indigenous N (N_i), is equal to the N_{mins} content in the rooted soil zone in spring [17,21,23]. The N_i affects N_{se} on the N-non fertilized plot. The second N source for WOSR plants during the growing season is equal to the N dose applied in fertilizer. The third N pool is the amount of N released from N soil resources during WOSR spring vegetation. The maximum N_{se} depends, however, on the optimum N supply to WOSR from all N pools (N_{intop}) or to its maximum (N_{intmax}) supply. These three N pools were used as criteria for a site evaluation with respect to N_{se}:

(1) N_{se} response to N_i:

Do (102.5) ≥ Ko (93.1) > Wi (71.0) ≥ Go (66.7) ≥ Ve (66.1) > Bu (46.1 kg ha^{-1});

(2) N_{se} response to N_{in}, i.e., N_{seinop} or $N_{seinmax}$:

Do (153.2) ≥ Ve (145.8) > Ko (131.2) ≥ Wi (125.2) ≥ Bu (124.2 kg ha^{-1}) > Go (95.1, kg ha^{-1});

(2′) N_{se} net increase with respect to the N control:

Ve (+79.7) ≥ Bu (+78.1) > Wi (+54.2) ≥ Do (+50.7) > Ko (+38.1) > Go (+28.4 kg ha^{-1});

(1) N_{se} response to N_{int}, i.e., $N_{seintop}$ or $N_{seintmax}$:

Do (162.7) ≥ Ve (152.9) > Ko (135.1) ≥ Bu (123.0) ≥ Wi (116.4) > Go (91.9, kg ha^{-1});

(3′) N_{se} net increase with respect to N_{in}:

Do (+9.5) > Ve (+7.8) > Ko (+3.9) > Bu (−1.2) > Go (−3.2) > Wi (−8.8 kg ha^{-1}).

The first row clearly shows that fields at Ko and Do were naturally the most productive, as results from the highest N_{se} at harvest. On the opposite site are fields at Bu and Ve, which increased N_{se} the highest in response to N_f application during the growing season. The third group of sites is represented by the field at Go, which showed only a moderate productivity of N_i and low response to N_f. The highest productivity of N as recorded in 2011 was due to a high response of N_{se} to the amount of N released from the soil N resources during the spring growing season. The shortage of N supply to WOSR plants from the indigenous N pools during the spring growing season, as recorded for three sites, i.e., Bu, Go, and Wi, resulted in yield stagnation. The results obtained explain the observation presented by Grzebisz et al. [11]. According to these authors, the shortage of N during SFP, leads to an SD decrease, consequently resulting, as shown this study, in a decline of both, i.e., N_{se} and yield. This study corroborates the observation by Barłóg and Grzebisz [15], who documented that the N_c in leaves at the onset of pod growth (BBCH 71) is probably due to a net supply of N to the growing seeds, which explains 81% of yield variability.

The N_{semax}, irrespective of the N pool, was revealed as a significant discriminator of the studied sites with respect to yield (Table A5). The linear regression model obtained showed the same level of accuracy for N_{in} and N_{int} in yield prediction (Figure 4). The lowest WOSR yield as a result of low N_{semax} was characterized by the field located at Go. The main reason for low N_{semax} was the low efficiency of N in the soil/WOSR system (46% vs. 37% for N_{in} and $N_{int,}$ respectively). The second group, with a significantly higher N_{semax}, concomitant with a moderate yield level, comprised three sites, i.e., Ko, Bu, and Wi. This group was characterized by a high efficiency of N_{in} (60–70%), but a low efficiency of N_{int} (32–38%). The gap obtained indicates a low efficiency of N released from soil N resources during the growing season. The third group comprised two sites, i.e., Do and Ve, which yielded the highest due to both a much higher N_{semax}, and N_{int} efficiency. The key difference between these two sites resulted from differences in N efficiency. The field located at Ve was characterized by a moderate use efficiency of N_{in} (58%), and the field located at Do by the high efficiency of N_{int} (44%).

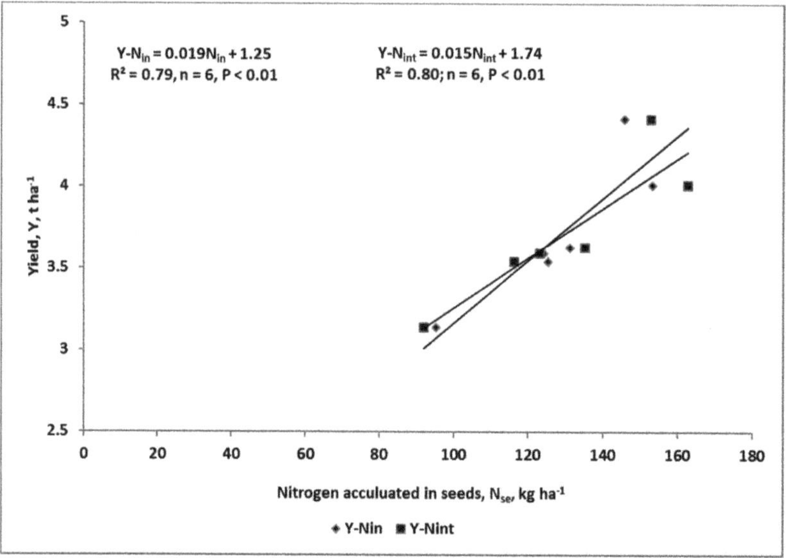

Figure 4. WOSR yield prognosis based on N_{semax} calculated for the N input to the soil/plant system at the beginning of WOSR spring regrowth and its total input.

The next question formulated during the study referred to the applicability of N_f, N_{in}, and N_{int} as yield prognostic tools. Excluding N_f, because its rates were the same in all experiments, the yield prognosis, based on N_{in} and N_{int}, was slightly better for the total N input:

$$(1)\ N_{in}\text{: } Y = -0.000034N_{in}^2 + 0.0178N_{in} + 1.324 \text{ for n} = 24, R^2 = 0.48, p \leq 0.05, \quad (19)$$

$$(2)\ N_{int}: Y = 0.009N_{inT} + 0.592 \text{ for } n = 24, R^2 = 0.61, p \leq 0.001. \quad (20)$$

The maximum yield, calculated based on equ. 19, for the N_{inop} of 261.8 kg ha^{-1} was 3.654 t ha^{-1}. The linear model for N_{int} corroborates the importance of N released during the SFP for exploitation of the WOSR yielding potential due the better supply of N to the growing seeds.

Nitrogen use efficiency (NUE) can also be used as a criterion for site discrimination with respect to the key yield driving factor, i.e., N_{se}. The NUE indices as shown in Table 5, in a major part corroborate the opinion expressed by Bouchet et al. [40] about the limited possibility of NUE improvement with respect to the increase of the seed yield of WOSR. The results obtained are in line with this opinion because the developed NUE indices, such as PFP_N, AE_N, PE_N, R decreased in accordance with the increasing N_f rate (Table 5). In spite of this, the best criterion for the study site discrimination was the PFP_N, but only for an N_f rate of 160 kg ha^{-1}. The order of sites obtained based on the PFP_{N160} was as follows:

Ve (27.9) > Do (24.5) ≥ Wi (23.6) ≥ Ko (23.0) > Go (18.4) ≥ Bu (18.3; kg seeds kg−1N).

The order obtained was significantly correlated with N_{semax}, as calculated, based on $N_{inop/max}$ or $N_{intop/max}$:

$$(1)\ N_{in}: N_{semax} = 4.279PFP_{N160} + 32.36 \text{ for } n = 6, R^2 = 0.61, p \leq 0.05, \quad (21)$$

$$(2)\ N_{int}: N_{semax} = 5.248PFP_{N160} + 11.65 \text{ for } n = 6, R^2 = 0.57, p \leq 0.05. \quad (22)$$

These two equations are contradictory to the opinion expressed by Bouchet et al. [40]. The relationships obtained clearly showed that N_{se} responded positively to the increasing PFP_N, including the site factor. This seemingly contradictory effect of N fertilizer on WOSR yield and NUE indices indirectly corroborates the hypothesis on N_{se} as the driving yield factor. This hypothesis is in accordance with the sink hypothesis presented by Körner [39]. This author clearly stated that sink strength in seed crops is a factor determining the source activity, and the consequent yield. On the other hand, too high N_c in seeds leads to decline in the crude oil concentration [41].

5. Conclusions

The study showed the occurrence of two different season-site N management strategies, resulting in a significant diversity in SD, the primary yield complement decisive for the final seed yield of WOSR. The first strategy was based on a high indigenous productivity of soil, as revealed in two of the six sites (Ko, Do). The second strategy was based on a high productivity of applied fertilizer N, which was revealed in two sites of the best quality of soil with respect to the soil agronomy class (Ve, Bu). The amount of N accumulated in WOSR seeds at harvest, i.e., seed nitrogen (N_{se}) was found to be the key yield driver factor. Its yield forming function is determined, however, by the SD and the N content in seeds, which in turn depends on N supply during SFP. The simultaneous increase of both these components is crucial for yield increase. A sufficiently high amount of N_{se} in WOSR seeds during the onset of pod and seed growth can be achieved provided there is a net N release from its soil resources during the growing season, but especially during SFP. The PFPN is a useful tool for discrimination of site productivity with respect to N_{semax}, treated as the prerequisite of WOSR yield evaluation.

Author Contributions: Conceptualization, R.Ł. and W.G.; methodology, R.Ł.; software, R.Ł.; validation, W.G.; formal analysis, W.G.; investigation, R.Ł.; resources, R.Ł.; data curation, W.G.; writing—original draft preparation, R.Ł.; writing—review and editing W.G.; visualization, R.Ł.; supervision, W.G.; project administration, R.Ł. All authors have read and agreed to the published version of the manuscript.

Funding: This publication was co-financed within the framework of Ministry of Science and Higher Education programme as Regional Initiative Excellence" in years 2019–2022. Project No. 005/RID/2018/19.

Acknowledgments: This research was in part supported by UCC Uralchem OJSC.

Conflicts of Interest: The authors declare no conflict of interest.

Appendix A

Table A1. Regression patterns of yield, seed density, and N accumulated in seeds response to the N fertilizer rate [1], n = 4.

Year	Site	Equation	R^2	*p*	N_{op}, kg ha^{-1}	$Y_{max}/SD_{max}/N_{semax}$
		Yield, Y, t ha^{-1}				
2009	Gostyń, Go	$Y = -0.00008N_f^2 + 0.017N_f + 2.38$	0.99	≤0.001	103.1	3.142
	Kołaczkowo, Ko	$Y = -0.00005N_f^2 + 0.014N_f + 2.65$	0.94	≤0.01	142.7	3.629
2010	Buszewo, Bu	$Y = -0.0002N_f^2 + 0.042N_f + 1.41$	0.99	≤0.001	104.3	3.589
	Wieszczyn, Wi	$Y = 0.009N_f + 2.1$	0.91	≤0.01	-	3.540
2011	Wenecja, Ve	$Y = 0.015N_f + 2.01$	0.98	≤0.001	-	4.410
	Donatowo, Do	$Y = -0.00007N_f^2 + 0.017N_f + 2.93$	0.99	≤0.001	128.0	4.012
		Seed density, SD, number m^{-2}				
2009	Gostyń, Go	$SD = -1.759N_f^2 + 343.3N_f + 49{,}562$	0.97	≤0.01	97.6	66,312
	Kołaczkowo, Ko	$SD = -0.988N_f^2 + 332N_f + 55{,}707$	0.87	≤0.05	168.1	83,612
2010	Buszewo, Bu	$SD = -3.09N_f^2 + 629N_f + 27{,}745$	0.99	≤0.001	101.8	59,755
	Wieszczyn, Wi	$SD = 138.9N_f + 34{,}135$	0.93	≤0.001	-	56,359
2011	Wenecja, Ve	$SD = 228.3N_f + 39{,}935$	0.98	≤0.001	-	74,463
	Donatowo, Do	$SD = -0.99N_f^2 + 268.2N_f + 49{,}646$	0.99	≤0.001	135.5	67,810
		Nitrogen accumulated in seeds, N_{se}, kg ha^{-1}				
2009	Gostyń, Go	$N_{se} = -0.0032N_f^2 + 0.59N_f + 68.9$	0.98	≤0.001	92,2	96.1
	Kołaczkowo, Ko	$N_{se} = 0.225N_f + 95.2$	0.87	≤0.05	-	131.2
	Buszewo, Bu	$N_{se} = -0.0064N_f^2 + 1.41N_f + 46.3$	0.99	≤0.001	110.3	124.2
	Wieszczyn, Wi	$N_{se} = 0.368N_f + 66.0$	0.85	≤0.05	-	124.9
	Wenecja, Ve	$N_{se} = 0.558N_f + 64.5$	0.95	≤0.001	-	153.8
	Donatowo, Do	$N_{se} = -0.0022N_f^2 + 0.67N_f + 102.8$	0.99	≤0.001	152.7	154.1

[1] N_f–fertilizer N, kg ha^{-1}.

Table A2. Matrix of Spearman's correlation coefficients between yield components and indices of N management by WOSR, n = 24.

Characteristics	TB	HI	PD	Se/Po	SD	TSW	N_c	N_{se}	N_r	TN	NHI	UNA	UNP
Y	0.56 **	0.19	0.65 **	−0.00	0.84 ***	0.19	5473	0.96 ***	0.47 *	0.83 ***	−0.00	0.18	−0.26
TB	1.00	−0.67 ***	0.63 **	−0.29	0.62 **	−0.06	0.24	0.49 *	0.91 ***	0.83 ***	−0.69 ***	0.74 ***	−0.70 ***
HI		1.00	−0.23	0.42 *	−0.01	0.24	0.16	0.25	−0.63 **	−0.24	0.84 ***	−0.70 ***	0.64 **
PD			1.00	−0.63 **	0.71 ***	−0.19	−0.06	0.47 *	0.36	0.48 *	−0.17	0.09	−0.12
Se/Po				1.00	0.06	−0.12	0.37	0.16	−0.07	0.05	0.24	−0.05	0.04
SD					1.00	−0.35	0.30	0.76 ***	0.49	0.73 ***	−0.09	0.17	−0.21
TSW						1.00	0.42 *	0.28	0.03	0.18	0.03	0.13	−0.19
N							1.00	0.75 ***	0.36	0.65 **	−0.01	0.43 *	−0.51 *
N_{se}								1.00	0.46 *	0.85 ***	0.03	0.25	−0.34
Nr									1.00	0.86 ***	−0.82 ***	0.91 ***	−0.87 ***
TN										1.00	−0.48 *	0.69 ***	−0.72 ***
NHI											1.00	−0.87 ***	0.86 ***
UNA												1.00	−0.95 ***

***, **, * indicate significant differences at $p < 0.001$, $p < 0.01$, and $p < 0.05$. Y—seed yield; TB–total biomass; HI–harvest index; PD–pod density; Se/Po—number of seeds per pod; SD—seed density; TSW—thousand seed weight; N_c—N concentration in seeds; N_{se}—amount of n accumulated in seeds; N_r—amount of n in WOSR post-harvest residues; TN—total N in WOSR biomass at harvest; NHI—nitrogen harvest index; UNA—unit N accumulation; UNP—unit N productivity.

Table A3. Spearman's correlation matrix between yield components, selected nitrogen variables and PCA factors.

Variables	F1	F2	F3	F4
Y	**0.95** [1]	−0.11	−0.05	−0.27
PD	0.60	**−0.77**	0.18	−0.05
Se/Po	0.02	0.66	**−0.74**	0.01
SD	**0.83**	−0.38	−0.39	0.03
TSW	0.16	0.55	0.68	−0.42
N	0.68	0.64	0.01	−0.05
N_{se}	**0.95**	0.14	−0.07	−0.25
TN	**0.95**	0.10	0.08	0.26
UNA	0.45	0.27	0.34	**0.78**

[1] bold = correlation coefficients for $R^2 \geq 0.50$. Y—seed; yield; PD—pod density; Se/Po—number of seeds per pod; SD—seed density; TSW—thousand seed weight; N—N content in seeds; N_{se}—amount of n accumulated in seeds; TN—total N in WOSR biomass at harvest; NHI—nitrogen harvest index; UNA—unit N accumulation.

Table A4. Regression patterns of key WOSR nitrogen management characteristics response to the N fertilizer rate, n = 4.

Year	Site	Equation	R^2	*p*	N_{op} kg ha^{-1}	N_{inmax}/ $N_{gainmax}$/ N_{intmax}
		N_{input}, N_{in}, kg ha^{-1}				
2009	Gostyń, Go	$N_{in} = -0.0033N_f^2 + 1.36N_f + 42.5$	0.98	≤0.001	206.2	182.8
	Kołaczkowo, Ko	$N_{in} = 0.225N_f + 88.1$	0.89	≤0.05	-	124.1
2010	Buszewo, Bu	$N_{in} = -0.006N_f^2 + 2.26N_f + 74.7$	0.99	≤0.001	176.3	273.7
	Wieszczyn, Wi	$N_{in} = 0.37N_f + 47.9$	0.85	≤0.05	-	107.1
2011	Wenecja, Ve	$N_{in} = 0.57N_f - 5.27$	0.95	≤0.001	-	85.9
	Donatowo, Do	$N_{se} = -0.0022N_f^2 + 20.91N_f + 59.1$	0.99	≤0.001	207.5	154.1
		N_{gain}, kg ha^{-1}				
2009	Gostyń, Go	$N_{gain} = -0.73N_f^2 + 0.78N_f + 31.5$	0.99	≤0.001	53.5	52.4
	Kołaczkowo, Ko	$N_{gain} = -0.62N_f + 229.6$	0.96	≤0.001	-	129.6
2010	Buszewo, Bu	$N_{gain} = -0.014N_f^2 + 1.75N_f + 104.8$	0.97	≤0.001	64.8	161.6
	Wieszczyn, Wi	$N_{gain} = -0.0057N_f^2 + 0.62N_f + 186.2$	0.52	≤0.05	54.4	169.3
2011	Wenecja, Ve	$N_{gain} = -0.111N_f + 135$	0.52	≤0.05	-	117.2
	Donatowo, Do	$N_{gain} = -0.381N_f + 203.3$	0.89	≤0.05	-	142.3
		$N_{input\ total}$, N_{int}, kg ha^{-1}				
2009	Gostyń, Go	$N_{int} = -0.007N_f^2 + 1.78N_f + 140.7$	0.99	≤0.001	122.0	249.4
	Kołaczkowo, Ko	$N_{int} = 0.375N_f + 289$	0.89	≤0.05	-	349.0
2010	Buszewo, Bu	$N_{int} = -0.0135N_f^2 + 2.75N_f + 171$	0.99	≤0.001	101.9	311.0
	Wieszczyn, Wi	$N_{int} = -0.006N_f^2 + 1.62N_f + 235$	0.83	≤0.05	142.4	344.6
2011	Wenecja, Ve	$N_{int} = 0.89N_f + 240$	0.99	≤0.001	-	382.4
	Donatowo, Do	$N_{int} = 0.619N_f + 258.3$	0.95	≤0.05	-	357.3

N_f—fertilizer N, kg ha^{-1}.

Table A5. Regression patterns of key WOSR N_{se} response to the N supply, n = 4.

Year	Site	Equation	R^2	p	N_{inop}/N_{inmax} kg ha^{-1}	N_{semax} kg ha^{-1}	NE [1] %
		N input, N_{in}, kg ha^{-1} as independent variable					NE_{in}
2009	Gostyń, Go	$N_{se} = -0.0033N_{in}^2 + 1.36N_{in} - 42.5$	0.98	≤0.001	206.2	95.1	46
	Kołaczkowo, Ko	$N_{se} = 0.225N_{in} + 88.1$	0.89	≤0.05	219.7	131.2	60
2010	Buszewo, Bu	$N_{se} = -0.0064N_{in}^2 + 2.257N_{in} - 74.7$	0.99	≤0.001	176.3	124.2	70
	Wieszczyn, Wi	$N_{se} = 0.37N_{in} + 47.9$	0.85	≤0.05	209.0	125.2	60
2011	Wenecja, Ve	$N_{se} = 0.56N_{in} + 5.89$	0.95	≤0.001	265.0	154.3	58
	Donatowo, Do	$N_{se} = -0.0022N_{in}^2 + 0.91N_{in} + 59.1$	0.99	≤0.001	206.8	153.2	74
		N input total, N_{int}, kg ha^{-1} as independent variable					NE_{int}
2009	Gostyń, Go	$N_{se} = 0.24N_{int} + 32.7$	0.79	≤0.01	246.7	91.9	37
	Kołaczkowo, K1	$N_{se} = 0.52N_{int} - 51.2$	0.72	≤0.05	358.3	135.1	38
2010	Buszewo, Bu	$N_{se} = -0.0033N_{int}^2 + 2.14N_{int} - 224$	0.99	≤0.001	342.0	123.0	36
	Wieszczyn, Wi	$N_{se} = 0.36N_{int} - 13.7$	0.61	≤0.05	361.3	116.4	32
2011	Wenecja, Ve	$N_{se} = 0.61N_{int} - 79.4$	0.90	≤0.01	380.9	152.9	40
	Donatowo, Do	$N_{se} = -0.0042N_{int}^2 + 3.11N_{int} - 413$	0.99	≤0.001	370.2	162.7	44

[1] NE—nitrogen efficiency.

References

1. Abbadi, A.; Leckband, G. Rapessed breeding for oil content, quality, and sustainability. *Eur. J. Lipid Sci. Technol.* **2011**, *113*, 1198–1206. [CrossRef]
2. Vinnichek, L.; Pogorelova, E.; Dergunov, A. Oilseed market: Global trends. *IOP Conf. Ser. Earth Environ. Sci.* **2019**, *274*, 0112030. [CrossRef]
3. FAOSTAT. Food and Agriculture Organization of the United Nations. Available online: http://faostat.fao.org/site/567/default.aspx#ancor (accessed on 25 June 2020).
4. Zając, T.; Klimek-Kopyra, A.; Oleksy, A.; Lorenc-Kozik, A.; Ratajczak, K. Analysis of yield and plant traits of oilseed rape (*Brassica napus* L.) cultivated in temperate region in light of the possibilities of sowing in arid areas. *Acta Agrobotanica* **2016**, *69*, 1696. [CrossRef]
5. Carre, P.; Pouzet, A. Rapeseed market, worldwide and in Europe. *Ocl* **2014**, *21*, D102. [CrossRef]
6. Pullens, J.W.M.; Sharif, B.; Trnka, M.; Balek, J.; Semenov, M.A.; Olesen, J.E. Risk factors for European winter oilseed rape production under climate change. *Agric. For. Meteorol.* **2019**, *272*, 30–39. [CrossRef]
7. Brown, J.K.; Beeby, R.; Penfield, S. Yield instability of winter oilseed rape modulated by early winter temperature. *Sci. Rep.* **2019**, *9*, 6953. [CrossRef]
8. Sieling, K.; Böttcher, U.; Kage, H. Effect of sowing method and N application on seed yield and N use efficiency of winter oilseed rape. *Agronomy* **2017**, *7*, 21. [CrossRef]
9. Diepenbrock, W. Yield analysis of winter oilseed rape (*Brassica napus* L.): A review. *Field Crops Res.* **2000**, *67*, 35–49. [CrossRef]
10. Habekotté, B. Quantitative analysis of pod formation, seed set and seed filling in winter oilseed rape (*Brassica napus* L.) under field conditions. *Field Crops Res.* **1993**, *35*, 21–33. [CrossRef]
11. Grzebisz, W.; Szczepaniak, W.; Grześ, S. Sources of nutrients for high-yielding winter oilseed rape (*Brassica napus* L.) during post-flowering growth. *Agronomy* **2020**, *10*, 626. [CrossRef]
12. Berry, P.M.; Spink, J.H. A physiological basis of oilseed rape yields: Past and future. *J. Agric. Sci.* **2006**, *144*, 381–392. [CrossRef]
13. Hoffmann, M.P.; Jacobs, A.; Whitbread, A.M. Crop modeling based analysis of site-specific production limitations of winter oilseed rape in northern Germany. *Field Crops Res.* **2015**, *178*, 49–62. [CrossRef]
14. Weymann, W.; Böttcher, U.; Sieling, K.; Kage, H. Effects of weather conditions during different growth phases on yield formation of winter oilseed rape. *Field Crops Res.* **2015**, *173*, 41–48. [CrossRef]

15. Barłóg, P.; Grzebisz, W. Effect of timing and nitrogen fertilizer application on winter oilseed rape (*Brassica napus L.*). II. Nitrogen uptake dynamics and fertilizer efficiency. *J. Agron. Crop. Sci.* **2004**, *190*, 314–323. [CrossRef]
16. Sieling, K.; Kage, H. Efficient N management using winter oilseed rape. A review. *Agron. Sustain. Dev.* **2010**, *30*, 271–279. [CrossRef]
17. Barraclough, P.B. Root growth, macro-nutrient uptake dynamics and soil fertility requirements of a high-yielding winter oilseed rape crop. *Plant Soil* **1989**, *119*, 59–70. [CrossRef]
18. Li, H.; Cong, R.; Ren, T.; Li, X.; Ma, C.; Zheng, L.; Zhang, Z.; Lu, J. Yield response to n fertilizer and optimum N rate of winter oilseed rape under different soil indigenous N supplies. *Field Crops Res.* **2015**, *181*, 52–59. [CrossRef]
19. Fordoński, G.; Pszczółkowska, A.; Okorski, A.; Olszewski, J.; Załuski, D.; Gorzkowska, A. The yield and chemical composition of winter oilseed rape seeds depending on different nitrogen fertilization rates and preceding crop. *J. Elem.* **2016**, *21*, 1225–1234.
20. Olfs, H.-W.; Blankenau, K.; Brentrup, F.; Jasper, J.; Link, A.; Lammel, J. Soil- and plant-based nitrogen-fertilizer recommendations in arable farming. *J. Plant Nutr. Soil Sci.* **2005**, *168*, 414–431. [CrossRef]
21. Barłóg, P.; Łukowiak, R.; Grzebisz, W. Predicting the content of soil mineral nitrogen based on the content of calcium chloride-extractable nutrients. *J. Plant Nutr. Soil Sci.* **2017**, *180*, 624–635. [CrossRef]
22. Li, H.; Lu, J.; Ren, T.; Li, X.; Cong, R. Nutrient efficiency of winter oilseed rape in an intensive cropping system: A regional analysis. *Pedosphere* **2017**, *27*, 364–370. [CrossRef]
23. Łukowiak, R.; Barłóg, P.; Grzebisz, W. Soil mineral nitrogen and the rating of $CaCl_2$ extractable nutrients. *Plant Soil Environ.* **2017**, *63*, 177–183.
24. Grzebisz, W.; Szczepaniak, W.; Barłóg, P.; Przygocka-Cyna, K.; Potarzycki, J. Phosphorus sources for winter oilseed rape (*Brassica napus* L.)—Magnesium sulfate management impact on P use efficiency. *Arch. Agron. Soil Sci.* **2018**, *12*, 1646–1662. [CrossRef]
25. Szczepaniak, W.; Grzebisz, W.; Potarzycki, J.; Łukowiak, R.; Przygocka-Cyna, K. Nutritional status of winter oilseed rape in cardinal stages of growth as yield indicator. *Plant Soil Environ.* **2015**, *61*, 291–296. [CrossRef]
26. Grzebisz, W.; Łukowiak, R.; Sassenrath, G. Virtual nitrogen as a tool for assessment of nitrogen at the field scale. *Field Crops Res.* **2018**, *218*, 182–184. [CrossRef]
27. Bardsley, C.E.; Lancaster, J.D. Determination of reserve sulfur and soluble sulfates in soils. *Soil Sci. Soc. Am. Proc.* **1980**, *24*, 265–268. [CrossRef]
28. Fotyma, E.; Fotyma, M.; Pietruch, C. The content of mineral N in arable soils in Poland. *Fertil. Fertil.* **2004**, *VI*, 11–54.
29. Cassman, G.; Dobermann, A.; Walters, D. Agro-ecosystems, nitrogen-use efficiency, and nitrogen Management. *AMBIO* **2002**, *31*, 132–140. [CrossRef]
30. Rathke, G.W.; Christen, O.; Diepenbrock, W. Effects of nitrogen source and rate on productivity and quality of winter oilseed rape (*Brassica napus* L.) grown in different crop rotations. *Field Crops Res.* **2005**, *94*, 103–113. [CrossRef]
31. Wang, X.; Mathieu, A.; Cournede, P.-H.; Allirand, J.-M.; Jullien, A.; de Reffye, P.; Zhang, B.G. Variability and regulation of the number of ovules, seeds, and pods according to assimilate availability in winter oilseed rape (*Brassica napus* L.). *Field Crops Res.* **2011**, *122*, 60–69. [CrossRef]
32. Wang, Y.; Liu, T.; Li, X.; Ren, T.; Cong, R.; Lu, J. Nutrient deficiency limits population development, yield formation and nutrient uptake of direct sown winter oilseed rape. *J. Integr. Agric.* **2015**, *14*, 670–680. [CrossRef]
33. Szczepaniak, W. A mineral profile of winter oilseed rape in critical stages of growth—Nitrogen. *J. Elem.* **2014**, *19*, 759–778. [CrossRef]
34. Gomez, N.; Miralles, D.J. Factors that modify early and late reproductive phases in oilseed rape (*Brassica napus* L.): Its impact on seed yield and oil content. *Ind. Crops Prod.* **2011**, *34*, 1277–1285. [CrossRef]
35. Jankowski, K.J.; Hulanicki, P.S.; Krzebietke, S.; Żarczyński, P.; Hulanicki, P.; Sokólski, M. Yield and quality of winter oilseed rape in response to different systems of foliar fertilization. *J. Elem.* **2016**, *21*, 1017–1027. [CrossRef]
36. Pan, Y.; Lu, Z.; Lu, J.; Li, X.; Cong, R. Effects of low sink demand on leaf photosynthesis under potassium deficiency. *Plant Physiol. Biochem.* **2017**, *113*, 110–121. [CrossRef] [PubMed]

37. Dreccer, M.F.; Schapendonk, A.H.; Slafer, G.A.; Rabbinge, R. Comparative response of wheat and oilseed rape to nitrogen supply: Absorption and utilization efficiency of radiation and nitrogen during the reproductive stages of growth. *Plant Soil* **2000**, *220*, 189–205. [CrossRef]
38. Grzebisz, W.; Łukowiak, R.; Biber, M.; Przygocka-Cyna, K. Effect of multi-micronutrient fertilizers applied to foliage on nutritional status of winter oilseed rape and development of yield forming elements. *J. Elem.* **2010**, *15*, 477–491. [CrossRef]
39. Körner, C. Paradigm shift in plant growth control. *Curr. Opin. Plant Biol.* **2015**, *25*, 107–114. [CrossRef]
40. Bouchet, A.-S.; Laperche, A.; Bissuel-Belaygue, C.; Snowdon, R.; Nesi, N.; Stahl, A. Nitrogen use eficiency in rapeseed. A review. *Agron. Sustain. Dev.* **2016**, *36*, 38. [CrossRef]
41. Szczepaniak, W.; Grzebisz, W.; Barłóg, P.; Przygocka-Cyna, K. Mineral composition of winter oilseed rape (*Brassica napus* L.) seeds as a tool for oil seed prognosis. *J. Cen. Eur. Agric.* **2017**, *18*, 196–213. [CrossRef]

Article

A Remote Sensing-Based Approach to Management Zone Delineation in Small Scale Farming Systems

Davide Cammarano [1,*,†], Hainie Zha [2], Lucy Wilson [3], Yue Li [2], William D. Batchelor [4] and Yuxin Miao [5,*]

1 James Hutton Institute, Errol Road, Invergowrie DD25BQ, UK
2 College of Resource and Environmental Sciences, China Agricultural University, 2 Yuanmingyuan W. Rd, Beijing 100193, China; zhahn@cau.edu.cn (H.Z.); yli@cau.edu.cn (Y.L.)
3 RSK ADAS Ltd., 172 Chester Road, Helsby WA60AR, UK; Lucy.Wilson@adas.co.uk
4 Department of Biosystems Engineering, Auburn University, 200 Tom Corley Building, Auburn, AL 36849, USA; wdb0007@auburn.edu
5 Precision Agriculture Center, Department of Soil, Water and Climate, University of Minnesota, 1991 Upper Buford Circle, St. Paul, MN 55108, USA
* Correspondence: dcammar@purdue.edu (D.C.); ymiao@umn.edu (Y.M.)
† Current address: Department of Agronomy, Purdue University, West Lafayette, IN 47906, USA.

Received: 18 September 2020; Accepted: 7 November 2020; Published: 12 November 2020

Abstract: Small-scale farms represent about 80% of the farming area of China, in a context where they need to produce economic and environmentally sustainable food. The objective of this work was to define management zone (MZs) for a village by comparing the use of crop yield proxies derived from historical satellite images with soil information derived from remote sensing, and the integration of these two data sources. The village chosen for the study was Wangzhuang village in Quzhou County in the North China Plain (NCP) (30°51′55″ N; 115°02′06″ E). The village was comprised of 540 fields covering approximately 177 ha. The subdivision of the village into three or four zones was considered to be the most practical for the NCP villages because it is easier to manage many fields within a few zones rather than individually in situations where low mechanization is the norm. Management zones defined using Landsat satellite data for estimation of the Green Normalized Vegetation Index (GNDVI) was a reasonable predictor (up to 45%) of measured variation in soil nitrogen (N) and organic carbon (OC). The approach used in this study works reasonably well with minimum data but, in order to improve crop management (e.g., sowing dates, fertilization), a simple decision support system (DSS) should be developed in order to integrate MZs and agronomic prescriptions.

Keywords: site-specific nutrient management; soil brightness; satellite remote sensing; crop yield; soil fertility; spatial variability

---

1. Introduction

Small scale farms represent about 80% of the farming area of China. Given the need to produce more food on the same amount (or less) of land while also reducing environmental pollution, such areas are faced with tough challenges. Farmers manage their fields by experience and need science-based evidence to make the system more efficient. The mismanagement of nitrogen (N) fertilizer is a known problem in the smallholder farming systems of China [1].

Precision nutrient management can be achieved either through a sensor-based or a map-based approach. The former uses sensors to guide site-specific N management based on the quantification of crop reflectance. However, a sensor-based approach is affected by the inter-annual interactions between soil and weather. Colaço & Bramley [2] demonstrated how the sensor-based approach could be improved by also considering the impact of other environmental covariates. The latter approach consists of using multiple images (e.g., soil, remote sensing, yield monitor) with the aim of dividing the fields into management zones (MZs). MZs are defined as sub-regions within the field that have similar combinations of yield-limiting factors and are managed accordingly [3,4]. Basso et al. [5] stated that understanding the factors that lead to the spatial and temporal variability of a crop within the field is the first step for optimal agronomic management. In addition, the delineation of fields into MZs helps with obtaining soil/crop samples cost-effectively and applying site-specific agronomic input [4–7].

Several approaches have been proposed to define MZs at the field level. One approach is based on gathering soil or landscape information, such as sampling the soil using electrical conductivity (EC) sensor, sampling for soil physical and chemical properties or using remotely sensed data for estimating soil and landscape properties [8–12]. Another approach uses yield maps or remotely sensed images to reconstruct spatial and temporal patterns of crop growth within the field to define a given number of MZs [13–16]. Finally, the integrated approach uses both soil-landscape and crop information to define MZs at field level [4,17].

However, the small size of single farms in the North China Plain (NCP) does not allow for cost-effective on-farm management using farm-scale MZs. Moreover, yield maps are not available in most small-scale farming systems and, in such systems, farmers do not measure field-level yield at harvest [18], but measure total grain that they sell from all of their fields. Jin et al. [19] combined a crop simulation model with remotely sensed data to map yield heterogeneity on smallholder farms in East Africa. High-resolution satellite images were used to define a smallholder farming system, but the prevalence of small field size was one of the challenges in improving the performance of the approach proposed [19]. In addition, high-resolution satellite data usually has to be purchased from a private provider (e.g., RapidEye) or obtained from a contemporary open-source sensor such as Sentinel-2, which does not yet have enough years available to capture the long-term inter-annual variability. A practical strategy is to divide fields in a village into several MZs disregarding the current field management structure. Some of the common open-source satellites, such as Landsat that has a long time-series of data, may allow the mapping of fields in a village for MZ delineation purposes; therefore, each field can be classified in a given MZ, maintaining the existing boundary structure. Fields in the same MZ could then use similar management practices or inputs optimised for their particular conditions and requirements regardless of their geographical proximity.

The objective of this work was to integrate soil information derived from remote sensing with crop yield proxies derived from historical satellite images to define MZs at the village scale.

2. Materials and Methods

2.1. Site Description

The village chosen for the study was Wangzhuang village in Quzhou County in the NCP (36°51′55″ N; 115°02′06″ E). The cropped area of the village covered approximately 177 ha and comprised a total of 540 fields, with an average farm size of 0.33 ha (Figure 1). The village is part of the Science and Technology Backyards (STB) initiative. The STB was established originally in Quzhou County in 2009 by China Agricultural University to carry out specific research and extension services aimed at transferring research technology to smallholder farmers [18]. The main cropping system of the village was winter wheat (*Triticum aestivum* L.) (sown in October) and summer maize (*Zea mays* L.) (sown in mid-June), with irrigation used by farmers on both crops. The soil was classified as silty-loam [20] and no significant slopes were present. The 2008–2018 weather had a mean, maximum and minimum air temperature during the growing season of 9.7, 15, and 4.3 °C, respectively.

Figure 1. Location of the Wangzhaung village in China (red mark), the area occupied by the village (white polygon), and an image of the typical field layout in Wangzhuang village, China. The fields were separated only by a thin bare patch of soil. The similar growth stages of the wheat plants suggest that planting occurred at the same time in the different fields.

2.2. Overall Methodological Approach

Satellite data for the village were obtained from the Landsat satellites (30 m pixels) for 2008–2018 using the USGS explorer website [21]. Level 2 processed imagery was requested and analyzed following the protocols described on the Landsat Explorer [21]. The Landsat data came from 2 different sensors having a different wavelength range in the near-infrared bands; one was used from 2008 to 2011 (Landsat 5 with 0.76–0.90 μm with Landsat Thematic Mapper - TM sensor [21]) and another from 2013 to 2018 (Landsat 8 with 0.85–0.88 μm with Operational Land Imager - OLI sensor [21]). Atmospherically-corrected cloud-free scenes from the end of April/beginning of May were selected to coincide with flowering in winter wheat. The choice of using images at that particular developmental stage was justified by the strong correlation between remote sensing data and grain yield [22–24]. In addition, in Northern China, the use of proximal and/or remote sensing as a proxy of canopy N and yield has been validated on several crops and in different geographical regions [25–28]. The Green Normalized Difference Vegetation Index (GNDVI) [29] was calculated for each scene. The GNDVI is more closely related to the photosynthetically absorbed radiation than NDVI and has shown a linear correlation with the Leaf Area Index (LAI) and biomass [29]. This makes the GNDVI more sensitive to changes in biomass and chlorophyll concentration compared to the NDVI. It was calculated as follows:

$$GNDVI = \frac{\rho_{NIR} - \rho_{Green}}{\rho_{NIR} + \rho_{Green}} \quad (1)$$

where ρ_{NIR} and ρ_{Green} are the reflectance in the Near Infrared and Green bands, respectively.

Soil Brightness (SOB) data at 3 m spatial resolution were purchased from Courtyard Agriculture Ltd. in the UK and were derived from RapidEye optical satellite images at a time in the season when the land was un-cropped (bare soil). Most of the fields were contiguous, with 'boundaries' being narrow ridges of bare soil. There was little other vegetation or paths/roads present. Therefore, most of the pixels corresponded to cropland (Figure 1).

Prior to sowing of wheat in 2015, soil sampling was carried out in the village. For each field, 10 samples were collected at 0–20 cm depth following "S" patterns. The samples were pooled together, mixed thoroughly and divided into four subsamples. One subsample of about 1–2 kg was kept for determining inorganic soil N and soil organic carbon (OC). The soil samples were later dried, ground to powder and analyzed for total soil N concentration using the Kjeldahl digestion method [30]. Soil OC was analyzed following the wet combustion method [31]. Soil texture information was also available at selected points and used for this study and additional information is available elsewhere [32].

Spatial and temporal variability in GNDVI were quantified following the method of Basso et al. (2012). The spatial variability of GNDVI from 2008 to 2018 was calculated from the relative percentage difference of GNDVI at each 30 m pixel from the average GNDVI obtained for the whole cropped area of the village, according to Equation (2) [13]:

$$\overline{\sigma}^2_{si} = \frac{1}{n}\sum_{k=1}^{n}\left[\frac{y_{i,k} - \overline{y}_k}{\overline{y}_k} \times 100\right] \quad (2)$$

where n is the total number of available years, $k = 1, \ldots, n$ is the integer corresponding to every year, $\overline{\sigma}^2_{si}$ is the average percentage difference at location i, $\overline{y}_k$ is the average of the variable obtained for the whole village at year k, $y_{i,k}$ is the variable monitored at location i at year k. Pixels that have high values of $\overline{\sigma}^2_{si}$ are associated with high yield (under the assumption that GNDVI is directly proportional to yield) and pixels with low values are lower yielding.

The temporal variability, which is a measure of stability, was calculated as temporal variance to overcome the issues with varying stability over time [33]. The temporal variance of patterns in the GNDVI data was calculated using Equation (3):

$$\overline{\sigma}^2_{ti} = \frac{1}{n}\sum_{k=1}^{n}\left(y_{i,k} - \overline{y}_{i,n}\right)^2 \quad (3)$$

where $\overline{\sigma}^2_{ti}$ is the temporal variance value at location i, $y_{i,k}$ is the value of the variable monitored at location i at year k, $\overline{y}_{i,n}$ is the average variable value at location i over the n years. The higher the temporal variance, the more unstable the GNDVI measurement (and thus the yield) at that particular location over time. Threshold values have previously been used to identify stable zones within fields, however, the choice of the threshold for determining the stable zone can affect the result considerably [13]. To overcome this problem, a clustering algorithm was applied to the spatial and temporal variability layers, which is further described below. The GNDVI variability metrics and the soil brightness images were summarized at the field scale for consistency with the measured soil variables that were used for evaluation purposes.

For site-specific agronomic management of any input, there is the need to develop MZs that will be subject to a unique combination of potential yield-limiting factors [12]. Management zones are most effective if the variation (of the factors under consideration) within them is minimized, and there are a manageable number of zones. Clustering is an important method in precision agriculture [34–38]. In all these studies it was found that the *k*-means was not among the best methods to define MZs, but the best algorithm differed. For example, in smallholder farming systems, Possiblistic Fuzzy C Mean worked best but in other conditions, the McQuitty method seemed to give more consistent results [36,37]. Although the *k*-means is still widely adopted, it was decided to use the partitioning around medoids (PAM) method [39] to derive the MZs. The PAM is a clustering algorithm that, like *k*-means clustering, aims to minimize the distances between points within a cluster and the point at the center of the cluster. It is a more robust alternative to *k*-means clustering when noisy data are expected, as is the case in this study due to unmeasured variation in growing conditions due to the soil spatial variability. The term medoid refers to an observation within a cluster for which the sum of the distances between it and all the other members of the cluster is a minimum. PAM requires a priori information of the number of clusters, but it does more computation than the more commonly used *k*-means clustering to ensure that the medoids that it finds are truly representative of the observations within a given cluster. It has been found that the PAM showed better results than the *k*-means in terms of execution time, sensitivity towards outliers, reduction of noise in the data due to minimization of the sum of dissimilarities within the dataset [36]. Field-scale variables for (i) spatial and temporal variability in GNDVI; and (ii) mean soil brightness were normalised and used as variables with the PAM method, using R software, to define clusters.

Soil N and soil OC were used for MZ evaluation as reported in [4]. Relative variance (RV) of measured soil properties at the field scale was used to evaluate the accuracy of this approach for delineating MZs given by Equation (4):

$$RV = 1 - \frac{S_w^2}{S_T^2} \times 100 \tag{4}$$

where S_w^2 is the total within-zone variance of the soil property and S_T^2 is the total village-level variance of the corresponding property. RV approximates the amount of variability explained by the MZ delineation and can be interpreted similarly to the R^2 value in regression analysis in terms of the percentage of variation explained. RV was calculated on a per-field basis for measured total N and OC because they best characterise areas that would benefit from differential management.

3. Results

The maximum and minimum air temperatures from Jan to mid-May from 2008 to 2018 are shown in Figure 2. Overall, there is inter-annual variability in amplitudes and patterns of maximum and minimum air temperature. For example, in 2011 the minimum air temperatures were the lowest recorded during the period of study. The spatial patterns of the Landsat GNDVI images for a period of 10 years, each around the time of flowering, are shown in Figures 3 and 4. Overall, GNDVI was generally more variable for the earlier years, with values ranging between 0.1 and 0.8 for the 2008–2011 period (Landsat 5) and the majority of values over 0.6 from 2013 to 2018 (Landsat 8). The years 2010 and 2011 showed higher spatial variability than other years.

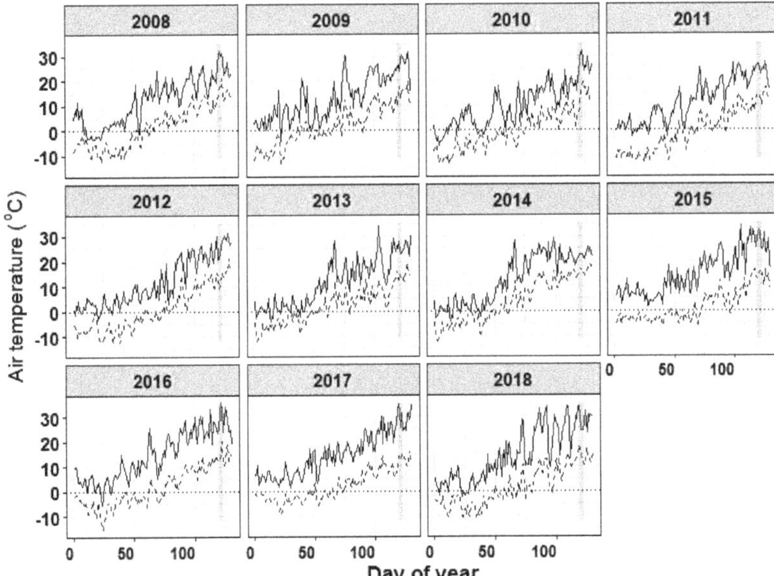

Figure 2. Maximum (full black line) and minimum (dotted black line) air temperature recorded at the Wangzhaung village from 2008 to 2018 from January to mid-May. The horizontal dotted line represents the 0 °C threshold while the vertical grey box represents the time when satellite images were acquired.

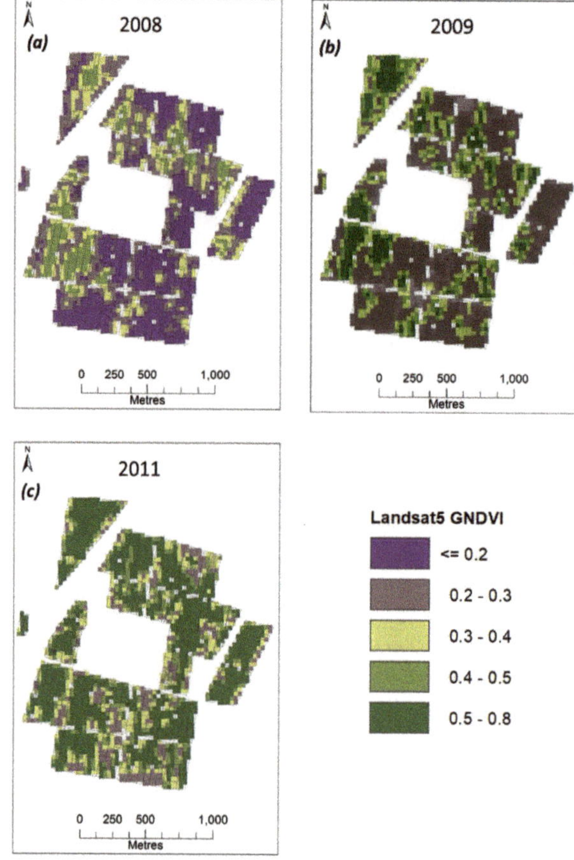

Figure 3. Calculated Green Normalized Vegetation Index (GNDVI) for LANDSAT 5 scenes for the growing season (**a**) 2008; (**b**) 2009; and (**c**) 2011. LANDSAT 5 scenes were obtained for late April/early May every year from 2008 to 2011.

The soil brightness resampled to 30 m resolution to match the Landsat data is shown in Figure 5a and the original soil brightness data at a 3 m resolution is shown in Figure 5b. The mean soil brightness was lower in the North-East portion of the village where the values ranged between 3 and 7. Areas with the highest soil brightness values were in the north-western and south-western portions of the village.

The spatial distributions of measured soil N and OC throughout the village are shown in Figure 6a,b respectively. Soil N varied between 0.68 and 1.37 g/kg (Figure 6a). Low values of soil N were located mainly on the North-West portion of the village and the highest values were located across the village without any clear spatial patterns (Figure 6a). The soil OC varied between 0.52 and 1.2% and its spatial variability matched the soil N content (Figure 6b). Therefore, there was not any clear spatial clustering of its value across the village.

The spatial and temporal decomposition of the GNDVI is shown in Figure 7. No obvious spatial clustering was found in the spatial variability metric (Figure 7a). The temporal variability of the GNDVI was higher in the eastern half of the village and also tended to coincide with areas of low GNDVI (Figure 7b).

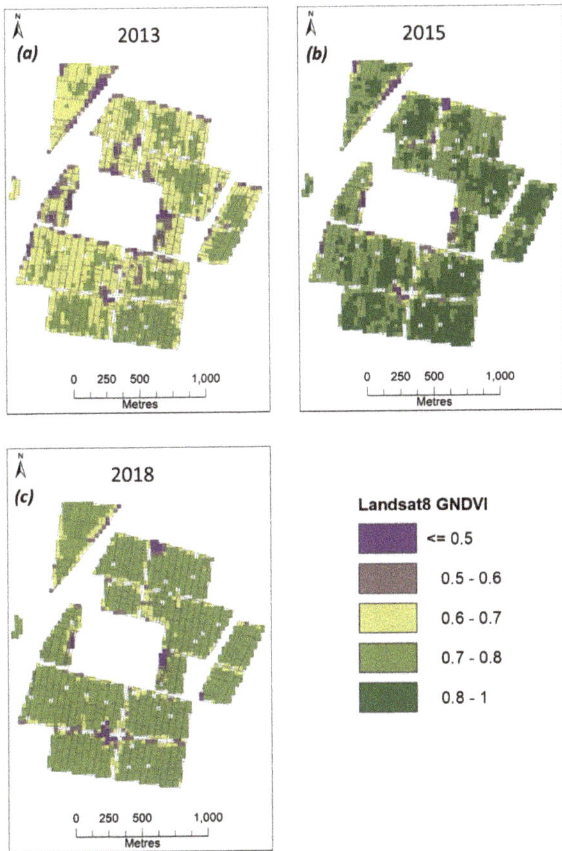

Figure 4. Calculated GNDVI for LANDSAT 8 scenes for growing season (**a**) 2013; (**b**) 2015; and (**c**) 2018. LANDSAT 8 scenes were obtained for late April/early May from 2013 to 2018. There were no scenes available from 2012.

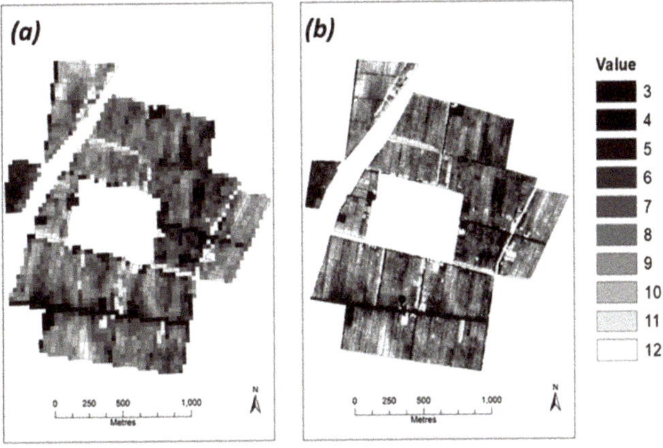

Figure 5. Soil brightness image (**a**) for 2017 resampled to match the LandSat8 spatial resolution (30 m) and (**b**) the original soil brightness image (3 m resolution).

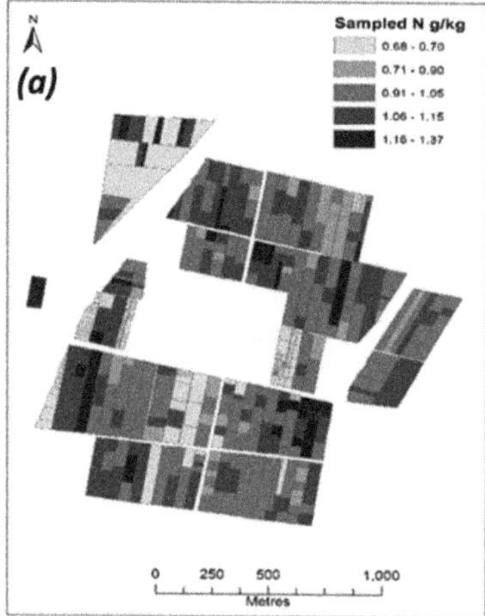

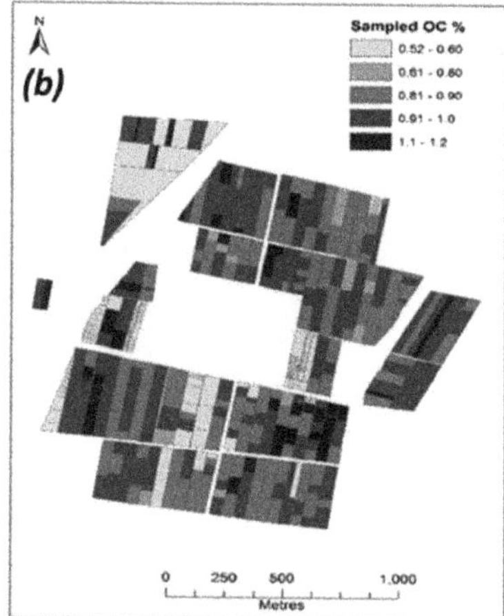

Figure 6. Measured values of (**a**) soil nitrogen content (g/kg) and (**b**) soil organic carbon (%) at the field scale.

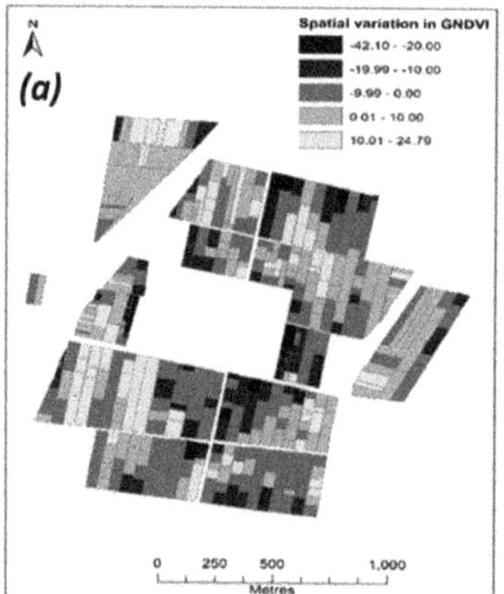

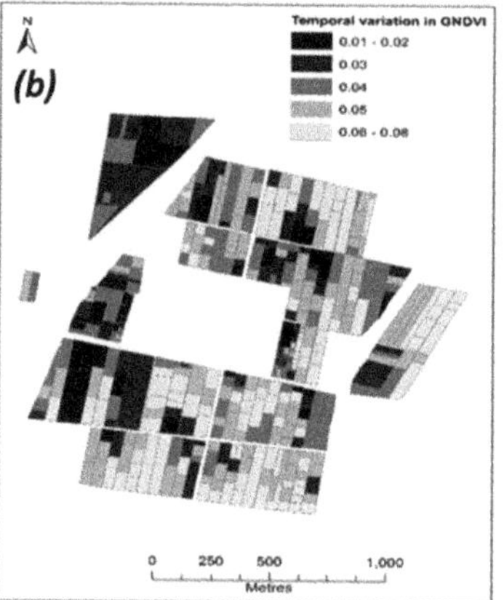

Figure 7. Ten-year time-series of GNDVI decomposed into (**a**) spatial variance and (**b**) temporal variance.

Using soil brightness to define the zones resulted in the three and four zones shown in Figure 8a,b. Using the GNDVI variability metrics to define the zones resulted in the three and four zones shown in Figure 8c,d. Using a combination of soil brightness and GNDVI variability metrics to define the zones resulted in the three and four zones shown in Figure 8e,f.

The RV of measured soil variables total N and OC within these clusters compared to the village mean is shown in Table 1. Soil brightness alone was a fairly poor predictor of measured N and OC in this study, explaining only up to 9% of the variability. The GNDVI variability metrics were a reasonable predictor (up to 39%) of variability in measured N and OC. The three-cluster solution in the combined model (soil brightness + GNDVI) was the best predictor of N and OC variability at up to 45%.

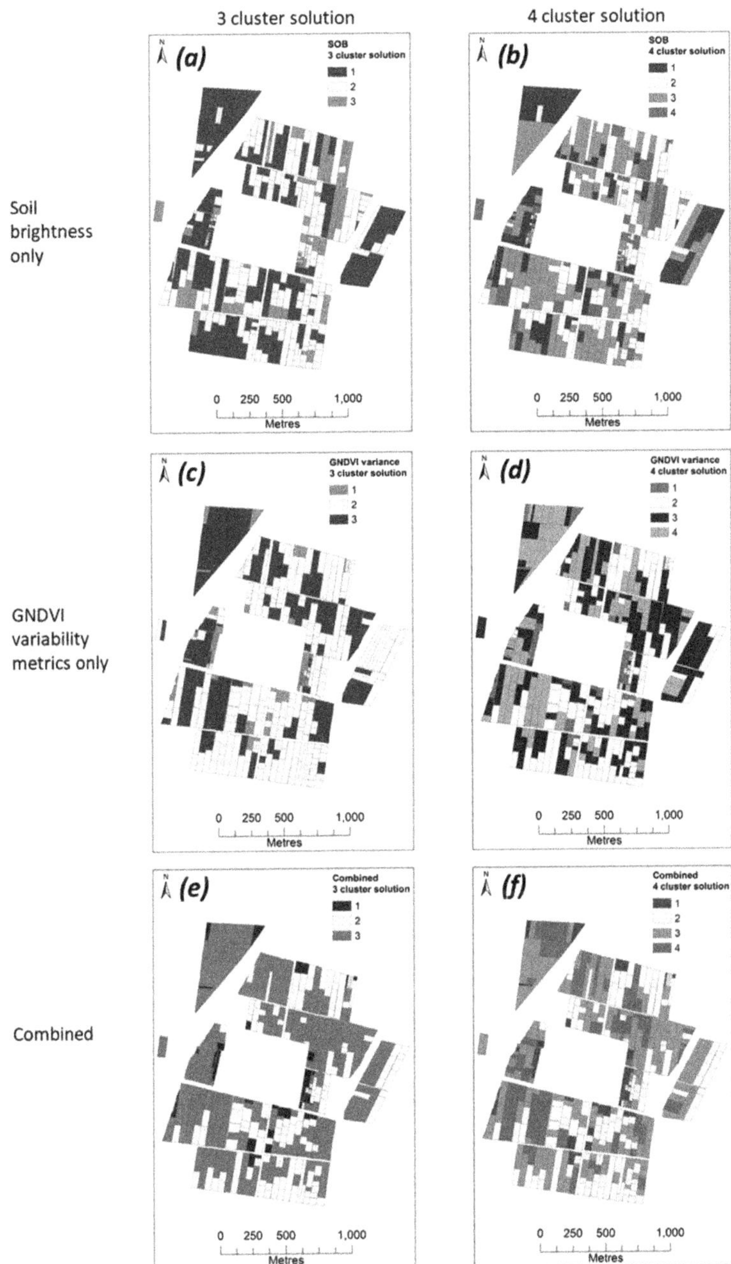

Figure 8. Results of clustering at field scale using (**a**,**b**) soil brightness alone; (**c**,**d**) the GNDVI variability metrics and (**e**,**f**) both combined, testing three and four cluster solutions.

Table 1. Relative variance of measured soil variables total N and organic carbon at the field scale for three and four cluster solutions using (i) soil brightness; (ii) the spatial and temporal variability in GNDVI and (iii) a combination of both.

Clustering Approach	Relative Variance (RV)	
	Total N	Organic Carbon
Soil brightness 3 zones	7.5	7.3
Soil brightness 4 zones	9.3	9.5
GNDVI variability metrics 3 zones	35.8	37.3
GNDVI variability metrics 4 zones	35.8	38.1
Combined 3 zones	43.9	44.9
Combined 4 zones	36.1	37.9

4. Discussion

Overall, the GNDVI derived from remote sensing allowed for the discrimination of zones within the village, with a reasonably good explanation of the variability in measured N and OC. The GNDVI values were low in years in which the minimum growing season temperatures were higher than in other years (e.g., 2008 c.f. 2011). This may have resulted in m the crop developing faster, meaning that by the time the satellite data were acquired (grey box in Figure 2) the crops might have been at an advanced developmental stage i.e.past flowering, meaning higher senescence rates causing lower GNDVI values. For years when minimum growing season temperatures were lower, meaning a longer wheat growing season, by the time our images were acquired, the wheat would still be at the flowering stage and therefore with less senesced material. In addition, there were differences in GNDVI index values between TM and OLI because of the differences in the wavelengths for each of the band that the sensor collected. The image data were not corrected for this. There was a very narrow range of the measured N and OC within the village along with the unmeasured variability in field management and inputs, both of which would have affected the relationships. From a statistical point of view, the soil N and OC showed a narrow range of values. However, from an agronomic point of view, the values of OC ranging between 0.5 and 1.2% have big differences in terms of impact on soil chemical processes and impacts on yield. In fact, soil organic matter is a reserve for nutrients and an agent that improves soil structure. It is a storage pool of plant nutrients. In addition, the humus (which is the stable fraction of the soil organic matter) adsorbs and holds nutrients in a plant-available form. Soil organic matter also releases nutrients in a plant-available form upon decomposition [40].

Satellite images of crops provide an indirect tool to obtain spatial information of crop growing conditions for a given year and therefore are a good tool to quantify spatial field variability [26,41]. The use of a longer remotely sensed time-series enabled the quantification of temporal stability within the spatial context of the village. Satellite images of the bare soil also provide an indirect method to obtain spatial information about the variability in soil conditions, in particular soil moisture, which affects crop growth. However, one limitation of the soil brightness is that it is a weak proxy of soil moisture because the soil–plant relationship is deeper than the first few cm of soil. In addition, whilst there was a significant correlation between soil brightness and measured soil OC, this relationship was not strong (0.19). The soil brightness was reconstructed by a private company and it is not a good proxy of the soil samples measured in the field during the study. This is because the resulting soil brightness image is captured by the company on a given date at a time when the soil is bare (it could have been many months before the soil sampling), and given the high temporal variability in soil nitrogen concentration the time when soil samples were collected do not reflect the spatial variability of the brightness map.

The subdivision of the village into 3–4 zones improved the explanation of the variability in measured soil parameters. However, there is a trade-off between the number of zones and site-specific management. The spatial coherence of the zones also needs to be considered, since it is likely to be more economically and practically efficient to zone the village fields into "blocks" rather than by individual field. It has been found in the literature that three zones are a common number adopted in commercial precision agriculture solutions regardless of the size of the field [42]. More zones could have been defined, but this would translate into more management recommendations and organisation of farmers into more co-operative "clusters", which would add to the complexity and time commitment for co-operative leads.

The results of this study can be considered as a preliminary method based on the integration of different remotely sensed data to delineate MZs at the village scale. More studies are needed to further refine them for guiding site-specific management in small scale farming systems. In addition, incorporating measurements of field level yields would aid in validating this approach in the future.

The main limitation of this study is in the spatial resolution of the satellite data. If a higher resolution and consistently measured dataset had been available over a similar timescale, a more accurate measurement of spatial and temporal variability at the field scale would have been detected. However, recent advancements in sensors on free satellite products (e.g., Sentinel-2) will make the acquisition of a long temporal series of images with higher resolution easier in the future. In addition, the MZ approach could be improved with data on historic management (e.g., timing of key operations) and crop yields. The lack of mechanization and extension services at the village level might hamper the application of modern technologies. Even the subdivision of a village into zones might be of no help if it is not coupled with additional information on how to translate this information into agronomic management.

The approach used in this study was developed considering the limited data availability in small scale farming systems used in the NCP. Therefore, it may not be the best approach if more data are available. In order to improve crop management (e.g., sowing dates, fertilizer amount and timing), a decision support system (DSS) should be developed in order to integrate MZs and agronomic prescriptions. The design of a DSS should provide a science-based approach to quantify the optimal practices that can evaluate the trade-off between economic and environmental benefits. The DSS should be system-based in order to take into account the dynamic interactions between soil–plant–atmosphere–agronomy continuum [5,16]. Future research is therefore needed to overcome the limits highlighted in [19] for a better link between the remotely-sensed approach to define zones and crop models. It is likely that the development of simplified crop growth models will be a step forward for better integration. In this regard, [43] developed simplified models that are less data-intensive. In particular, [44] developed a simple scalable and satellite-based yield model to predict yield for canola (Brassica Napus L.) and wheat (Triticum spp.) at different regional scales. The results of this study could be integrated within the modelling approach highlighted in [44] to provide the system-based DSS approach. Future research would also be needed to concentrate the efforts to consider the impacts of other agronomic practices such as crop rotation (wheat–maize), the use of manure and tillage on the zoning.

5. Conclusions

Spatial and temporal data of remotely estimated crop growth or soil variation measurements from satellites can be used to delineate zones at the village level that explain a reasonable amount of the variation in measured soil variables. In this study on winter wheat, the GNDVI was collected around flowering for 10 years in order to identify spatial and temporal patterns of crop yield. The subdivision of the village into three zones will be optimal for the NCP villages because it tends to cluster many fields into few zones that are easy to manage in situations where low mechanization is the norm. The next step is to develop a system-based DSS that will help to translate the zoning into site-specific agronomic management prescriptions in terms of planting dates and fertilizer amount and timing.

Author Contributions: Conceptualization, Y.M. and D.C.; methodology, L.W., Y.M., D.C., W.D.B.; data analysis: L.W., H.Z.; image processing, L.W., soil data sampling: H.Z., Y.L.; writing—original draft preparation, D.C.; writing—review and editing, Y.M., W.D.B., L.W., H.Z., Y.L., D.C. All authors have read and agreed to the published version of the manuscript.

Funding: This project was funded by the Agri Tech in China Newton Network+ (ATCNN) and Norwegian Ministry of Foreign Affairs (SINOGRAIN II, CHN-17/0019).

Acknowledgments: We thank Hu Kelin and Li Baoguo from China Agricultural University and Yong He from the Chinese Academy of Agricultural Science for providing the spatial soil texture information of the Village. We would like to thank Shuhua Zhang and other students at the Wangzhuang Science and Technology Backyard for soil sampling and analysis. We also thank the three anonymous reviewers that helped improving the manuscript.

Conflicts of Interest: The authors declare no conflict of interest. The funders had no role in the design of the study; in the collection, analyses, or interpretation of data; in the writing of the manuscript, or in the decision to publish the results.

References

1. Miao, Y.; Stewart, B.A.; Zhang, F. Long-term experiments for sustainable nutrient management in China. A review. *Agron. Sustain. Dev.* **2011**, *31*, 397–414. [CrossRef]
2. Colaço, A.F.; Bramley, R.G.V. Site–Year Characteristics Have a Critical Impact on Crop Sensor Calibrations for Nitrogen Recommendations. *Agron. J.* **2019**, *111*, 2047–2059. [CrossRef]
3. Nawar, S.; Corstanje, R.; Halcro, G.; Mulla, D.; Mouazen, A.M. Chapter Four—Delineation of Soil Management Zones for Variable-Rate Fertilization: A Review. *Adv. Agron.* **2017**, *143*, 175–245.
4. Miao, Y.; Mulla, D.J.; Robert, P.C. An integrated approach to site-specific management zone delineation. *Front. Agric. Sci. Eng.* **2018**, *5*, 432–441. [CrossRef]
5. Basso, B.; Ritchie, J.T.; Cammarano, D.; Sartori, L. A strategic and tactical management approach to select optimal N fertilizer rates for wheat in a spatially variable field. *Eur. J. Agron.* **2011**, *35*, 215–222. [CrossRef]
6. Basso, B.; Ritchie, J.T.; Pierce, F.J.; Braga, R.P.; Jones, J.W. Spatial validation of crop models for precision agriculture. *Agric. Syst.* **2001**, *68*, 97–112. [CrossRef]
7. Batchelor, W.D.; Basso, B.; Paz, J.O. Examples of strategies to analyze spatial and temporal yield variability using crop models. *Eur. J. Agron.* **2002**, *18*, 141–158. [CrossRef]
8. Franzen, D.W.; Hopkins, D.H.; Sweeney, M.D.; Ulmer, M.K.; Halvorson, A.D. Evaluation of Soil Survey Scale for Zone Development of Site-Specific Nitrogen Management. *Agron. J.* **2002**, *94*, 381–389. [CrossRef]
9. Mulla, D. Using geostatistics and GIS to manage spatial patterns in soil fertility. In *Automated Agriculture for the 21st Century*; Kranzler, G., Ed.; Am. Soc. Ag. Eng.: St. Joseph, MI, USA, 1991.
10. Arshad, M.; Li, N.; Zhao, D.; Sefton, M.; Triantafilis, J. Comparing management zone maps to address infertility and sodicity in sugarcane fields. *Soil Tillage Res.* **2019**, *193*, 122–132. [CrossRef]
11. Moral, F.J.; Serrano, J.M. Using low-cost geophysical survey to map soil properties and delineate management zones on grazed permanent pastures. *Precis. Agric.* **2019**, *20*, 1000–1014. [CrossRef]
12. Fraisse, C.; Sudduth, K.; Kitchen, N. Delineation of Site-Specific Management Zones by Unsupervised Classification of Topographic Attributes and Soil Electrical Conductivity. *Trans. ASAE* **2001**, *44*, 155–166. [CrossRef]
13. Blackmore, S. The interpretation of trends from multiple yield maps. *Comput. Electron. Agric.* **2000**, *26*, 37–51. [CrossRef]
14. Burke, M.; Lobell, D.B. Satellite-based assessment of yield variation and its determinants in smallholder African systems. *Proc. Natl. Acad. Sci. USA* **2017**, *114*, 2189–2194. [CrossRef] [PubMed]
15. Maestrini, B.; Basso, B. Drivers of within-field spatial and temporal variability of crop yield across the US Midwest. *Sci. Rep.* **2018**, *8*, 14833. [CrossRef] [PubMed]
16. Cammarano, D.; Holland, J.; Ronga, D. Spatial and Temporal Variability of Spring Barley Yield and Quality Quantified by Crop Simulation Model. *Agronomy* **2020**, *10*, 393. [CrossRef]
17. Hornung, A.; Khosla, R.; Reich, R.; Inman, D.; Westfall, D.G. Comparison of Site-Specific Management Zones. *Agron. J.* **2006**, *98*, 407–415. [CrossRef]
18. Zhang, W.; Cao, G.; Li, X.; Zhang, H.; Wang, C.; Liu, Q.; Chen, X.; Cui, Z.; Shen, J.; Jiang, R.; et al. Closing yield gaps in China by empowering smallholder farmers. *Nature* **2016**, *537*, 671. [CrossRef]
19. Jin, Z.; Azzari, G.; Burke, M.; Aston, S.; Lobell, D.B. Mapping Smallholder Yield Heterogeneity at Multiple Scales in Eastern Africa. *Remote Sens.* **2017**, *9*, 931. [CrossRef]
20. United States Department of Agriculture (USDA). Soil Survey Staff, Natural Resources Conservation Service. Available online: https://websoilsurvey.sc.egov.usda.gov/App/HomePage.htm (accessed on 18 September 2020).

21. USGS. Earth Explorer. Available online: https://www.usgs.gov/core-science-systems/nli/landsat/landsat-collection-2-level-2-science-products/ (accessed on 18 September 2020).
22. Tucker, C.J.; Holben, B.H.; Elgin, J.H.J.; McMurtrey, J.E., III. Relationship of Spectral Data to Grain Yield Variation. *Photogramm. Eng. Remote Sens.* **1980**, *46*, 9.
23. Doraiswamy, P.C.; Moulin, S.; Cook, P.W.; Stern, A. Crop Yield Assessment from Remote Sensing. *Photogramm. Eng. Remote Sens.* **2003**, *69*, 9. [CrossRef]
24. Rasmussen, M.S. Advances in crop yield assessment in the Sahel using remote sensing data. *Dan. J. Geogr.* **1999**, *1*, 5.
25. Koppe, W.; Li, F.; Gnyp, M.L.; Miao, Y.; Jia, L.; Chen, X.; Zhang, F.; Bareth, G. Evaluating multispectral and yperspectral satellite remote sensing data for estimating winter wheat growth parameters at regional scale in the North China Plain. *Photogramm. Fernerkund. Geoinf.* **2010**, *3*, 11.
26. Chen, Z.; Miao, Y.; Lu, J.; Zhou, L.; Li, Y.; Zhang, H.; Lou, W.; Zhang, Z.; Kusnierek, K.; Changhua, L. In-Season Diagnosis of Winter Wheat Nitrogen Status in Smallholder Farmer Fields Across a Village Using Unmanned Aerial Vehicle-Based Remote Sensing. *Agronomy* **2019**, *9*, 619. [CrossRef]
27. Huang, S.; Miao, Y.; Zhao, G.; Yuan, F.; Ma, X.; Tan, C.; Yu, W.; Gnyp, M.L.; Lenz-Wiedemann, V.I.S.; Rascher, U.; et al. Satellite Remote Sensing-Based In-Season Diagnosis of Rice Nitrogen Status in Northeast China. *Remote Sens.* **2015**, *7*, 10646–10667. [CrossRef]
28. Xia, T.; Miao, Y.; Wu, D.; Shao, H.; Khosla, R.; Mi, G. Active Optical Sensing of Spring Maize for In-Season Diagnosis of Nitrogen Status Based on Nitrogen Nutrition Index. *Remote Sens.* **2016**, *8*, 605. [CrossRef]
29. Gitelson, A.A.; Kaufman, Y.J.; Merzlyak, M.N. Use of a green channel in remote sensing of global vegetation from EOS-MODIS. *Remote Sens. Environ.* **1996**, *58*, 289–298. [CrossRef]
30. Nelson, D.W.; Sommers, L.E. Determination of total nitrogen in plant material. *Agron. J.* **1962**, *65*, 423–425. [CrossRef]
31. Nelson, D.W.; Sommers, L.E. Total Carbon, Organic Carbon, and Organic Matter. Methods of Soil Analysis Part 3. *Chem. Methods Soil Sci. Soc. Am.* **1996**, *5*, 961–1010.
32. He, Y.; Chen, D.; Li, B.G.; Huang, Y.F.; Hu, K.L.; Li, Y.; Willett, I.R. Sequential indicator simulation and indicator kriging estimation of 3-dimensional soil textures. *Soil Res.* **2009**, *47*, 622–631. [CrossRef]
33. Pringle, M.J.; McBratney, A.B.; Whelan, B.M.; Taylor, J.A. A preliminary approach to assessing the opportunity for site-specific crop management in a field, using yield monitor data. *Agric. Syst.* **2003**, *76*, 273–292. [CrossRef]
34. Castrignanò, A.; De Benedetto, D.; Girone, G.; Guastaferro, F.; Sollitto, D. Characterization, Delineation and Visualization of Agro-Ecozones Using Multivariate Geographical Clustering. *Ital. J. Agron.* **2010**, *5*, 11. [CrossRef]
35. Guastaferro, F.; Castrignanò, A.; De Benedetto, D.; Sollitto, D.; Troccoli, A.; Cafarelli, B. A comparison of different algorithms for the delineation of management zones. *Precis. Agric.* **2010**, *11*, 600–620. [CrossRef]
36. Gavioli, A.; de Souza, E.G.; Bazzi, C.L.; Schenatto, K.; Betzek, N.M. Identification of management zones in precision agriculture: An evaluation of alternative cluster analysis methods. *Biosyst. Eng.* **2019**, *181*, 86–102. [CrossRef]
37. Janrao, P.; Mishra, D.; Bharadi, V. Clustering Approaches for Management Zone Delineation in Precision Agriculture for Small Farms. In Proceedings of the International Conference on Sustainable Computing in Science, Technology & Management (SUSCOM-2019), Jaipur, India, 26–28 February 2019.
38. Soni, K.G.; Patel, A. Comparative Analysis of K-means and K-medoids Algorithm on IRIS Data. *Int. J. Comput. Intell. Res.* **2017**, *5*, 7.
39. Kaufman, L.; Rousseeuw, P. Partitioning Around Medoids. In *An Introduction to Cluster Analysis*; Kaufman, L., Rousseeuw, P., Eds.; John Wiley & Sons, Inc.: Hoboken, NJ, USA, 1990.
40. Johnston, A.E. Soil organic matter, effects on soils and crops. *Soil Use Manag.* **1986**, *2*, 97–105. [CrossRef]
41. Pinter, P.J.J.; Hatfield, J.; Schepers, J.S.; Barnes, E.M.; Moran, S.M.; Daughtry, C.S.T.; Upchurch, D.R. Remote Sensing for Crop Management. *Photogramm. Eng. Remote Sens.* **2003**, *69*, 17. [CrossRef]
42. Basso, B.; Fiorentino, C.; Cammarano, D.; Cafiero, G.; Dardanelli, J. Analysis of rainfall distribution on spatial and temporal patterns of wheat yield in Mediterranean environment. *Eur. J. Agron.* **2012**, *41*, 52–65. [CrossRef]

43. Zhao, C.; Liu, B.; Xiao, L.; Hoogenboom, G.; Boote, K.J.; Kassie, B.T.; Pavan, W.; Shelia, V.; Kim, K.S.; Hernandez-Ochoa, I.M.; et al. A SIMPLE crop model. *Eur. J. Agron.* **2019**, *104*, 97–106. [CrossRef]
44. Donohue, R.J.; Lawes, R.A.; Mata, G.; Gobbett, D.; Ouzman, J. Towards a national, remote-sensing-based model for predicting field-scale crop yield. *Field Crop Res.* **2018**, *227*, 79–90. [CrossRef]

Publisher's Note: MDPI stays neutral with regard to jurisdictional claims in published maps and institutional affiliations.

Article

Within-Field Relationships between Satellite-Derived Vegetation Indices, Grain Yield and Spike Number of Winter Wheat and Triticale

Ewa Panek [1,*], Dariusz Gozdowski [1], Michał Stępień [1,2], Stanisław Samborski [3], Dominik Ruciński [4] and Bartosz Buszke [4]

1 Department of Biometry, Warsaw University of Life Sciences, Nowoursynowska 159, 02-776 Warsaw, Poland; dariusz_gozdowski@sggw.edu.pl (D.G.); michal1966@gmail.com (M.S.)
2 Complex of Agricultural Schools, Grzybno 48, 63-112 Brodnica, Poland
3 Department of Agronomy, Warsaw University of Life Sciences, Nowoursynowska 159, 02-776 Warsaw, Poland; stanislaw_samborski@sggw.edu.pl
4 Wasat Sp. z o.o., Office in Gdansk, Science and Technology Park, Trzy Lipy 3, 80-172 Gdansk, Poland; dominik.rucinski@wasat.pl (D.R.); office@wasat.pl (B.B.)
* Correspondence: ewa_panek@sggw.edu.pl

Received: 14 October 2020; Accepted: 19 November 2020; Published: 23 November 2020

Abstract: The aims of this study were to: (i) evaluate the relationships between vegetation indices (VIs) derived from Sentinel-2 imagery and grain yield (GY) and the number of spikes per square meter (SN) of winter wheat and triticale; (ii) determine the dates and plant growth stages when the above relationships were the strongest at individual field scale, thus allowing for accurate yield prediction. Observations of GY and SN were performed at harvest on six fields (three locations in two seasons: 2017 and 2018) in three regions of Poland, i.e., northeastern (A—Brożówka), central (B—Zdziechów) and southeastern Poland (C—Kryłów). Vegetation indices (Normalized Difference Vegetation Index (NDVI), Soil-Adjusted Vegetation Index (SAVI), modified SAVI (mSAVI), modified SAVI 2 (mSAVI2), Infrared Percentage Vegetation Index (IPVI), Global Environmental Monitoring Index (GEMI), and Ratio Vegetation Index (RVI)) calculated for sampling points from mid-March until mid-July, covering within-field soil and topographical variability, were included in the analysis. Depending on the location, the highest correlation coefficients (of about 0.6–0.9) for most of VIs with GY and SN were obtained about 4–6 weeks before harvest (from the beginning of shooting to milk maturity). Therefore, satellite-derived VIs are useful for the prediction of within-field cereal GY as well as SN variability. Information on GY, predicted together with the results for soil nutrient availability, is the basis for the formulation of variable fertilize rates in precision agriculture. All examined VIs were similarly correlated with GY and SN via the commonly used NDVI. The increase in NDVI by 0.1 unit was related to an average increase in GY by about 2 t ha^{-1}.

Keywords: winter wheat; winter triticale; vegetation indices; NDVI; grain yield; number of spikes

1. Introduction

Satellite remote sensing (RS) helps in the mapping of current crop status and the assessment of biophysical parameters. Currently, RS data are publicly accessible due to the availability of an unprecedented amount of Free Sentinel data from the Copernicus Program, established by the European Space Agency [1]. Constellations of the Sentinel-2 satellites (2A and 2B) can be also used for precision farming applications [2]. Thanks to the high spatial resolution (pixel size of 10 m), relatively short revisit time (about 5 days) and multispectral sensors, it is possible to observe and analyze the crop status using vegetation indices (VIs). For the purpose of such an evaluation, satellite Sentinel-2 images of two radiation bands, red (650–680 nm) and NIR (785–900 nm), are usually used [2].

Recently, many research results on yield evaluation based on data from the Sentinel-2 satellites [3–8] as well as from other satellite sources has been available. However, studies in which significant relationships between NDVI and the grain yield of cereals were proved have been carried out since the 1980s [3–6]. Most of these studies were done at a regional level and low-resolution satellite imagery was used (mainly using AVHRR/NOAA - Advanced Very High Resolution Radiometer/National Oceanic and Atmospheric Administration sensors with pixel sizes of 1 km).

Research on the evaluation of the relationships between VIs from high-resolution satellite sensors (e.g., QuickBird) and the grain yield and crop status of cereals has become more common since the year 2000 [7–10], due to the availability of new sources of satellite data. Such of studies were conducted at different spatial scales from one field scale (evaluation of within-field yield variability) to a whole country level. Many studies refer to Vis in relation to wheat grain yield because of the high importance of this crop. In most studies, Normalized Difference Vegetation Index (NDVI) [5] has been used as a predictor of grain yield. The highest accuracy of grain yield forecast was possible at different growth stages, even 2–3 months before harvest [11–16]. Recently, other VIs besides NDVI have been used as grain yield predictors. Ali et al. (2019) [17] studied the relationship between six vegetation indices (NDVI, EVI—Enhanced Vegetation Index, SAVI—Soil-Adjusted Vegetation Index, GNDVI—Green Normalized Vegetation Index, GCI—Green Chlorophyll Index and SR—Simple Ratio), derived from Landsat 5, 7 and 8 satellite imagery and a grain yield of durum and bread wheat in one 11 ha field in Italy, during five consecutive years (2010–2014). Most frequently, NDVI and SR were characterized as having the strongest relationship with yields.

Most of the studies that examined relationships between satellite-derived VIs and the grain yield of cereals were conducted at a regional level and did not analyze the relationships of VIs with other yield components such as the number of spikes. Nowadays, the common availability of satellite images of relatively high spatial resolution allows us to evaluate these relationships even at the within-field level. The results of such an evaluation can be useful for site-specific crop management, including variable fertilization, which should be based, besides soil nutrient availability, on the prediction of grain yields.

The aim of this study was to evaluate the relationships between vegetation indices derived from Sentinel-2 imagery and grain yield and the number of spikes per square meter of winter wheat and triticale. These two crops were selected because of their high cropping area among cereals in Poland. Recently, winter wheat has been grown on about 1.9 M hectares and winter triticale on 1.1 M hectares [18]. Moreover, the aim was to determine the dates and plant growth stages when the above relationships were the strongest at the individual field scale, thus allowing an accurate yield prediction.

2. Materials and Methods

2.1. Study Area

Three research sites cropped with winter wheat (*Triticum aestivum* L.) or winter triticale (*Triticosecale* Witt.) (Table 1) were located in Brożówka, northern Poland, in the land of the Great Masurian Lakes (location A), in Zdziechów, central Poland, in the Mazowieckie Voivodeship (location B), and in Kryłów, southeast Poland, close to the border with Ukraine (location C) (Figure 1).

Table 1. Characteristics of the studied fields in three locations and in two years.

		Location A—Brożówka		Location B—Zdziechów		Location C—Kryłów	
Year		2017	2018	2017	2018	2017	2018
Crop		winter wheat	winter triticale	winter wheat	winter triticale	winter wheat	winter wheat
Sowing date		20 September 2016	27 September 2017	27 September 2016	28 September 2017	22 September 2016	23 September 2017
Area of the field (ha)		14.9	14.9	9.7	7.8	9.7	9.6
Geographical coordinates		54°6′36″ N, 22°0′3.6″ E	54°6′36″ N, 22°0′3.6″ E	51°24′57.6″ N, 21°3′7.2″ E	51°25′1.2″ N, 21°3′10.8″ E	50°41′27.6″ N, 24°1′58.8″ E	50°42′32.4″ N, 24°3′10.8″ E
Soil WRB 2015 Reference Group-dominant (associated) *		Luvisols (Phaeozems, Histosols, Gleysols, Cambisols, Regosols)	Luvisols (Phaeozems, Histosols, Gelysols, Cambisols, Regosols)	Phaeozems (Luvisols, Arenosols)	Luvisols (Arenosols)	Gleysols	Luvisols
USDA soil texture class dominant (associated) *		sandy loam (loam, clay loam)	sandy loam (loam, clay loam)	sandy loam (loam, clay loam, loamy sand, sand)	loamy sand (sandy loam, sand)	silt loam (silty clay loam, silty clay)	silt loam (silty clay loam)
Number of sampling points of soil in spring/plots used at harvest for grain yield and spike number evaluation		16/18	18/36	10/12	12/24	10/12	12/21
Available elements in soil in a layer of 0–30 cm ($mg \cdot kg^{-1}$) ***	P	71.5 (92.0) **	52.2 (18.4)	55.4 (16.6)	118.0 (42.8)	119.5 (31.8)	79.8 (41.8)
	K	160.2 (104.6)	143.6 (44.6)	112.1 (27.4)	106.8 (19.4)	175.1 (32.4)	193.7 (38.6)
	Mg	109.7 (206.2)	150.8 (135.0)	67.5 (32.0)	69.7 (47.3)	54.9 (13.3)	78.8 (22.1)
pH		6.2 (0.8)	6.5 (0.7)	5.7 (0.5)	5.8 (0.5)	6.0 (0.4)	6.7 (0.3)

* Approximated soil data based on early spring field examination obtained during soil augerings done in locations of spike sampling (see Figure 3). **—means and standard deviations in parentheses. ***—using Mehlich 3 extract.

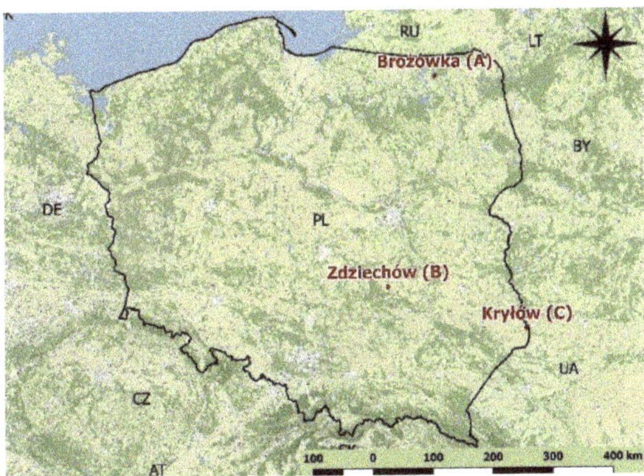

Figure 1. Location of the three research sites in Poland (Open Geospatial Consortium (OGC) Web Map Service).

According to the Köppen classification [19], the majority of Poland lies in the warm temperate climate zone with fully humid continental climate and hot summers (Cfb). The locations of Kryłów (C) and Zdziechów (B) are situated within this zone, while the location of Brożówka (A) is placed in the transitional area between the Cfb zone and the snowy climate zone, with fully humid and warm summers (Dfb).

The sum of precipitation in two studied seasons (from the beginning of September to the end of August) ranged from about 520 mm (location C, season 2016/2017) to 740 mm (location A, season 2016/2017). The lowest air temperatures were observed for location A, while locations B and C were characterized by similar air temperatures. Growing degree days calculated at a base air temperature of 5 °C for the period from the sowing date to the end of July were much lower in 2017 (for locations: A—1364, B—1531, and C—1492) in comparison to 2018 (for locations: A—1795, B—1975, and C—1930). Monthly sums of precipitation and mean air temperatures for the nearest meteorological stations from the Institute of Meteorology and Water Management National Research Institute IMGW-PIB (https://danepubliczne.imgw.pl/) are presented in the Figure 2.

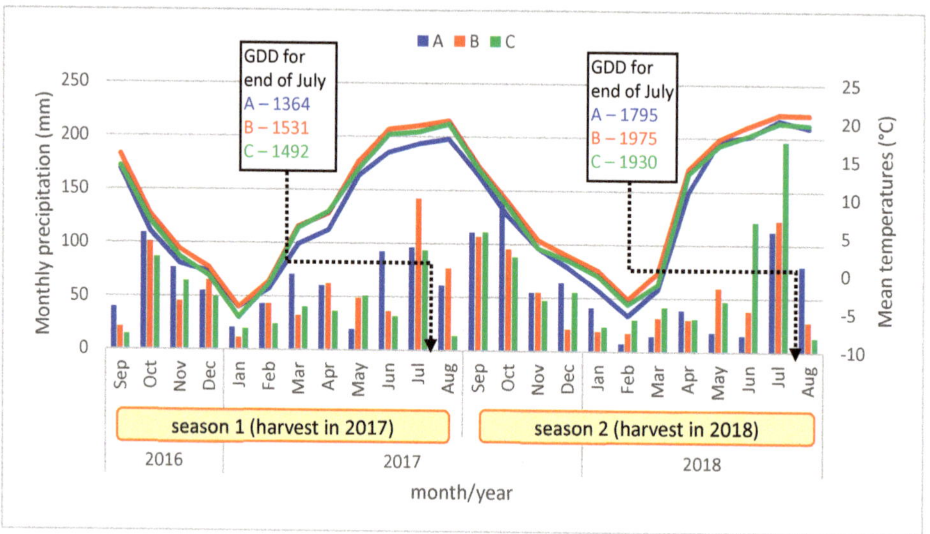

Figure 2. Monthly sums of precipitation, mean air temperature and growing degree days (GDD) for meteorological stations located closest to the research sites (for location A—Olecko, for location B—Puławy, for location C—Strzyżów) from the IMGW-PIB national institute (source of data: https://danepubliczne.imgw.pl/).

2.2. Soil and Plant Sampling

Within each of the fields, sampling points of soil and plants were georeferenced using a Global Positioning System (GPS) receiver. The locations of the sampling points were selected to obtain large variability in the studied traits, i.e., soil properties as well grain yield and ear number (Figure 3). The characteristics of each of the three research sites are presented in Table 1.

Soil samples were collected from a layer of 0–90 cm at the end of winter or in early spring before the starting of the vegetation. Grain yield and number of ears per square meter were determined at the full maturity of plants in plots of 2 m^2 (2 × 1 m^2 subplots, each located very closely to the soil and plant sampling points). The following statistical parameters were calculated: average, standard deviation, minimum, and maximum for each field in a given year.

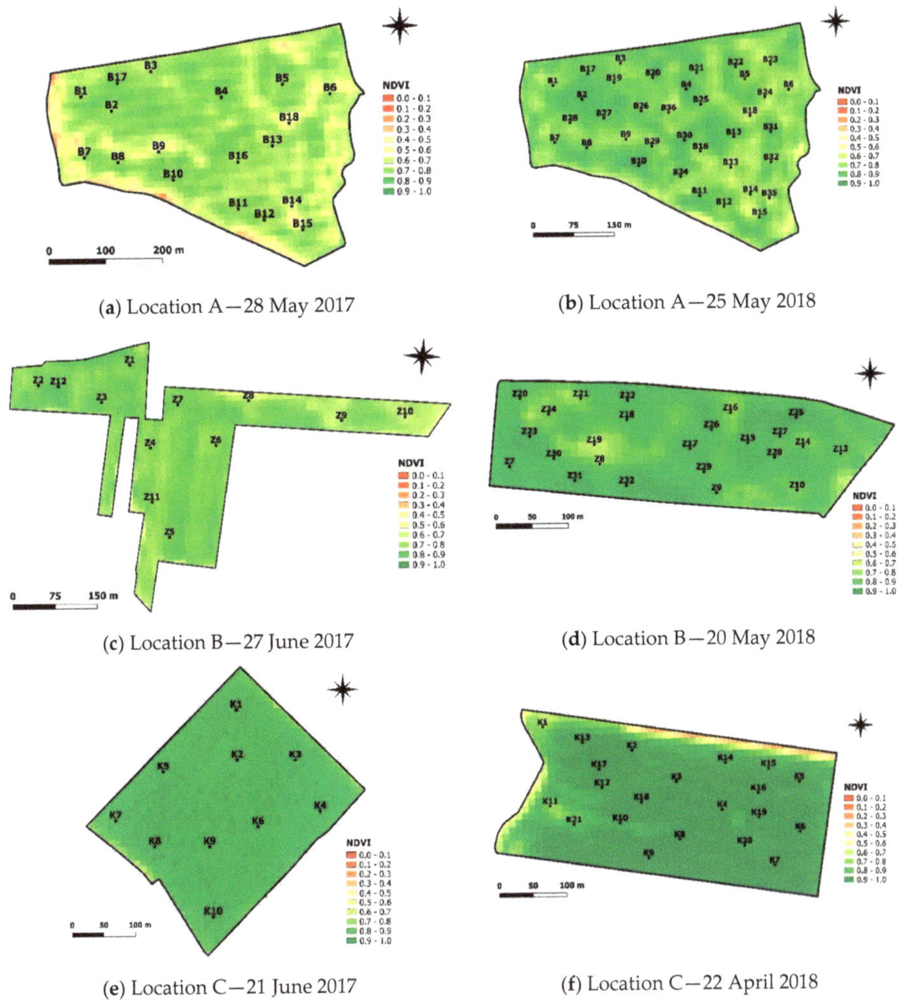

(a) Location A—28 May 2017
(b) Location A—25 May 2018
(c) Location B—27 June 2017
(d) Location B—20 May 2018
(e) Location C—21 June 2017
(f) Location C—22 April 2018

Figure 3. Normalized Difference Vegetation Index (NDVI) maps of the experimental fields in three locations ((**a**,**b**): Brożówka—A, (**c**,**d**): Zdziechów—B, and (**e**,**f**): Kryłów—C) in two seasons: 2017 (on the left) and 2018 (on the right), from dates when maximum average value of this index was achieved, with soil and plant sampling points superimposed. Explanation of color representation of NDVI values and the dates of NDVI acquisition are given.

2.3. Satellite Data

Satellite Sentinel-2 (2A and 2B) images of Level-1C with a spatial resolution of 10 to 20 m in the range of visible and near-infrared (NIR—band 8 at central wavelength 833 nm) light were downloaded from the Copernicus Open Access Hub [20]. Level-1C processing includes radiometric and geometric corrections, namely ortho-rectification and spatial registration, on a global reference system with a sub-pixel accuracy. Then, the acquired Level 1C products were subjected to atmospheric correction using the Dark Object Subtraction (DOS1) method in Semi-Automatic Classification Plugin documentation (SCP), developed by Luca Congedo (2016) [21].

Vegetation indices were derived for the representative soil and plant sampling points established within fields and imported together with the coordinates of these points. Then, they were compiled into tables and subjected to regression analyses. Depending on the location, despite the revisit time of about 5 days, it was only possible to acquire cloudless satellite images during the intensive growth of cereals in spring and early summer for 4–6 dates (Table 2).

Table 2. Dates for which satellite-based vegetation indices were derived from Sentinel images for three locations and two seasons.

Location A Brożówka		Location B Zdziechów		Location C Kryłów	
2017	**2018**	**2017**	**2018**	**2017**	**2018**
16 March	19 March	29 March	26 March	2 April	7 April
8 April	8 April	18 May	20 May	22 April	22 April
25 April	23 April	4 June	4 June	21 June	21 June
28 May	25 May	27 June	29 June	11 July	11 July
27 June	27 June	12 July	12 July	-	-
14 July	19 July	-	-	-	-

2.4. Spectral Vegetation Indices

The selection of the vegetation indices (VIs) used for the winter wheat and triticale yield analysis depended on several factors. First of all, we only considered the same satellite data spatial resolution. Another factor determining our choices was the prevalence in the literature and the comparability with the yield of cereal crops. The VIs were calculated on the basis of the reflectance of wavelength bands registered by Sentinel-2 at spatial resolution 10 m (squared pixels 10 × 10 m). The Normalized Differential Vegetation Index (NDVI) allows us to determine the indirect absorption of photosynthetic radiation on a landscape scale. NDVI calculation is based on the contrast between the largest reflection in the near-infrared band and the absorption in the red band. In the case of this index, the difference in reflection in the near-infrared and red bands is divided by their sum. This approach compensates for the differences in the radiation amounts in both bands. It is believed that NDVI is more sensitive to small amounts of vegetation [22,23].

$$\mathrm{NDVI} = \frac{\mathrm{NIR} - \mathrm{RED}}{\mathrm{NIR} + \mathrm{RED}}$$

Another VI used in this study was the Soil-Adjusted Vegetation Index (SAVI). This indicator was designed to minimize the effect of soil reflection on red and near-infrared radiation by adding an estimated background correction factor [24].

$$\mathrm{SAVI} = \frac{(1+\mathrm{L})(\mathrm{NIR} - \mathrm{RED})}{\mathrm{NIR} + \mathrm{RED} + \mathrm{L}}$$

where L is a canopy background adjustment factor. An L value of 0.5 was adopted to minimize soil brightness variations and eliminate the need for additional calibration for different soils.

In the literature, modifications of the SAVI index, known as modified SAVI (mSAVI) and modified SAVI 2 (mSAVI2), are often used. These two VIs also include the soil background correction factor [25]:

$$\mathrm{mSAVI} = \frac{(1+\mathrm{L})(\mathrm{NIR} - \mathrm{RED})}{\mathrm{NIR} + \mathrm{RED} + \mathrm{L}}$$

where L is:

$$\mathrm{L} = 1 - \frac{2 \times s \times (\mathrm{NIR} - s \times \mathrm{RED})(\mathrm{NIR} - \mathrm{RED})}{\mathrm{NIR} + \mathrm{RED}},$$

and:

$$\mathrm{mSAVI2} = \frac{(2 \times \mathrm{NIR} + 1 - \sqrt{(2 \times \mathrm{NIR} + 1)^2 - 8 \times (\mathrm{NIR} - \mathrm{RED})}}{2}$$

Crippen (1990) proposed the Infrared Percentage Vegetation Index (IPVI), which simplifies the calculation by eliminating the subtraction of the red radiation value in the NDVI indicator counter. This simplification of a new VI calculation proved to be important for fast image processing [26].

$$\text{IPVI} = \frac{\text{NIR}}{\text{NIR} + \text{RED}}$$

In 1992, Pinty and Verstraete proposed the Global Environmental Monitoring Index (GEMI), a new nonlinear crop status indicator, which includes a correction factor adjusted by the effect of soil reflectance and atmospheric background [27].

$$\text{GEMI} = n(1 - 0.25n) - \frac{\text{RED} - 0.125}{1 - \text{RED}}$$

where:

$$n = \frac{\left[2 \times \left(\text{NIR}^2 - \text{R}^2\right) + 1.5\text{NIR} + 0.5\text{R}\right]}{\text{NIR} + \text{RED} + 0.5}$$

The last VI used in this study for a comparison with NVDI was the Ratio Vegetation Index (RVI), sometimes also referred to as the Simple Ratio (SR). This VI is a ratio of NIR and red reflectance [28].

$$\text{RVI} = \frac{\text{NIR}}{\text{RED}}$$

The investigated Vis, except RVI, are normalized and placed on a comparative scale. These indices have values from −1 to 1, where values close to −1 indicate water, or any inanimate matter, values from 0 to 0.20 relate to barren rock, sand, barren soil and plants at an early stage of growth, values of 0.20–0.50 relate to sparse vegetation such as shrubs and grasslands, while values above 0.60 indicate crops at their peak growth stages.

2.5. Statistical Analysis

The Orfeo Toolbox (OTB Development Team, 2018) [29] was used for the calculation of vegetation indices for pixels located nearest to the soil and plant sampling points in 2017 and 2018. The indices was subjected to statistical analysis and the mean, standard deviation, coefficient of variation (CV), minimum, maximum were determined (Tables 3 and 4). Relationships between VI values and grain yield as well as the number of spikes per square meter were evaluated using Pearson's correlation and a linear regression analysis. These analyses were conducted for the dates from mid-March to mid-July (Table 2). The choice of dates remained strictly dependent on the availability of cloudless Sentinel-2 satellite images. Depending on location, there were 4 to 6 dates (obtained, on average, every 3–4 weeks) in which the images were available without clouds.

Table 3. Grain yield (t ha^{-1}) and number of spikes per m^2 m in sampling points at harvest in Brożówka (A), Zdziechów (B), and Kryłów (C) in 2017 and 2018.

	Grain Yield (t ha^{-1})					Spikes Number Per m^2				
Location and Year	**Mean**	**SD**	**CV**	**Min.**	**Max.**	**Mean**	**SD**	**CV**	**Min.**	**Max.**
A 2017	8.33	1.77	21.2%	3.62	10.1	458	91.5	20.0%	240	589
A 2018	5.10	1.63	32.0%	1.80	8.26	317	43.3	13.7%	230	397
B 2017	5.07	1.99	39.3%	0.93	7.08	415	95.4	23.0%	205	533
B 2018	5.01	1.86	37.1%	0.98	8.79	389	77.5	19.9%	231	582
C 2017	10.5	0.63	6.0%	9.43	11.6	739	34.0	4.6%	676	801
C 2018	8.63	1.38	16.0%	4.67	10.4	514	62.0	12.1%	348	593

Table 4. Mean values and ±SD of vegetation indices for the dates when the highest NDVI values were observed.

Location and Date	NDVI	SAVI	mSAVI	mSAVI2	IPVI	GEMI	RVI
A 28 May 2017	0.84 ± 0.06	0.51 ± 0.05	0.51 ± 0.05	0.59 ± 0.07	0.92 ± 0.03	0.83 ± 0.04	12.95 ± 3.93
A 25 May 2018	0.74 ± 0.07	0.49 ± 0.05	0.44 ± 0.05	0.49 ± 0.07	0.87 ± 0.03	0.78 ± 0.04	7.31 ± 2.25
B 27 June 2017	0.90 ± 0.01	0.42 ± 0.03	0.65 ± 0.03	0.35 ± 0.03	0.39 ± 0.04	9.38 ± 1.30	0.80 ± 0.03
B 20 May 2018	0.80 ± 0.06	0.52 ± 0.06	0.47 ± 0.07	0.53 ± 0.08	0.90 ± 0.03	0.80 ± 0.06	9.88 ± 2.78
C 21 June 2017	0.86 ± 0.02	0.62 ± 0.08	0.57 ± 0.09	0.63 ± 0.01	0.90 ± 0.08	0.85 ± 0.09	11.52 ± 3.39
C 22 April 2018	0.86 ± 0.07	0.60 ± 0.06	0.55 ± 0.07	0.64 ± 0.09	0.93 ± 0.03	0.87 ± 0.05	15.39 ± 3.99

The correlation coefficient values for each date are presented over time in a graphical form for better visualization of the relationship change between NDVI and grain yield (Tables 5 and 6). For periods when maximal values of the correlation coefficients were observed, the diagrams of linear regression are presented (Figures 5 and 6), together with regression equations and coefficients of determination (R^2).

Spatial analyses were conducted, and maps were prepared using QGIS 2.18 software (QGIS Development Team, Gossau, Switzerland) [30], while the statistical analyses were performed using GNU R and TIBCO Statistica software.

Table 5. Correlation coefficients (r) between NDVI values and grain yield ($t{\cdot}ha^{-1}$) for three locations in 2017 and 2018. The background color of the cells with the correlation coefficient values indicates the strength of the relationship (significant correlations at $\alpha = 0.05$ are given in bold).

Location	Date (2017)	Growth Stage *	r	Location	Date (2018)	Growth Stage	r
A	16-March	tillering	0.421	A	19-March	tillering	0.254
	8-April	tillering	**0.493**		8-April	tillering	0.241
	25-April	tillering/shooting	0.341		23-April	tillering/shooting	**0.389**
	28-May	shooting/heading	**0.587**		25-May	shooting/heading	**0.622**
	27-June	milk maturity	0.441		27-June	milk maturity	**0.570**
	14-July	dough maturity	0.239		19-July	dough maturity	−0.012
B	29-March	tillering	−0.457	B	26-March	tillering	0.064
	18-May	shooting/heading	0.154		20-May	shooting/heading	**0.859**
	4-June	heading/flowering	**0.889**		4-June	heading/flowering	**0.790**
	27-June	milk maturity	**0.840**		29-June	milk maturity	**0.804**
	12-July	dough maturity	**0.693**		12-July	dough maturity	−0.152
C	2-April	tillering	−0.423	C	7-April	tillering	**0.713**
	22-April	tillering/shooting	0.067		22-April	tillering/shooting	**0.795**
	21-June	milk maturity	0.594		21-June	milk maturity	**0.899**
	11-July	dough maturity	0.774		11-July	dough maturity	0.361

−1.0 −0.9 −0.8 −0.7 −0.6 −0.5 −0.4 −0.3 −0.2 −0.1 0.1 0.2 0.3 0.4 0.5 0.6 0.7 0.8 0.9 1.0

very strong negative correlation — very strong positive correlation

*—approximated growth stages of the whole field crop.

Table 6. Correlation coefficients (r) between NDVI values and number of spikes per square meter for three locations in 2017 and 2018. The background color of the cells with the correlation coefficient values indicates the strength of the relationship (significant correlations at $\alpha = 0.05$ are given in bold).

Location	Date (2017)	Growth Stage *	r	Location	Date (2018)	Growth Stage	r
A	16-March	tillering	0.452	A	19-March	tillering	0.102
	8-April	tillering	**0.535**		8-April	tillering	0.120
	25-April	tillering/shooting	0.372		23-April	tillering/shooting	0.167
	28-May	shooting/heading	**0.603**		25-May	shooting/heading	**0.508**
	27-June	milk maturity	**0.471**		27-June	milk maturity	**0.485**
	14-July	dough maturity	0.288		19-July	dough maturity	−0.096
B	29-March	tillering	−0.341	B	26-March	tillering	0.068
	18-May	shooting/heading	0.159		20-May	shooting/heading	**0.685**
	4-June	heading/flowering	**0.822**		4-June	heading/flowering	**0.625**
	27-June	milk maturity	**0.786**		29-June	milk maturity	**0.728**
	12-July	dough maturity	0.529		12-July	dough maturity	−0.064
C	2-April	tillering	0.025	C	7-April	tillering	**0.731**
	22-April	tillering/shooting	−0.162		22-April	tillering/shooting	**0.785**
	21-June	milk maturity	**0.649**		21-June	milk maturity	**0.647**
	11-July	dough maturity	**0.592**		11-July	dough maturity	0.275

−1.0	−0.9	−0.8	−0.7	−0.6	−0.5	−0.4	−0.3	−0.2	−0.1	0.1	0.2	0.3	0.4	0.5	0.6	0.7	0.8	0.9	1.0
very strong negative correlation																			very strong positive correlation

*—approximated growth stages of the whole field crop.

3. Results

3.1. Grain Yield and Spike Number

The highest grain yield as well the highest number of spikes per m^2 was observed for location C (Kryłów) in both years (Table 3). This was mainly due to the very favorable soil conditions where the research sites were located in both years in (Table 1). In 2017, soil conditions were very good and uniform within the very flat field (denivelation of less than 2 m). This caused very low yield variability (SD = 0.63) within this field. In 2018, soil conditions were slightly less favorable and more variable, but still quite uniform in comparison to soil conditions in the other two locations (A and B). In location A (Brożówka), the average grain yield of winter wheat was much higher (8.33 t·ha^{-1}) in 2017 than the yield of winter triticale in 2018 (5.10 t·ha^{-1}). The main reason for the much lower grain yield as well as the lower spike number per m^2 (458 spikes in 2017 versus 317 in 2018) was the shorter tillering time in 2017/2018, related to the later sowing date of winter triticale when compared to winter wheat (2016/2017). Moreover, we must mention that winter wheat is usually sown later than winter triticale. In both years in location A, yield variability within this field was quite high, mainly due to the undulated surface of the field (denivelation of about 20 m), which determined the soil variability. In location B, in both years, a grain yield of about 5 t ha^{-1} and a spike density of about 400 spikes per m^2 were relatively low and very variable within the field. The main reason for such high within-field variability of the grain yield traits was the variable soil texture (Table 1), which caused varied water availability for plants. A shortage of water usually occurred in later growth stages, especially during heading and grain filling. This caused very low-weight grains and, despite quite large numbers of spikes, grain yield was very low. Such a situation was especially visible in the sandy parts of the fields in location B.

3.2. Changes in NDVI over the Vegetation Season and in Research Locations

Due to cloud cover, the availability of useful satellite images from Sentinel-2 was limited to 4–6 scenes per season during intensive growth of winter cereals, i.e., from half of March to half of July. Despite this limitation, it was possible to evaluate changes in VI values during the growing seasons. The most substantial differences between the average and a range of NDVI values for the three locations were observed from the end of March to the beginning of April. For location A, mean values of NDVI in both years were very low (0.3–0.4) until the beginning of April. The main cause of these low NDVI values in location A were lower air temperatures and consequently growing degree days (from sowing date to the end of April in 2017: location A—283, B—298, and C—451; in 2018: A—415; B—501, and C—605) during early spring than in the other two locations (Figure 2). The highest values of NDVI (of about 0.7 at the beginning of April) were observed in location C. This was caused by warmer conditions in both autumn and early spring as well as by more favorable soil conditions, which caused more intensive plant growth. The very high values of NDVI (0.7 or higher) in location C were observed for a much longer time, even until the end of June, in comparison to the other two locations. Consequently, this longer and more intensive crop growth in location C allowed us to obtain higher grain yields. In location B, NDVI values in early spring were at a medium level (higher than in location A and lower in comparison to location C) and very variable depending on the season at the later growth stages. In all locations, in 2017, a maximum NDVI value was achieved in very late June and much earlier in 2018, at the end of April (Figure 4, Table 4).

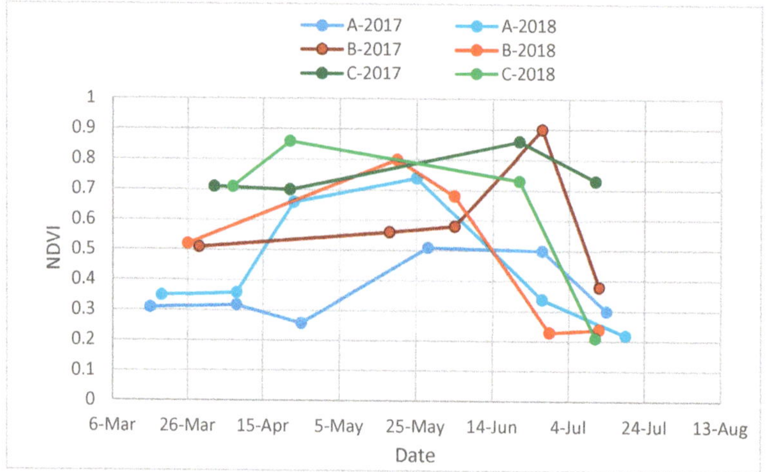

Figure 4. Mean values of NDVI values for six fields (three locations: A—Brożówka, B—Zdziechów, and C—Kryłów in two years: 2017 and 2018).

3.3. Relationships between NDVI and Grain Yield and Spike Number

One of the aims of the study was to determine the dates and plant growth stages when the relationships between VI values and grain yield, as well as number of spikes, were the strongest. The strongest correlations were achieved at various growth stages (Tables 5 and 6).

For fields located in northeastern Poland, the highest correlation coefficients between grain yield and as well as number of spikes and NDVI were observed at the end of May (approximately shortly before or during the heading stage) in both years. On the basis of the regression slope values, it can be concluded that an increase in NDVI by 0.1 unit corresponded to the increase in grain yield by about 1.5–1.6 t·ha^{-1} in both years (Figures 5 and 6).

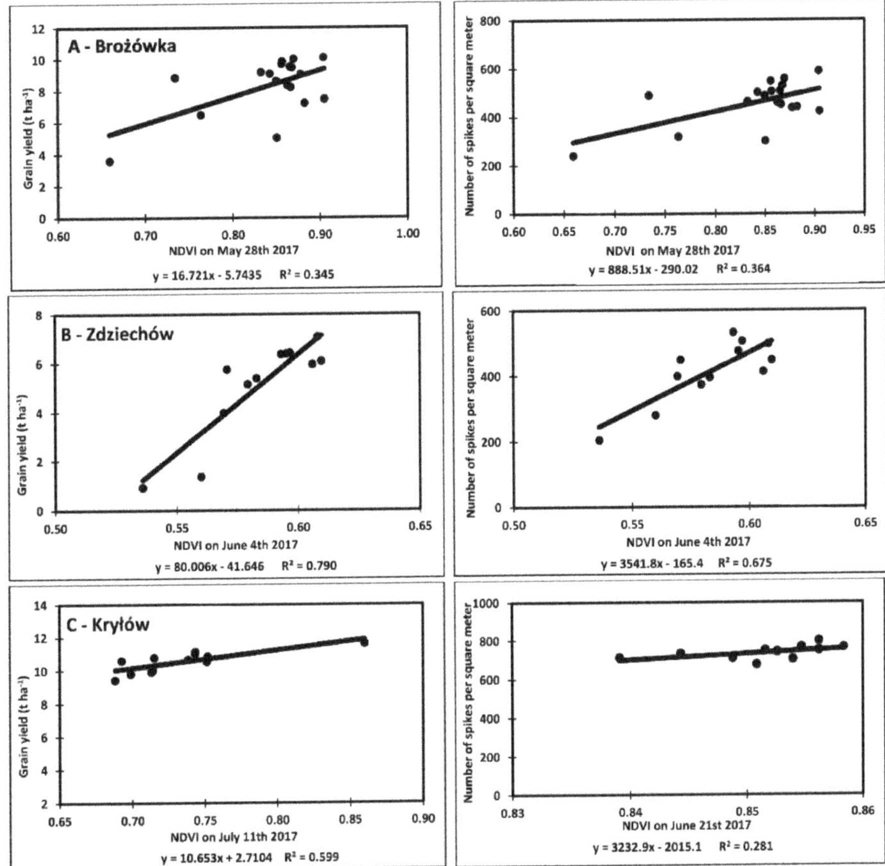

Figure 5. Relationships between grain yield (charts in the left side), number of spikes per square meter (charts in the right side) and NDVI values for three locations (A—Brożówka, B—Zdziechów, C—Kryłów) for the dates when the correlations reached maximum values in 2017. Regression equations and the value of the R^2 coefficient are given for each relationship.

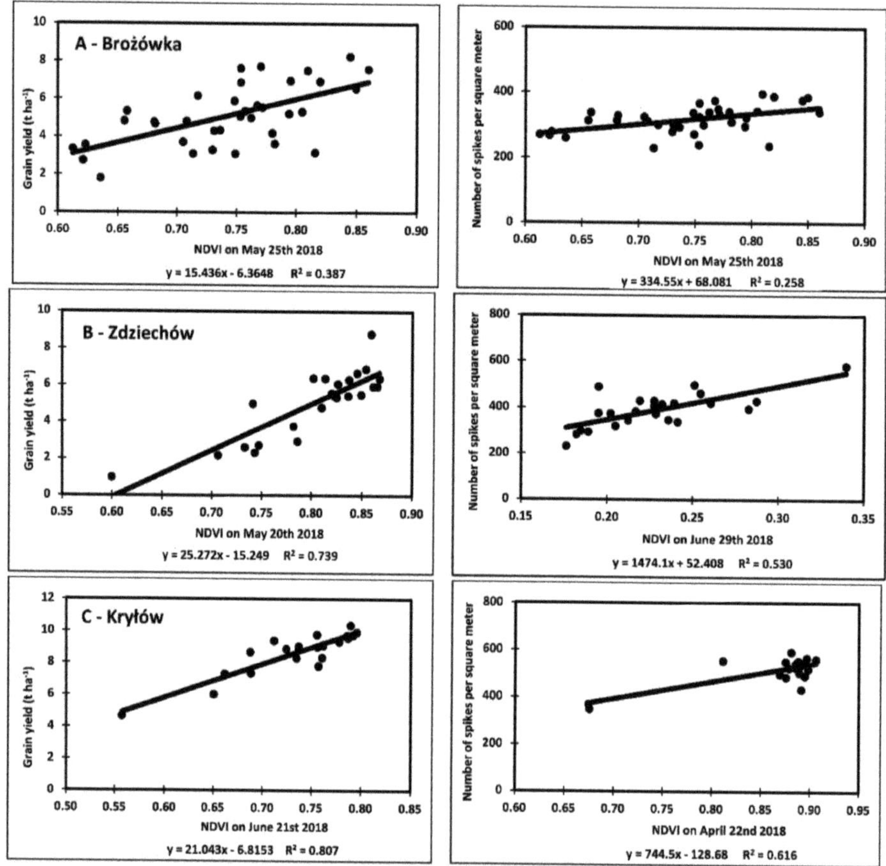

Figure 6. Relationships between grain yield (charts in the left side), number of spikes per square meter (charts in the right side) versus NDVI for three locations (A—Brożówka, B—Zdziechów, C—Kryłów) for the date when their correlations reached maximum values in 2018. Regression equations and the value of the R^2 coefficient are given for each relationship.

In location B, the strongest relationship between NDVI and grain yield was registered in late June of 2017 (end of milk maturity of winter wheat) and in the second half of May 2018 (shooting and heading of winter triticale). In this research site, the relationships were the strongest among the three examined locations, i.e., the R^2 value was about 0.70–0.75 (Figures 6 and 7). Moreover, the relationship between NDVI and the number of spikes was slightly weaker. The increase in NDVI by 0.1 unit was related to an increase in grain yield of about 8.0 t·ha^{-1} in 2017 and about 2.5 t·ha^{-1} in 2018.

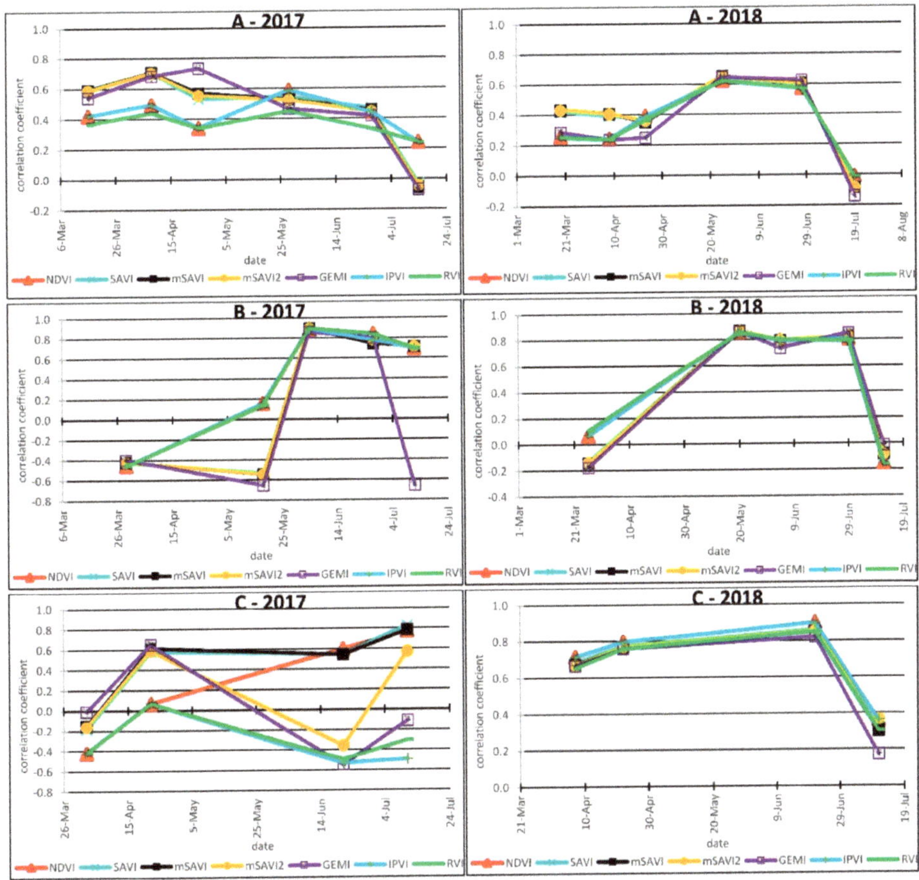

Figure 7. Values of correlation coefficients between vegetation indices and grain yield (t·ha^{-1}) in three locations (A—Brożówka, B—Zdziechów, C—Kryłów) on the selected dates in years 2017 (charts in the left side) and 2018 (charts in the right side).

In location C in 2017, the strongest correlation between NDVI and both grain yield and number of spikes was observed, respectively, on 21 June and 11 July (milk and dough maturity). In 2018, the strongest relationships between NDVI and both grain yield and spike number per square meter were obtained at the end of April and June. Due to a lack of data (clouds), it was not possible to verify this relationship at the end of May. The relationship was much stronger in 2018 than in 2017, mainly due to the considerably higher within-field variability of soil and consequently grain yield variability (Coefficient of Variation—CV) for grain yield was 6% and 16%, respectively, for 2017 and 2018). Based on the regression equations (Figures 5 and 6), it was evaluated that an increase in NDVI by 0.1 unit was related to the increase in grain yield by about 1.0 t·ha^{-1} in 2017 and about 2.1 t·ha^{-1} in 2018. For all six site years, the strongest correlation between NDVI and grain yield and number of spikes was observed on similar dates and at similar growth stages. This is because the number of spikes per square meter usually very strongly affects grain yield.

3.4. Relationships between the Other Vegetation Indices and Grain Yield and Spike Number

Regression analyses performed to evaluate the strength of the relationship between the vegetation indices (SAVI, mSAVI, mSAVI2, GEMI, IPVI, RVI) and grain for each crop in 2017 and 2018 (Figure 7) indicated that the strongest relationships were observed for most of the VIs on the same dates. In location A, the highest correlations were observed at the end of May (end of shooting stage) in 2017, and from late May to mid-June (heading stage) in 2018. In location B, in 2017, the value of the correlation coefficients differed for individual VIs on several dates, while, in 2018, the strongest relationships were achieved in mid-May (beginning of heading stage) and at the end of June (milk maturity stage) for all VIs. The highest differentiation in the correlation coefficient values for the relationship between various VIs and yields was observed at the end of June and beginning of July (milk maturity stage) of 2017 in location C. In 2018, in all locations, the correlation coefficients had similar values for all VIs across the whole season.

The vegetation indices (VI) are ranked by order of decreasing correlation (data not shown) with yield as follows: SAVI, MSAVI, NDVI, MSAVI2, IPVI, RVI, GEMI. This ranking was carried out by averaging the R^2 values from all of the measurement dates and research areas. The strength of the correlation of VIs with the number of ears was evaluated in the same way as above, and the order of decreasing correlation was as follows: NDVI, SAVI, MSAVI, MSAVI2, IPVI, RVI, GEMI.

4. Discussion

Constant field monitoring using RS methods is important not only for grain yield prediction, but also for the assessment of the site-specific conditions of plant development during the growing season [31,32]. This may allow us to introduce site-specific management of the field, and thus increase yields or save inputs.

4.1. Relationships between NDVI and Grain Yield

Our results for winter cereals, which constitute the majority of crop production in Central Europe, proved moderate and strong relationships (R^2 in range of 0.35–0.81 for all fields in two seasons, Figures 5 and 6) between the NDVI and grain yield of winter wheat and triticale for all of the studied fields. At location A, the correlation coefficient grew from mid-March to the end of May, when it reached the highest value. From the beginning of June to mid-July, we noticed a decrease in this correlation. The highest correlation coefficients are observed from the beginning of June to mid-July due to the growth stages such as heading, milk maturity, or dough maturity for location B. At the site C, we can see high rates of milk maturity at the end of June. A similar tendency, though with lower correlation values, can be observed when comparing the number of spikes per square meter. Similar results were achieved by Benedetti and Rossini (1993) [4] in Italy and by Smith et al. (1995) [5] in Western Australia for wheat. In both studies, AVHRR/NOAA satellite data were used for the calculation of NDVI. In Italy, R^2 coefficient values for the relationship between NDVI and grain yield ranged from 0.24 to 0.749, and for Australia they ranged from 0.46 to 0.72. Ali et al. (2019) [17] observed higher R^2 values of 0.878 and 0.926 for the relationship between the grain yield of winter and bread durum wheat and NDVI when the values of this index were derived from Landsat images. In turn, Labus et al. (2002) [6] achieved low R^2 values ranging from 0.00 to 0.69 when wheat yield prediction was based on NDVI derived from AVHRR/NOAA satellite data. In our study, very low R^2 values were obtained in April and May, and higher R^2 values were mainly close to the heading stage and at the end of the growing season (Tables 5 and 6). Therefore, while comparing many different studies, we can expect that, regardless of the research region, the relationship between NDVI and the grain yield remains at a similar level depending on the growth stage of the studied cereal.

4.2. Determination of Dates and Plant Growth Stages When Relationship between NDVI and Grain Yield Was the Strongest

Veloso et al. (2017) [10] found, for southeastern France, that NDVI reached maximum values in the second half of May, while, in Poland, depending on the research location, the highest NDVI values were reached from the end of April to the end of June (Figure 4). The site-related differences in the absolute NDVI values and the date at which they reached their maximum may primarily be influenced by the environmental conditions, mainly weather and soil conditions. This, in turn, closely depends on the climate in a specific region (amount of rainfall, growing degree days) and the amount of nitrogen applied. In the study of Naser et al. (2020) [33], much stronger correlations between NDVI, measured using a ground sensor, and the grain yield of winter wheat were observed for dryland than for irrigated conditions. The reason for such results was the greater range of absolute NDVI values for dryland in comparison to irrigated conditions. In that study, the highest correlations were observed after anthesis for both types of conditions. Very strong relationships between NDVI, measured using a ground spectrometer, and wheat yield were observed in a study where the effect of various nitrogen doses (from 0 to 210 kg/ha) was investigated [34]. The strongest correlations (R^2 up to 0.96) were achieved at the heading stage. According to Satir and Berberoglu (2016) [35], the strength of the relationship between satellite-derived NDVI and the yields of wheat, corn and cotton was modified by soil conditions, e.g., soil salinity. This is because soil surface and soil water content can affect the values of vegetation indices, as well as the fact that water availability for plants is limited in high-salinity conditions. A better prediction of grain yield in the variable soil conditions of south Turkey can be achieved not only using prediction models such as NDVI, but also other spectral indices such as Normalized Difference Water Index (NDWI), Soil-Adjusted Vegetation Index (SAVI) or Tasseled Cap Wetness Index (WETNESS). The highest prediction accuracy for wheat was using the model which includes NDVI, NDWI and WETNESS. Prediction of wheat yield based on satellite-derived NDVI was improved if auxiliary data such as grain yield from previous seasons were included in models [36]. Dempewolf et al. (2013) [12] obtained the highest R^2 values of 0.964 for NDVI and the yield of wheat grown in Pakistan six weeks before harvest. In our studies, the highest R^2 value was observed at the end of April, i.e., about 12 weeks before harvest in the case of the field located in southeastern Poland. These results are similar to those of Kussul et al. [13] (2014), obtained in Ukraine, where a good prediction of winter wheat yields was possible even 2–3 months before harvest. The latest date (beginning of July) when the correlation coefficient for the relationship yield versus NDVI was the highest was observed in location B (Zdziechów) in central Poland in 2017. Similar results were obtained in Hungary by Nagy et al. (2018) [15] and Lopresti et al. (2015) [11] in Argentina, where good yield prediction was possible six weeks before harvest. Ali et al. (2019) [17] observed the highest correlation between NDVI and yield for winter bread wheat from stem elongation to the milk maturity of the grain, i.e., 6–17 weeks before harvest. In summary, the choice of the date of the yield estimation with the best accuracy does not depend only on long-term field and satellite measurements. Environmental, weather, and soil conditions should also be taken into account. As shown above, a yield estimation can be properly performed even between 3 months and 6 weeks before harvest, depending on the research region and the field-specific conditions.

4.3. Relationship between Other VIs and Grain Yield

According to Purevdorj et al. (1998) [37], SAVI and its modifications usually show higher correlations with grain at the beginning of the growing season compared to NDVI. According to Ren et al. (2018) [38], SAVI, with its modifications, should be used to estimate vegetation coverage at low vegetation density on arable land. In our study, the value of the correlation coefficient for the relationship between SAVI and grain yield was similar to the relationship with NDVI for both seasons and locations. Moreover, only in locations A and C in 2017 were the correlations between SAVI and grain yield stronger in early spring. For location A, during two vegetative seasons, we saw an increase in the correlation coefficient at the beginning of the growing season and a decrease in the cereals' subsequent growth stages. For location B, the correlation coefficient increases until the beginning of June and slightly decreases until the end of July. In tested area C, an increase in the correlation coefficient value is noticed until mid-June, followed by a sharp decline by the end of July.

Ali et al. (2019) [17] selected NDVI and SR (referred to RVI in our study) as better VIs in comparison to EVI, SAVI, GNDVI, and GCI. Their conclusion was that SAVI never exhibited the strongest correlation with yield in comparison with other VIs, although, its correlation with grain yields was, most frequently, significant and strong. On the contrary, our research results proved that SAVI (and its modifications) and NDVI are ranked as VIs with the highest correlations with yield and number of spikes.

Moreover, Ali et al. (2019) [17] obtained R^2 values of about 0.8 for durum and bread wheat yield and NDVI, EVI, and SR. We achieved Pearson's correlation values of 0.4 between winter wheat and triticale yield and SAVI, but 0.3 with mSAVI, NDVI, IPVI, RVI (SR), and GEMI. These differences most probably arise from a different method being used for the collection of yield data.

5. Conclusions

Various vegetation indices (SAVI, MSAVI, NDVI, MSAVI2, IPVI, RVI and GEMI), their calculation based on the red and near-infrared radiation derived from the Sentinel−2 imagery, showed similar relationships with grain yield and number of spikes per square meter of winter wheat and triticale as the commonly used NDVI. While comparing the grain yield and the number of spikes from NDVI for all three locations during two seasons, it can be seen that the values of R^2 for the number of spikes per square meter are lower than the values of R^2 for the grain yield. Consequently, the grain yield is a better yield prediction parameter than number of spikes per square meter. In general, the relationship between NDVI values and grain yield was stronger at more advanced growth stages. Depending on the region of Poland, the strongest correlations between NDVI and yield and its main component were obtained for NDVI derived from images obtained at the end of April (beginning of shooting) in southeastern Poland, at the end of May (beginning of heading) in the northeastern part of the country and only at the end of June (at milk maturity) in the central region of Poland. The divergence in these periods may result from the sowing date, but also from different climatic conditions, especially GDD values, in the three micro-climate regions of Poland. The different strengths of the relationships may be caused by the soil water deficiency due to variable weather and soil conditions, depending on the research location, at different growth stages. Within-field water variability was higher, especially on sandy soils (which were common in some parts of the field in location B) at later growth stages when air temperatures were higher and negative balance between rainfall and evapotranspiration was more common than in locations A and C [39]. Moreover, the presence of clouds, which made the registration of useful satellite images impossible on some dates, also contributed to the results obtained in this study. For example, in location C in southeastern Poland, there were no cloudless images available in May in both seasons and, because of this, it was not possible to evaluate the relationships between VIs and yield for some important growth stages of the crop. The proposed estimated time for accurate yield prediction is about 4–6 weeks before harvest (from the beginning of shooting to milk maturity).

Author Contributions: Conceptualization, E.P., D.G.; methodology, E.P., D.G., M.S., formal analysis, E.P., D.G., M.S.; investigation, E.P., D.G., M.S., S.S.; data curation, E.P., D.G.; writing—original draft preparation, E.P., D.G., M.S., S.S.; writing—review and editing, E.P., D.G., M.S., D.R., B.B., S.S.; visualization, E.P., D.G.; supervision, D.G.; project administration, D.R., B.B.; funding acquisition, D.R., B.B. All authors have read and agreed to the published version of the manuscript.

Funding: This research was funded by the European Space Agency, FERTISAT—Satellite-based Service for Variable Rate Nitrogen Application in Cereal Production, grant number 4000118613/16/NL/EM.

Conflicts of Interest: The authors declare no conflict of interest.

References

1. Szantoi, Z.; Strobl, P. Copernicus Sentinel-2 Calibration and Validation. *Eur. J. Remote Sens.* **2019**, *52*, 253–255. [CrossRef]
2. Sentinel-2 User Handbook-Document Library-Sentinel Online. Available online: https://sentinels.copernicus.eu/web/sentinel/user-guides/document-library/-/asset_publisher/xlslt4309D5h/content/sentinel-2-user-handbook (accessed on 17 August 2020).
3. Asrar, G.; Fuchs, M.; Kanemasu, E.T.; Hatfield, J.L. Estimating absorbed photosynthetic radiation and Leaf Area Index from spectral reflectance in wheat. *Agron. J.* **1984**, *76*, 300–306. [CrossRef]

4. Benedetti, R.; Rossini, P. On the use of NDVI profiles as a tool for agricultural statistics: The case study of wheat yield estimate and forecast in Emilia Romagna. *Remote Sens. Environ.* **1993**, *45*, 311–326. [CrossRef]
5. Smith, R.C.G. Forecasting wheat yield in a Mediterranean-type environment from the NOAA satellite. *Aust. J. Agric. Res.* **1995**, *46*, 113–125.
6. Labus, M.P.; Nielsen, G.A.; Lawrence, R.L.; Engel, R.; Long, D.S. Wheat yield estimates using multi-temporal NDVI satellite imagery. *Int. J. Remote Sens.* **2002**, *23*, 4169–4180. [CrossRef]
7. Shou, L.; Jia, L.; Cui, Z.; Chen, X.; Zhang, F. Using high-resolution satellite imaging to evaluate nitrogen status of winter wheat. *J. Plant Nutr.* **2007**, *30*, 1669–1680. [CrossRef]
8. Jeppesen, J.H.; Jacobsen, R.H.; Jørgensen, R.N.; Halberg, A.; Toftegaard, T.S. Identification of high-variation fields based on open satellite imagery. *Adv. Anim. Biosci.* **2017**, *8*, 388–393. [CrossRef]
9. Delegido, J.; Verrelst, J.; Alonso, L.; Moreno, J. Evaluation of Sentinel-2 red-edge bands for empirical estimation of green LAI and chlorophyll content. *Sensors* **2011**, *11*, 7063–7081. [CrossRef]
10. Veloso, A.; Mermoz, S.; Bouvet, A.; Le Toan, T.; Planells, M.; Dejoux, J.F.; Ceschia, E. Understanding the temporal behavior of crops using Sentinel-1 and Sentinel-2-like data for agricultural applications. *Remote Sens. Environ.* **2017**, *199*, 415–426. [CrossRef]
11. Lopresti, M.F.; Di Bella, C.M.; Degioanni, A.J. Relationship between MODIS-NDVI data and wheat yield: A case study in Northern Buenos Aires province, Argentina. *Inf. Process. Agric.* **2015**, *2*, 73–84. [CrossRef]
12. Dempewolf, J.; Adusei, B.; Becker-Reshef, I.; Barker, B.; Potapov, P.; Hansen, M.; Justice, C. Wheat Production Forecasting for Pakistan from Satellite Data. In *2013 IEEE International Geoscience and Remote Sensing Symposium-IGARSS*; IEEE: Melbourne, Australia, 2013; pp. 3239–3242.
13. Kussul, N.; Kolotii, A.; Skakun, S.; Shelestov, A.; Kussul, O.; Oliynuk, T. Efficiency Estimation of Different Satellite Data Usage for Winter Wheat Yield Forecasting in Ukraine. In *2014 IEEE Geoscience and Remote Sensing Symposium*; IEEE: Quebec City, QC, Canada, 2014; pp. 5080–5082.
14. Bu, H.; Sharma, L.K.; Denton, A.; Franzen, D.W. Comparison of satellite imagery and ground-based active optical sensors as yield predictors in sugar beet, spring wheat, corn, and sunflower. *Agron. J.* **2017**, *109*, 299–308. [CrossRef]
15. Nagy, A.; Fehér, J.; Tamás, J. Wheat and maize yield forecasting for the Tisza river catchment using MODIS NDVI time series and reported crop statistics. *Comput. Electron. Agric.* **2018**, *151*, 41–49. [CrossRef]
16. Yang, A.; Zhong, B.; Wu, J. Monitoring Winter Wheat in ShanDong Province Using Sentinel Data and Google Earth Engine Platform. In *2019 10th International Workshop on the Analysis of Multitemporal Remote Sensing Images*; IEEE: Shanghai, China, 2019; pp. 1–4.
17. Ali, A.; Martelli, R.; Lupia, F.; Barbanti, L. Assessing multiple years' spatial variability of crop yields using satellite vegetation indices. *Remote Sens.* **2019**, *11*, 2384. [CrossRef]
18. GUS Statistical Yearbook of Agriculture. 2019. Available online: https://stat.gov.pl/en/topics/statistical-year books/statistical-yearbooks/statistical-yearbook-of-agriculture-2019,6,14.html (accessed on 16 August 2020).
19. Kottek, M.; Grieser, J.; Beck, C.; Rudolf, B.; Rubel, F. World Map of the Köppen-Geiger climate classification updated. *Meteorol. Z.* **2006**, *15*, 259–263. [CrossRef]
20. Open Access Hub. Available online: https://scihub.copernicus.eu/ (accessed on 16 August 2020).
21. Brief Introduction to Remote Sensing—Semi-Automatic Classification Plugin 6.4.0.2-Documentation. Available online: https://semiautomaticclassificationmanual.readthedocs.io/pl/latest/remote_sensing.html#d os1-correction. (accessed on 16 August 2020).
22. Rouse, J.W., Jr.; Haas, R.H.; Deering, D.W.; Schell, J.A.; Harlan, J.C. Monitoring the vernal advancement and retrogradation (Green Wave Effect) of natural vegetation. In *Great Plains Corridor*; Final Rep. RSC 1978–4; Remote Sensing Center, Texas A&M Univ.: College Station, TX, USA, 1974.
23. Wang, J.; Rich, P.M.; Price, K.P.; Kettle, W.D. Relations between NDVI and tree productivity in the central Great Plains. *Int. J. Remote Sens.* **2004**, *25*, 3127–3138. [CrossRef]
24. Huete, A.R. A soil-adjusted vegetation index (SAVI). *Remote Sens. Environ.* **1988**, *25*, 295–309. [CrossRef]
25. Jiang, Z. Interpretation of the modified soil-adjusted vegetation index isolines in red-NIR reflectance space. *J. Appl. Remote Sens.* **2007**, *1*, 013503. [CrossRef]
26. Crippen, R. Calculating the vegetation index faster. *Remote Sens. Environ.* **1990**, *34*, 71–73. [CrossRef]
27. Pinty, B.; Verstraete, M.M. GEMI: A non-linear index to monitor global vegetation from satellites. *Vegetatio* **1992**, *101*, 15–20. [CrossRef]

28. Jordan, C.F. Derivation of Leaf-Area Index from quality of light on the forest floor. *Ecology* **1969**, *50*, 663–666. [CrossRef]
29. Orfeo ToolBox 7.1.0 Documentation. Available online: https://www.orfeo-toolbox.org/CookBook/ (accessed on 17 August 2020).
30. Welcome to the QGIS Project! Available online: https://www.qgis.org/en/site/ (accessed on 16 August 2020).
31. Mengmeng, D.; Noboru, N.; Atsushi, I.; Yukinori, S. Japan multi-temporal monitoring of wheat growth by using images from satellite and unmanned aerial vehicle. *Int. J. Agric. Biol. Eng.* **2017**, *10*, 1–13. [CrossRef]
32. Toscano, P.; Castrignanò, A.; Di Gennaro, S.F.; Vonella, A.V.; Ventrella, D.; Matese, A.A. Precision agriculture approach for durum wheat yield assessment using remote sensing data and yield mapping. *Agronomy* **2019**, *9*, 437. [CrossRef]
33. Naser, M.A.; Khosla, R.; Longchamps, L.; Dahal, S. Using NDVI to differentiate wheat genotypes productivity under dryland and irrigated conditions. *Remote Sens.* **2020**, *12*, 824. [CrossRef]
34. Chandel, N.S.; Tiwari, P.S.; Singh, K.P.; Jat, D.; Gaikwad, B.B.; Tripathi, H.; Golhani, K. Yield prediction in wheat (*Triticum aestivum* L.) using spectral reflectance indices. *Curr. Sci.* **2019**, *116*, 272. [CrossRef]
35. Satir, O.; Berberoglu, S. Crop yield prediction under soil salinity using satellite derived vegetation indices. *Field Crops Resolut.* **2016**, *192*, 134–143. [CrossRef]
36. Fieuzal, R.; Bustillo, V.; Collado, D.; Dedieu, G. Combined use of multi-temporal Landsat-8 and Sentinel-2 images for wheat yield estimates at the intra-plot spatial scale. *Agronomy* **2020**, *10*, 327. [CrossRef]
37. Purevdorj, T.S.; Tateishi, R.; Ishiyama, T.; Honda, Y. Relationships between percent vegetation cover and vegetation indices. *Int. J. Remote Sens.* **1998**, *19*, 3519–3535. [CrossRef]
38. Ren, H.; Zhou, G.; Zhang, F. Using negative soil adjustment factor in soil-adjusted vegetation index (SAVI) for aboveground living biomass estimation in arid grasslands. *Remote Sens. Environ.* **2018**, *209*, 439–445. [CrossRef]
39. The Agricultural Drought Monitoring System. Available online: http://susza.iung.pl/en/ (accessed on 16 August 2020).

Article

Effects of Landscape, Soils, and Weather on Yields, Nitrogen Use, and Profitability with Sensor-Based Variable Rate Nitrogen Management in Cotton

James A. Larson [1,*], Melissa Stefanini [1], Xinhua Yin [2], Christopher N. Boyer [1], Dayton M. Lambert [3], Xia Vivian Zhou [1], Brenda S. Tubaña [4], Peter Scharf [5], Jac J. Varco [6], David J. Dunn [5], Hubert J. Savoy [7] and Michael J. Buschermohle [7]

1 Department of Agricultural & Resource Economics, University of Tennessee, Knoxville, TN 37996, USA; reynoldsmo@gmail.com (M.S.); cboyer3@utk.edu (C.N.B.); xzhou11@utk.edu (X.V.Z.)
2 Department of Plant Sciences, West Tennessee Research & Education Center, University of Tennessee, Jackson, TN 38301, USA; xyin2@utk.edu
3 Department of Agricultural Economics, Oklahoma State University, Stillwater, OK 74078, USA; dayton.lambert@okstate.edu
4 School of Plant, Environmental & Soil Sciences, Louisiana State University, Baton Rouge, LA 70803, USA; BTubana@agcenter.lsu.edu
5 Division of Plant Science, University of Missouri, Columbia, MO 65211, USA; ScharfP@missouri.edu (P.S.); DunnD@missouri.edu (D.J.D.)
6 Department of Plant & Soil Science, Mississippi State University, Mississippi State, MS 39762, USA; jjv3@msstate.edu
7 Department of Biosystems Engineering & Soil Science, University of Tennessee, Knoxville, TN 37996, USA; hsavoy@utk.edu (H.J.S.); mbuscher@utk.edu (M.J.B.)
* Correspondence: jlarson2@utk.edu

Received: 15 October 2020; Accepted: 20 November 2020; Published: 25 November 2020

Abstract: Farmers may be reluctant to adopt variable rate nitrogen (VRN) management because of uncertain profits. This study assessed field landscape, soil, and weather effects on optical sensing (OS)-based VRN on cotton (*Gossypium hirsutum* L.) N rates, yields, and net returns (NRs). Field data were collected from 21 locations in Louisiana, Mississippi, Missouri, and Tennessee, USA, between 2011 and 2014. Data included yields, N rates, and NRs for the farmer practice (FP), OS-based VRN, and OS-based VRN supplemented with other information. Production data were augmented with landscape, soils, and weather data, and ANOVA and logistic regressions were used to identify field conditions where VRN was profitable, provided risk management benefits, and improved N efficiency. Key findings indicate that NRs were improved with VRN by applying additional N on more erodible soils. Higher organic matter soils also benefited from VRN through enhanced yields and NRs. VRN may also have provided risk management benefits by providing a lower probability of NRs below NRs for the FP on soils associated with greater water-holding capacity, higher organic matter levels, or deeper profiles. Results from this study may help identify farm fields with similar characteristics for adoption of VRN management.

Keywords: economics; normalized difference vegetation index (NDVI); on-the-go sensors; site-specific nutrient management

---

1. Introduction

Upland cotton (*Gossypium hirsutum* L.) is an important crop in the lower Mississippi River Basin (MRB) of the United States (US) that includes the states of Louisiana, Mississippi, Missouri, and Tennessee [1].

Cotton area planted in the four states was 700,405 ha in 2019 [2]. Nitrogen (N) is the plant nutrient most often applied in the largest amounts by farmers growing upland cotton [3,4]. Nitrogen is especially important for lint yield formation after the cotton plant's first bloom [5]. Under application of fertilizer N reduces lint yield and profit. However, over application of fertilizer N in cotton increases fertilizer costs and can also cause excessive vegetative plant growth rather than increased production of cotton bolls that contribute to lint yield and profit [6]. Excessive vegetative growth can decrease lint yield due to boll rot and insects, reduce lint fiber quality, and cause increased expenses due to additional applications of pesticides and plant growth regulators to prevent lint yield losses [6].

Over application of fertilizer N can also negatively affect water quality. Nitrogen, especially in the form of nitrates, can leach from farm fields into surface and ground water [7]. This may be especially true if farmers apply a uniform N rate across individual fields. Efficient N management on fields in the lower MRB is an important priority for the US Department of Agriculture (USDA) Natural Resources Conservation Service (NRCS) [1]. The goal of the USDA NRCS is to reduce nutrient and sediment loading to local and regional water bodies and to improve water use efficiency. The USDA NRCS promotes the use of variable rate N (VRN) management to apply different rates of fertilizer N across farm fields based upon soil N and crop needs through the Environmental Quality Incentives Program (EQIP) [8]. However, by 2017, only 9.5% of upland cotton growers adopted VRN [9]. Grower uncertainty about the profitability of managing soil N spatial and temporal variability may be an important factor influencing VRN adoption by farmers [10].

Optimal fertilizer N management depends on the amount of available N derived from the soil and fertilizer [11]. Complex interactions between land use, crop management, landscape characteristics, soil properties, and weather influence N soil availability to plants [12]. Soil properties and soil N can vary substantially within farm fields [11]. Alluvial soils in the floodplains of the lower MRB (USDA NRCS Major Land Resource Area 131) frequently exhibit significant variation in texture and N availability [6]. Loess soils are common on cotton fields located in the lower MRB (USDA NRCS Major Land Resource Area 134) and are subject to water-induced soil erosion because of the rolling landscapes upon which these soils occur in the region [13]. Soils redistributed by water-induced soil erosion cause variation in a field's soil properties and, consequently, field soil N [14].

Rainfall and temperature also interact with soil and landscape attributes to cause spatial and temporal variability of soil N that complicates cotton N management [15]. Soil testing for N is unreliable in the warm, humid climate of the lower MRB because soil N varies greatly with soil organic matter, soil texture, tillage, and other factors [16]. Consequently, lower MRB cotton growers generally do not completely rely on soil test information to manage N [6]. Growers and their crop consultants develop a single (uniform) rate for the field using Land Grant University fertility recommendations, their experience, and other considerations, including cotton variety, soil texture, and crop rotation.

Given the unreliability of N soil tests, plant-based measurements can be used to determine crop demand for N. For example, in-season N status can be assessed using visual inspection of plants for N deficiency symptoms, petiole NO^3-N or leaf tissue sampling, or chlorophyll meters to determine N status in the growing cotton for in-season fertilizer N applications up to the early bloom stage [16,17]. However, assessing plant N status using hand-held devices is labor intensive and may not provide sufficient information to determine N rates for VRN management. Ground-based optical sensing (OS) of the growing crop canopy facilitates assessment of crop N status throughout the field and provides growing spatial plant canopy data useful for determining VRN rates that vary across the field [11,17].

Most of the studies evaluating OS-based VRN reported crop yields similar to yields for the uniform fertilizer N rate (i.e., conventional or farmer practice) [18–25]. Thus, an important factor driving the profitability of OS-based VRN is lower fertilizer N rates relative to the uniform rate. Researchers have reported fertilizer N savings with OS-based VRN of as much as 69 kg ha^{-1} [22]. However, other studies have reported increased applications of fertilizer N relative to the uniform rate of as much as 84 kg ha^{-1} with OS-based VRN [24]. Researchers evaluating the economic feasibility of OS-based VRN have found mixed profitability results. Studies that reported a lack of profitability found similar yields but did not find sufficient fertilizer N cost savings to provide a profit [18–20,24]. Research reporting positive profitability through enhanced yields and N cost savings did not include the costs of OS information and VRN application [22,23,25]. Costs of information used to produce the VRN prescription and VRN application costs are also important factors influencing the profitability of the technology [26]. VRN management may also mitigate yield and profit risk compared to uniform N management by reducing the probability of yield or profit below a threshold level [27,28].

Farmers are often unwilling to adopt technologies such as OS-based VRN unless they see the potential for positive profits [26]. This is especially a problem for N management in the lower MRB because plant response to N is influenced by landscape, soil, and weather characteristics. Quantifying how spatial and temporal factors affect yields, N use, profitability, and the risk management potential of VRN may be useful to cotton farmers in the lower MRB interested in adopting OS-based VRN. The objective of this research was to determine how landscape, soil, and weather influence fertilizer N use, lint yields, and profitability of OS-based VRN for cotton in the lower MRB.

2. Materials and Methods

2.1. Lint Yield and Fertilizer N Data

Lint yield and fertilizer N application rate data for the farmer practice (FP) and OS-based VRN management were from 21 study locations (Table 1). Farmers participating in the trials were eligible to receive payments to adopt VRN through USDA NRCS EQIP [8]. Stefanini et al. [24] previously reported differences in field-level fertilizer N use, lint yields, and profitability. The on-farm field trials were conducted between 2011 and 2014 at six locations in Tennessee, four locations in Mississippi, five locations in Louisiana, and six locations in Missouri. Most locations had only one year of data. However, several locations had two to three years of trials. Within each of the locations with multiyear trials, different fields were used for each year. A total of 29 site-years of data were collected in the study.

The field trial experimental design for each site-year was a randomized complete block design with three fertilizer N treatments and three replications. A strip-plot running the entire field length was used as the plot for each treatment in each replicate. Each strip-plot was further divided into sub-plots. The sub-plots were used to implement the two VRN treatments evaluated in this study. Cotton was planted on the nine strip-plots at each site, each with 8 to 10 sub-plots, that measured approximately 30.5 by 11.6 m. A different field on each participating farm was used for each site-year. While researchers attempted to choose similar field sizes in each year of the study, variation in field sizes resulted in different numbers of sub-plots within the strip-plots among the site-years (Table 1). However, the same number of sub-plots for each strip-plot within each site-year was maintained during the study.

The trials evaluated FP N management versus two OS-based VRN management regimes. The FP treatment was N application based on the farmer's current practice. Cotton farmers and their crop consultants often formulate their fertilizer N rate for the field using University recommendations, their experience, and agronomic and soil considerations [6]. Optical sensing-based VRN treatment 1 (VRN 1) was VRN management calculated using the normalized difference vegetation index (NDVI) readings collected with the GreenSeeker™ Crop Sensing System (Trimble, Sunnyvale, CA, USA) or Yara™ N-Sensor (Yara North America, Tampa, FL, USA) canopy optical-sensing. The configurations of sensor arrays were different in each state where the field trials took place. In Tennessee, a GreenSeeker™ RT200 system with six sensors (1.93 m apart and 0.76 m above the cotton canopy) covering 12 rows of cotton (11.58 m wide) was used to collect about two scans s^{-1} at a field speed of about 7.64 km h^{-1}. The second OS-based VRN treatment (VRN 2) was VRN management based on NDVI readings but augmented with additional information.

Table 1. State and county/parish locations of the farm fields.

State	County/Parish Field Locations	Years (Number of Field Sub-Plots) [A]		
Louisiana	Tensas Parish Res Station	2012 (89)		
Louisiana	Tensas Parish Middle	2012 (90)	2013 (90)	
Louisiana	Tensas Parish Middle Low	2014 (90)		
Louisiana	Tensas North	2012 (90)	2013 (100)	
Louisiana	Tensas Parish South	2012 (90)	2013 (90)	2014 (80)
Missouri	Dunklin	2013 (12)		
Missouri	New Madrid East	2012 (24)		
Missouri	New Madrid North	2012 (33)		
Missouri	New Madrid South	2012 (12)		
Missouri	Pemiscot North	2013 (6)		
Missouri	Pemiscot South	2013 (6)		
Mississippi	Adams	2012 (107)		
Mississippi	Leflore East	2014 (35)		
Mississippi	Leflore North	2013 (60)		
Mississippi	Leflore South	2013 (48)		
Tennessee	Carroll	2014 (72)		
Tennessee	Gibson	2011 (72)	2012 (88)	
Tennessee	Lauderdale	2012 (90)	2013 (90)	2014 (90)
Tennessee	Madison North	2012 (72)	2013 (72)	
Tennessee	Madison South	2014 (72)		
Tennessee	Tipton	2012 (72)		

[A] The number in parentheses indicates the total number of sub-plots at each field site. The field trials were conducted using a randomized complete block design at each site. Two variable rate nitrogen treatments were compared against the existing farmer practice. Each treatment was replicated three times in three strip-plots at each site. Strip-plots were divided into 8–10 sub-plots to implement the two variable rate nitrogen treatments in the field trials. Different fields with dissimilar sizes were used on each farm in each year of the study and resulted in a variable number of sub-plots at each site. However, the same number of sub-plots for each strip-plot within each site-year was maintained during the study. Yields were measured in Missouri at the strip-plot rather than the sub-plot level.

Two split applications of fertilizer N were made for the three fertilizer N management regimes. Starter fertilizer was applied at or before the planting of cotton and was determined by each farmer participating in the study. A uniform blanket rate of fertilizer N was applied to the entire field (covering all three treatment areas) with rates ranging from 33.6 to 78.4 kg N ha^{-1}, depending on the farm field location. A second side dress application of fertilizer N for the FP was made at approximately the early bloom stage. For the two VRN treatments, crop N status was determined using canopy optical-sensing at about the early bloom stage for each site-year of the trial and fertilizer N was side dressed variably on the sub-plots, thereafter based on the NDVI readings for the VRN 1 treatment and NDVI readings and either digital yield maps (Mississippi and Tennessee), soil productivity zones (Louisiana), or soil zones (Missouri) for VRN 2. Each state used different algorithms for the VRN 1 and VRN 2 treatments because each state has different soils, climates, and management practices for cotton. The unpublished algorithms were developed based on multiple-year and multiple-location data from previous research in each state.

The other production practices used to grow cotton on each field trial site were determined by the farmer cooperators in the study. Data collected for each sub-plot (strip-plot in Missouri) included harvested seed cotton yield, lint yields, applied fertilizer N rates, and latitude and longitudes for every field site, except in Missouri, where yield data were collected at the strip-plot level rather than by sub-plot (Table 2). In Louisiana, Mississippi, and Tennessee, cotton pickers with yield monitors were used to harvest cotton and determine sub-plot seed cotton and lint yields. Yield monitors were not available on cotton pickers at the Missouri sites so strip-plot yields rather than sub-plot yields were measured using a weigh wagon. A measure of nitrogen use efficiency (NEFF), defined as lint yield divided by fertilizer N rate, was also calculated for each N management regime (Table 2) [24].

Table 2. Field trial sub-plot mean, maximum, and minimum values and the number of sub-plot observations for lint yields, fertilizer nitrogen (N) rates, N efficiency (lint yield/fertilizer N rate), and net returns for the three fertilizer N treatments that were collected from the 2011–2014 field trials.

Variable Name/ Summary Statistics	Fertilizer N Treatment		
	FP [a]	VRN 1 [b]	VRN 2 [c]
Lint yield (kg ha^{-1})			
Mean	1332	1360	1349
Maximum	2397	2585	2565
Minimum	226	133	204
Observations	649	658	635
Fertilizer N rate (kg ha^{-1})			
Mean	107	109	114
Maximum	244	226	253
Minimum	34	54	34
Observations	660	659	635
Nitrogen efficiency (lint yield/fertilizer N rate, index)			
Mean	18	14	14
Maximum	120	54	40
Minimum	1	1	1
Observations	649	658	635
Net return (USD ha^{-1})			
Mean	2226	2315	2264
Maximum	4081	4233	4167
Minimum	481	239	333
Observations	649	658	635

[a] FP, farmer practice nitrogen management on each field in the study. [b] VRN 1, variable rate nitrogen management calculated using normalized difference vegetative index readings. [c] VRN 2, variable rate nitrogen management based on normalized difference vegetative index readings and either digital yield maps (Mississippi and Tennessee), soil productivity zones (Louisiana), or soil zones (Missouri).

2.2. Landscape, Soil, and Weather Data

Landscape, soil, and weather data were collected to determine differences within and between fields for each location-year. Georeferenced landscape, soil, and weather data were assembled from the center point of each sub-plot (strip-plot for Missouri locations) using ArcGIS 10.1 (ESRI, Redlands CA, USA). Sub-plot elevations (m above sea level) were collected from the National Elevation Dataset [29]. Soil water-holding capacity (volume fraction), soil organic matter (%), soil texture, soil depth (cm), field slope, and soil erosion factors were gathered from the Soil Survey Geographic (SSURGO) database [30]. Soil texture data in SSURGO were used to rank textures by coarseness—clay (finest), silt, loam, and sand (coarsest)—using the USDA soil texture calculator [31].

A soil erosion index (SEI) was created using SSURGO [30] data, USDA Revised Universal Soil Loss Equation, version 2 (RUSLE2) data [32], and a modified universal soil loss equation to account for the physical factors of the fields [24]:

$$SEI = (KF \times LS \times R)/TF \quad (1)$$

where *KF* is an erodibility factor due to water, *LS* is a soil length (*L*) and slope steepness (*S*) factor, *R* is the rainfall and runoff factor from USDA RUSLE2 version 2.5.2.11 [32]; and *TF* is a soil tolerance factor.

Weather was measured by temperature [33], expressed as seasonal growing degree days. To calculate seasonal growing degree days, the positive values of daily average temperature minus 15.6 °C was summed over 1 April through 31 October for each site-year.

2.3. Fertilizer N Management Net Returns

Net returns for the FP, VRN 1, and VRN 2 treatments were estimated using sub-plot lint yields, fertilizer N rates, lint and N fertilizer prices, and partial budgeting costs for OS and VRN technologies (Table 2). Price and budget data are in real 2013 US dollars indexed using the annual Gross Domestic Product Price Deflator Index [34]. Crop revenues were estimated by multiplying lint yields for each N management treatment by the national average marketing year cotton lint price of USD 1.86 kg^{-1} received for 2011 through 2014 [35]. EQIP cost-share payments (NRCS precision nutrient management practice code number 590) for each state for 2011 through 2014 were also added to crop revenues. Estimated payments were USD 68.21 ha^{-1} in Mississippi [36], USD 68.46 ha^{-1} in Louisiana [37], USD 65.85 ha^{-1} in Tennessee [38], and USD 32.64 ha^{-1} in Missouri [39].

Fertilizer N cost of USD 0.93 kg^{-1} was multiplied by the fertilizer N rate to determine fertilizer N cost for each N management regime. The fertilizer N price is the national average marketing year fertilizer N prices received for 2011 through 2014 [40]. Following Stefanini et al. [24], budgeted skilled operator labor and equipment operating and ownership costs of USD 2.14 ha^{-1} and USD 2.45 ha^{-1}, respectively, for OS of the crop canopy was assumed for GreenSeeker™ sensors retrofitted to a boom sprayer measuring 24.7 m wide. The cost of yield monitoring data identifying yield productivity zones in the field was assumed to be used to augment OS information for the VRN 2 prescription and had a budgeted cost of USD 2.73 ha^{-1}. In addition, the budgeted costs of a computer to manage yield monitor data of USD 0.31 ha^{-1} and reported cost of technical advice for incorporating yield monitor with OS information of USD 12.63 ha^{-1} [41], respectively, were included in the total cost for VRN 2. The cost of VRN application was estimated to be USD 6.60 ha^{-1} more than for the FP [41].

2.4. Statistical Analysis

Two statistical models were used to evaluate OS-based VRN in-field fertilizer N rate, lint yield, and net return (NR) relationships with farm field characteristics. The first is a general linear model for the fertilizer N management mean differences. The sub-plot lint yields (YLD), fertilizer N rates (FNs), N efficiency (FNEFF), and NRs (FNRs) that are summarized in Table 2 were used to construct the regressions' dependent variables. The dependent variables were created using paired sub-plot observations in each strip-plot to measure differences between VRN 1 and the FP (VRN 1-FP) and VRN 2 and the FP (VRN 2-FP). For example, field 1, replication 1, and sub-plot 1 for the VRN 1 treatment versus field 1, replication 1, and sub-plot 1 for the FP treatment. This procedure resulted in 1263 observations available for each of the regressions (Table 3). Fixed effects included in the mean difference regressions are landscape, soil, and weather characteristics georeferenced to each sub-plot. To account for potential differences in landscape and soil characteristics between paired VRN and FP sub-plot observations within each replication, observations were omitted from the regressions if soil characteristics differed between the two sub-plots. For example, if soil texture differed across field 1, replication 1, and sub-plot 1 for the VRN treatment versus sub-plot 1 for the FP treatment, then the observation was omitted from the regressions; if not, the observation was retained for the estimation. The summary statistics for the landscape, soil, and weather variables used as fixed effects in the mean difference regressions are also presented in Table 3.

Table 3. Dependent and fixed-effect variable names, definitions, and statistics (mean, minimum, maximum, and number of available observations) for the mean difference and logit regression models.

Variable Name	Mean	Minimum	Maximum	Observations
Mean difference regression dependent variables				
ΔYLD [a]	37.05	−1941.20	2077.53	1263
ΔFN [b]	4.95	−67.59	125.36	1263
ΔFNEFF [c]	−3.21	−96.52	20.18	1263
ΔFNR [d]	102.37	−3630.32	3668.06	1263
Logit regression dependent variables				
YLDprob [e]	0.45	0	1	1263
FNprob [f]	0.55	0	1	1263
FNEFFprob [g]	0.47	0	1	1263
FNRprob [h]	0.37	0	1	1263
Fixed Effects				
Soil texture index [i]	2.14	1	4	1221
Elevation [j]	64.20	21.64	136.36	1262
WHC [k]	0.21	0.08	0.23	1168
SOM [l]	1.85	0.52	2.25	1166
SEI [m]	7.03	0.21	39.13	1158
Depth [n]	21.32	8.00	64.00	1152
GDD [o]	1574.3	1025.93	1943.27	1263
ν [p]	0.33	0	1	1263

[a] ΔYLD, difference in optical sensing-based variable rate nitrogen management and farmer practice nitrogen management lint yields (kg ha^{-1}). [b] ΔFN, difference in optical sensing-based variable rate nitrogen management and farmer practice nitrogen management fertilizer nitrogen rates (kg ha^{-1}). [c] ΔFNEFF, difference in optical sensing-based variable rate nitrogen management and farmer practice nitrogen management fertilizer nitrogen efficiency measured as lint yield divided by fertilizer nitrogen rate (index). [d] ΔFNR, difference in optical sensing-based variable rate nitrogen management and farmer practice nitrogen management net returns (USD ha^{-1}). [e] Yprob, if optical sensing-based variable rate nitrogen management lint yield is less than farmer practice nitrogen management lint yield, then 1; else 0. [f] Nprob, if optical sensing-based variable rate nitrogen management fertilizer nitrogen rate is less than farmer practice nitrogen management fertilizer nitrogen, then 1; else 0. [g] NEFFprob, if optical sensing-based variable rate nitrogen management nitrogen efficiency is less than farmer practice nitrogen management nitrogen efficiency, then 1; else 0. [h] NRprob, if optical sensing-based variable rate nitrogen management net return is less than farmer practice nitrogen management net return, then 1; else 0. [i] Soil texture index, 1 = Clay, 2 = Silt, 3 = Loam, and 4 = Sand. Sand is the reference variable in the regressions. Sources: [30,31]. [j] Elevation, vertical distance above sea level (m). Source: [29]. [k] WHC, water holding capacity (volume fraction). Source: [30]. [l] SOM, soil organic matter (%). Source: [30]. [m] SEI, soil erosion index. [n] Depth, soil depth (cm) from the top of the soil to the base of the soil horizon. Source: [30] [o] GDD, growing degree days, 1 April through 1 October, base 15.6 degrees Celsius: Source: [33]. ν [p], 0–1 variable indicating variable rate nitrogen treatment using normalized difference vegetative index and either digital yield maps (Mississippi and Tennessee), soil productivity zones (Louisiana), and soil zones (Missouri).

The general linear model for the fertilizer N management mean differences was:

$$\Delta Y_{ijklt} = \mu + X_{lt}\beta + v + \varphi_j + \varphi_{k(j)} + e_{ijklt}, \quad (2)$$

where $i = 1$ (VRN 1 − FP), 2 (VRN 2 − FP); $j = 1, \ldots, 21$ farm field locations; $k = 1$, 2, and 3 replications on fields; $l = 1, \ldots, 8$ to 10 replication sub-plots within each strip-plot; $t = 2011$, 2012, 2013, and 2014; $\Delta Y_{ijklt} = Y^{VRN_i} - Y^{FP}$ is defined as the mean difference in the response variable Y (lint yields (ΔYLD, kg ha^{-1}), fertilizer N rates (ΔFN, kg ha^{-1}), YLD/FN (ΔFNEFF, index), and NR (ΔFNR, USD ha^{-1})) for VRN 1 or VRN 2 compared to the FP; μ is the conditional mean; X includes sub-plot measurements on soil texture (clay, silt, loam, and sand), elevation above sea level (m), soil water-holding capacity (volume fraction), soil organic matter (%), soil depth (cm), soil erosion index, and seasonal growing degree days (degrees Celsius); β is a vector of the estimated average landscape, soil, and weather effects on ΔY; and v is a 0–1 variable indicating the VRN 2 treatment. The parameters φ_j and $\varphi_{k(j)}$ are the farm field location random effects and the nested random effects from replications in farm field locations, with $\varphi_j \sim N\left(0, \sigma^2_{\varphi_j}\right)$ and $\varphi_{k(j)} \sim N\left(0, \sigma^2_{\varphi_{k(j)}}\right)$. The model error is $e_{ijkt} \sim N\left(0, \sigma^2_e\right)$ [42].

The models using Equation (2) were estimated using the MIXED model procedure and restricted maximum likelihood in SAS 9.2 [43]. The sand soil texture 0–1 variable was dropped to estimate regressions and was included as the reference variable in the intercept term. The mean difference models were evaluated for multicollinearity using variance inflation factors (VIF) estimated using the REG model procedure in SAS 9.2 [43]. VIF exceeding 10 may indicate that multicollinearity is increasing the size of the parameters' standard errors [44]. Models estimated using Equation (2) tested the null hypotheses that mean yields, fertilizer N rates, NRs, and N efficiency were not different between VRN and FP, holding landscape, soil, and weather factors constant.

The second statistical model is estimated as a mixed logistic regression:

$$\Pr\left(VRN_{ijklt} > FP_{ijklt} \middle| X_{lt}\right) = \text{Logistic}\left(\mu + X_{lt}\beta + v + \varphi_j + \varphi_{k(j)} + e_{ijklt}\right) \quad (3)$$

where $\Pr(VRN_{ijklt} > FP_{ijklt}|X_{lt})$ is the probability that the response variable (lint yields (YLD, kg ha^{-1}), fertilizer N rates (FN, kg ha^{-1}), YLD/N (NEFF, index), and NRs (FNR, USD ha^{-1})) for VRN falls above or below the FP value. The sub-plot data summarized in Table 2 were used to construct the logit regressions' dependent variables and are presented in Table 3. The binary dependent variables using the paired sub-plot observations in each strip-plot were calculated as:

$$\text{If } YLD^{VRN_i} - YLD^{FP} < 0, \text{ then } YLDprob = 1; \text{ else, } YLDprob = 0; \quad (4)$$

$$\text{If } FN^{VRN_i} - FN^{FP} > 0, \text{ then } FNprob = 1; \text{ else, } FNprob = 0; \quad (5)$$

$$\text{If } FNEFF^{VRN_i} - FNEFF^{FP} < 0, \text{ then } FNEFFprob = 1; \text{ else, } FNEFFprob = 0, \text{ and} \quad (6)$$

$$\text{If } FNR^{VRN_i} - FNR^{FP} < 0, \text{ then } FNRprob = 1; \text{ else, } FNRprob = 0. \quad (7)$$

Equations (4)–(7) were estimated for each binary dependent variable with the same set of fixed effects summarized in Table 3 and the same random effects used for the mean difference regressions described above. The logit models were estimated using the GLIMIX model procedure and restricted maximum likelihood in SAS 9.2 [43]. Multicollinearity was also evaluated in the logit regressions with the same procedures used for the mean difference regressions [44]. The odds ratios calculated using the estimated coefficients β of these logistic regressions are used to test the hypotheses comparing FP and OS-based VRN. Each covariate's impact on the odds VRN < FP is $\exp(\beta)$. In percent terms, the change in the log odds probability that VRN lint yields, N rates, NEFF, or NRs exceeded those of the FP is $100 \times [\exp(\beta) - 1]$. The null hypotheses for Equations (4)–(7) was that the N management regime does not affect the probability that yields, N rates, N efficiency, and NRs differ for VRN versus the FP, holding soil, landscape, and weather variables constant.

3. Results and Discussion

3.1. VRN vs. FP Mean Differences

The VIFs were less than five for all covariates and all general linear regressions (lint yield, fertilizer N rate, N efficiency, and NR), suggesting that multicollinearity was not inflating the parameters' standard errors.

3.1.1. Lint Yields

The soil, landscape, and weather factors associated with lint yields in the estimated mean difference regressions were silt soil texture ($\Pr \le 0.01$), loam soil texture ($\Pr \le 0.05$), elevation ($\Pr \le 0.01$), organic matter ($\Pr \le 0.01$), soil depth ($\Pr \le 0.01$), soil erosion index ($\Pr \le 0.01$), and growing degree days (GDD) ($\Pr \le 0.01$) (Table 4). Soils classified as having a silty or loamy texture relative to sand (the intercept term) were negative in relation to VRN yields when compared to FP. Higher temperatures, as measured by seasonal GDD, or field sites at higher elevations also had a negative association with

VRN yields compared to FP. Therefore, soils with coarser textures, fields at higher elevations, or fields in locations with warmer temperatures were negatively related with VRN yields when compared to FP, all other factors equal. Thus, VRN management may not increase lint yields on fields with these conditions when compared to the FP. By contrast, soils with higher organic matter content, deeper profiles, or subject to more erosion were positively associated with VRN yields relative to FP, all else equal. Soils with more organic matter may have more natural N available to the plant [45]. Soils with a higher erosion index had a positive association with lint yields, potentially because more N was applied in areas of the field that were more eroded.

Table 4. Estimated average landscape, soil, and weather effects on lint yield, fertilizer nitrogen (N) rate, N efficiency (lint yield/fertilizer N rate), and net return.

Fixed Effect [a]	Lint Yield	N Fertilizer Rate	N Efficiency	Net Return
	(kg ha^{-1})	(kg ha^{-1})	(Index)	(USD ha^{1})
Intercept [b,c]	32.54	−7.77	−9.87	856.58
	(20.83)	(1.91) ***	(8.78)	(372.90) **
Clay [c]	−4.63	6.33	−12.41	−195.82
	(−6.61)	(0.61) ***	(2.77) ***	(118.31) *
Silt [c]	−24.45	−1.68	−2.67	−376.13
	(4.92) ***	(0.45) ***	(2.08)	(87.97) ***
Loam [c]	−14.00	−2.41	6.19	−190.16
	(6.15) **	(0.56) ***	(2.59) **	(110.02) *
Elevation	−1.98	−0.05	0.03	−3.80
	(0.53) ***	(0.05)	(0.02)	(0.95) ***
WHC [d]	6.74	1.92	6.54	929.64
	(−6.41)	(0.59) ***	(26.80)	(1148.78)
SOM [e]	1.71	−0.03	8.39	297.60
	(0.48) ***	−4.44	(2.01) ***	(86.15) ***
Depth	3.76	−0.27	0.08	6.58
	(1.26) ***	(0.12) **	(0.05)	(2.27) ***
SEI	5.89	0.90	−0.43	6.62
	(2.09) ***	(0.19) ***	(0.09) ***	(3.75) *
GDD	−0.34	0.04	−0.01	−0.71
	(0.05) ***	(0.00) ***	(0.00) ***	(0.09) ***
v	24.82	5.81	0.02	26.02
	(−20.59)	(1.91) ***	(0.86)	(36.90)
Observations	1140	1140	1140	1140

Note: Standard errors are in parentheses. * Significant at the 0.10 probability level, ** significant at the 0.05 probability level, *** significant at the 0.01 probability level. [a] Variable names are defined in Table 3. [b] Intercept contains sand soil texture. [c] Soil texture coefficients scaled by 10%. [d] WHC coefficients scaled by 100. [e] SOM coefficients scaled by 100%.

3.1.2. N Fertilizer Rates

Many field soil, landscape, and weather characteristics were significantly related with N rate differences between VRN-generated N rates and the FP (Table 4). Fertilizer N rate differences were negatively associated with sand (the intercept term, Pr ≤ 0.01), silt (Pr ≤ 0.01), or loam (Pr ≤ 0.01) soil textures but positively related with clay (Pr ≤ 0.01) when VRN was compared to FP. Finer-textured soils required more fertilizer N applied due to the higher yield potential while coarser soils needed less applied N. Soils with greater water-holding capacity (Pr ≤ 0.01), larger soil erosion indexes (Pr ≤ 0.01), and higher GDD (Pr ≤ 0.01) were positive in relation to VRN N rates. All else equal, more N was applied using VRN compared to FP on fields with a greater water-holding capacity, more erodible soils, or warmer temperatures. Soils with deeper profiles (Pr ≤ 0.05) had a negative association to VRN N rate compared to FP. The estimated dummy variable for VRN 2 showed significantly higher N rates, indicating that OS plus yield monitor information calculated higher mean fertilizer N rates than the FP. However, the higher fertilizer N rates generated with the additional information embodied in the N management regime were not associated with higher lint yields (Table 4) and, therefore, limit the profit potential of VRN 2. In addition, the higher cost of information utilizing more expensive map-based information with VRN 2 also impedes its profit potential [28].

3.1.3. N Efficiency

Differences in N efficiency for VRN and the FP were negatively associated with the clay soil texture ($Pr \leq 0.01$) compared to sand and positively associated with loam soil texture ($Pr \leq 0.05$) (Table 4). Soils that were richer in organic matter ($Pr \leq 0.01$) had a positive association with N efficiency for VRN compared to FP. More erodible soils ($Pr \leq 0.01$) or fields with warmer temperatures ($Pr \leq 0.01$) had negative associations with VRN efficiency. While higher organic matter content ($Pr \leq 0.01$) soils had a positive relation to N efficiency of VRN compared to FP, all else equal, fields with more erodible soils ($Pr \leq 0.01$) and warmer temperatures ($Pr \leq 0.01$) had negative associations to N use efficiency, likely due to the need for higher N rates.

3.1.4. Net Returns

Soil, landscape, and weather factors also had significant impacts on mean NR differences (Table 4). Silt soil textures ($Pr \leq 0.01$) had a negative impact on VRN NRs when compared to FP NRs. As noted above, the silt texture also had a negative association to VRN yields and VRN N rates. The N rates savings may not have been enough to increase NR for that soil type. The soil texture reference variable sand ($Pr \leq 0.05$), however, had positive associations with VRN NRs compared to FP. Soils with higher organic matter ($Pr \leq 0.01$) or deeper profiles ($Pr \leq 0.01$) were positively associated with VRN NRs compared to FP. Higher elevation ($Pr \leq 0.10$) fields had a negative association with VRN NRs compared to FP. Soils at higher elevations may be more exposed to erosion from wind and rain events. All else equal, warmer growing conditions as measured by GDDs were negatively associated with VRN yields compared to FP, positively with N rates, and, thus, negatively with NR. Warmer temperatures are correlated with dryer climates, particularly during the summer months in the United States [46], which may cause the need for higher N rates because of increased volatilization of N to the atmosphere. However, the higher N applied was not sufficient to increase yields such that VRN NRs were increased relative to the FP.

3.2. *VRN and Risk*

The VIFs were less than five for all covariates and all logit regressions (lint yield, fertilizer N rate, N efficiency, and NR), suggesting that multicollinearity was not inflating the parameters' standard errors.

3.2.1. Lint Yields

Soil, landscape, and weather factors associated with lint yields in the estimated logit model were silt soil texture ($Pr \leq 0.01$), loam soil texture ($Pr \leq 0.01$), water-holding capacity ($Pr \leq 0.10$), organic matter ($Pr \leq 0.05$), soil depth ($Pr \leq 0.05$), and growing degree days ($Pr \leq 0.01$) (Table 5). Silt- or loam-textured soils or soils on fields with warmer growing conditions are positively attributed with the probability of lower VRN yields than FP (Table 6). Soils with greater water-holding capacity, higher organic matter content, or deeper profiles are negatively associated with the probability of lower VRN yields than FP. All else equal, higher organic matter in soils could potentially lower the probability of yield loss enough to warrant VRN adoption for some farmers through lower fertilizer N rates.

Table 5. Estimated logit regression coefficients of landscape, soil, and weather effects on lint yield, fertilizer nitrogen (N) rate, N efficiency (lint yield/fertilizer N rate), and net return.

Fixed Effect [a]	Lint Yield	N Fertilizer Rate	N Efficiency	Net Return
Intercept [b,c]	−0.9069	−1.1892	−2.5115	−1.3077
	(1.2204)	(1.4787)	(1.3020) *	(1.3024)
Clay [c]	−0.2694	1.9090	4.1468	0.1044
	(0.4686)	(1.1793)	(1.0628) ***	(0.4491)
Silt [c]	1.7455	−3.8551	−0.1331	0.9767
	(0.3584) ***	(0.6889) ***	(0.2960)	(0.3522) ***
Loam [c]	1.2943	−3.7156	−0.9842	0.8824
	(0.3934) ***	(0.6505) ***	(0.3852) **	(0.4067) **
Elevation	0.0044	−0.0177	−0.0025	0.0050
	(0.0032)	(0.0041) ***	(0.00342)	(0.0033)
WHC [d]	−6.9961	17.7126	3.2916	−3.2100
	(3.9285) *	(5.3030) ***	(4.4964)	(4.0088)
SOM [e]	−0.6101	0.4142	−0.3834	−0.5137
	(0.2880) **	(0.3775)	(0.3052)	(0.2940) *
Depth	−0.01872	0.0040	−0.0174	−0.0121
	(0.0075) **	(0.0090)	(0.0080) **	(0.0077)
SEI	−0.0178	0.1221	0.0304	. 0.0003
	(0.01217)	(0.0166) ***	(0.0129) **	(0.0125)
GDD	0.0012	0.0002	0.0018	0.0010
	(0.0003) ***	(0.0003)	(0.0003) ***	(0.0003) ***
ν	0.1126	0.4640	0.0321	0.1378
	(0.1238)	(0.1398) ***	(0.1286)	(0.1262)
Observations	1140	1140	1140	1140

Note: Standard errors are in parentheses. * Significant at the 0.10 probability level. ** Significant at the 0.05 probability level. *** Significant at the 0.01 probability level. [a] Variable names defined in Table 3. [b] Intercept contains soil texture sand. [c] Soil texture coefficients scaled by 10%. [d] WHC coefficients scaled by 100. [e] SOM coefficients scaled by 100%.

Table 6. Odds ratios and percent changes in log odds probabilities for landscape, soil, and weather effects on lint yield, fertilizer nitrogen (N) rate, N efficiency (lint yield/fertilizer N rate), and net return calculated from logit regression estimated coefficients [a].

Fixed Effect [b]	Statistic	Lint Yield	N Fertilizer Rate	N Efficiency	Net Return
Intercept [c,d]	Odds ratio	NS	NS	0.0811 *	NS
	Percent change	NS	NS	−9.1885 *	NS
Clay [c]	Odds ratio	NS	NS	63.2313 ***	NS
	Percent change	NS	NS	622.3134 ***	NS
Silt [d]	Odds ratio	5.7287 ***	0.0212 ***	NS	2.6557 ***
	Percent change	47.2877 ***	−9.7883 ***	NS	16.5568 ***
Loam [d]	Odds ratio	3.6484 ***	0.0243 ***	0.3737 **	2.4167 **
	Percent change	26.4844 ***	−9.7566 ***	−6.2626 **	14.1669 **
Elevation	Odds ratio	NS	0.9825 ***	NS	NS
	Percent change	NS	−1.75443 ***	NS	NS
WHC [e]	Odds ratio	0.0009 *	4.9259×10^{7} ***	NS	NS
	Percent change	−0.999 1*	4.9259×10^{8} ***	NS	NS
SOM [f]	Odds ratio	0.5433 **	NS	NS	NS
	Percent change	−0.4567 **	NS	NS	NS
Depth	Odds ratio	0.9815 **	NS	0.9828 **	NS
	Percent change	−1.8546 **	NS	−1.7200 **	NS
SEI	Odds ratio	NS	1.1299	1.0309 **	NS
	Percent change	NS	12.9867 ***	3.0898 **	NS
GDD	Odds ratio	1.0012	NS	1.0018 ***	1.0010 ***
	Percent change	0.1248 ***	NS	0.1845 ***	0.0965 ***
ν	Odds ratio	NS	1.5904***	NS	NS
	Percent change	NS	59.0423 ***	NS	NS

*,**,*** 10, 5, and 1 percent significance for the estimated coefficient in the logit model, respectively. NS, not significant for the estimated coefficient in the logit model. [a] Odds ratios and the changes in the log odds probabilities were calculated using the estimated coefficients β of the logistic regressions reported in Table 5. [b] Variable names are defined in Table 3. [c] Intercept contains soil texture category sand. [d] Texture scaled by 10%. [e] WHC scaled by 100. [f] SOM scaled by 100%.

For the silt soil texture, the lint yield odds ratio indicated that VRN treatment yields were 5.73 ($e^{1.7455}$) times as likely to be lower than FP yields under these conditions. The percent change in the log odds of VRN yields lower than FP yields was 47.29%. A field with a silt soil texture had a high probability of lower yields with VRN and could potentially benefit from a keeping the current FP N rate instead of going with VRN management. Estimating the odds ratio for the loam soil texture indicated that VRN treatments on loam textured soils were 3.65 ($e^{1.2943}$) times likely to have lower yields than the FP. There was a 26.49% change in the log odds that VRN yields were lower than FP yields on loam textured soils. Loamy fields with the same mean landscape, soil, and weather characteristics would also likely benefit from keeping the FP instead of adopting VRN in terms of yields.

3.2.2. N Fertilizer Rates

The landscape, soil, and weather variables related with N rates in the estimated logit model were silt soil texture ($Pr \leq 0.01$), loam soil texture ($Pr \leq 0.01$), elevation ($Pr \leq 0.01$), water-holding capacity ($Pr \leq 0.01$), soil erosion index ($Pr \leq 0.01$), and VRN 2 treatment dummy variable ($Pr \leq 0.01$) (Table 5). Evaluating the percentage changes in the log odds probabilities of landscape, soil, and weather attributes indicated that silt- or loam-textured soils or soils at higher elevations are negatively associated with the probability that FP generates lower N rates than VRN (Table 6). Greater water-holding capacity, more erodible soils, or VRN 2 were positively associated with the probability that FP generates lower N rates than VRT.

The fertilizer N rate odds ratio for silt indicated that FP N rates were 0.0212 ($e^{-3.8551}$) times as likely to be lower than VRT N rates. There was a 9.79 percent change in the log odds that the FP N rates were lower than VRN N rates. Fields with silt textures with the mean soil conditions would likely benefit from VRN in terms of N cost savings and environmental benefits due to significant chances of VRN generating lower N rates than the FP. Evaluating the odds ratio at the loam soil texture indicated that the FP N rates were 0.0243 ($e^{-3.7156}$) times as likely to be lower than the VRT N rates. The percentage change in the log odds of FP N rates being lower than VRN N rates was 9.76 percent. Under these conditions, there was a relatively large chance that the VRN practice would be applied less N than the FP technology. A field with these conditions may benefit from VRN use for environmental benefits.

3.2.3. N Efficiency

Soil, landscape, and weather variables related with N efficiency were sand soil texture ($Pr \leq 0.10$), clay soil texture ($Pr \leq 0.01$), loam soil texture ($Pr \leq 0.05$), soil depth ($Pr \leq 0.05$), soil erosion index ($Pr \leq 0.05$), and growing degree days ($Pr \leq 0.01$) (Table 5). The percentage changes in the log odd probabilities of landscape, soil, and weather attributes in relation to NEFF indicated that finer soil textures or warmer temperatures were positively associated with the probability of a lower VRN N efficiency compared to FP (Table 6). Deeper soils or soil with coarser textures were negatively related to the probability of lower N efficiency of VRN compared to FP.

Fertilizer N efficiency on the clay soil texture indicated that VRN N efficiency was 63.2 ($e^{4.1468}$) times as likely to be lower than FP. There was a 622% change in the log odds of VRN N efficiency, lower than FP. Using VRN on clay fields, these may be inefficient in terms of N use relative to the FP. At the loam soil texture, the odds ratio indicated that VRN N efficiency was 0.3737 ($e^{-0.9842}$) times as likely to be lower than FP. Evaluating the odds ratio for the loam soil texture, there was a 6.26 percent change in the log odds of lower VRN N efficiency compared to the FP on loamy textured fields with these conditions. This finding indicates that there is a significant chance of obtaining higher N efficiency using VRN on loam soil textures.

3.2.4. Net Returns

The landscape, soil, and weather variables associated with NRs in the estimated logit model were silt soil texture ($Pr \leq 0.01$), loam soil texture ($Pr \leq 0.06$), organic matter ($Pr \leq 0.10$), and growing degree days ($Pr \leq 0.01$) (Table 5). Evaluating the aforementioned soil texture and weather attributes in relation to NRs indicated that coarser soil textures and warmer temperatures were positively associated with the probability of lower NRs using VRN compared to FP (Table 6). Evaluating the NR odds ratio for the silt soil texture indicated that VRN NRs were 2.66 ($e^{0.9767}$) times as likely to be lower than FP NRs. There was a 16.56% change in the log odds that had lower NRs than the FP. Fields with these conditions would likely not benefit from VRN adoption in terms of profits. The odds ratio evaluated at the loam soil texture indicated that VRN NRs were 2.42 ($e^{0.8824}$) times as likely to be lower than FP. The change in the log odds of lower NRs under these conditions was 14.17%. Fields with these soil conditions may be better suited, from a profitability standpoint, to continue using the current FP N management in place.

4. Conclusions

Many field landscape, soil, and weather factors impacted the performance of VRN in the farm MRB fields analyzed in this study. Soils with higher organic matter content, deeper profiles, and that are more erodible produced higher lint yields using VRN compared to the FP. In contrast, coarser soils, fields at higher elevations, or fields in locations with warmer temperatures were negatively associated with VRN yields compared to the FP. More N was applied using VRN compared to the FP on fields associated with greater water-holding capacity, more erodible soils, or warmer temperatures. In contrast, deeper profiled soils had a negative association to VRN N rate compared to FP. Soil, landscape, and weather had less of an impact on VRN NRs than on lint yields and fertilizer N rates. Most notable was the positive association with greater NRs with VRN relative to the FP for soils that were deeper and had higher organic matter. Soils with more organic matter had a positive relation to N efficiency of VRN compared to the FP. More erodible fields and warmer climates had negative associations to N efficiency, likely due to the need for higher N rates to account for lower available N in soils. Supplementing OS information with other map-based information resulted in higher VRN N rates but not yields and NRs. In addition, the additional expense of map-based information also impeded VRN profitability.

VRN may provide downside risk management benefits on fields with greater water-holding capacity, higher organic matter, or deeper profile soils by being associated with a smaller probability of low yields relative to the FP. Fields with silt and loam soils would likely benefit from VRN fertilizer N cost savings and environmental benefits because of a high probability of VRN resulting in lower N rates than the FP. In addition, the probability of enhanced N efficiency is more likely on a loam texture soil. However, the potential environmental benefits on these two soil textures may be obtained at the cost of a higher probability of lower NRs.

Key findings can be used by extension educators and cotton farmers to determine if adopting OS-based VRN on fields with certain characteristics would likely provide positive benefits. However, an important caveat of this study is that the profitability of OS and VRN were evaluated at the sub-field level to identify the conditions where the technology may provide an advantage over the FP. Notwithstanding the potential benefits of VRN, farmers are interested in the profitability of the technology at the field and farm levels. Future analyses should assess the profitability and risk management potential of the technology at the field level and farm levels as influenced by landscape, soil, and weather.

Author Contributions: Conceptualization, J.A.L., X.Y., B.S.T., P.S., J.J.V., D.J.D., H.J.S. and M.J.B.; data curation, J.A.L., M.S., X.Y. and X.V.Z.; formal analysis, J.A.L., M.S., X.Y., C.N.B. and D.M.L.; funding acquisition, J.A.L., X.Y., B.S.T., P.S., J.J.V., D.J.D., H.J.S. and M.J.B.; investigation, J.A.L., X.Y., B.S.T., P.S., J.J.V., D.J.D., H.J.S. and M.J.B.; methodology, J.A.L., X.Y., C.N.B. and D.M.L.; project administration, X.Y.; resources, J.A.L., B.S.T., P.S., J.J.V., D.J.D., H.J.S. and M.J.B.; software, J.A.L., M.S., D.M.L. and X.V.Z.; supervision, X.Y.; validation, J.A.L., M.S. and X.V.Z.; writing—Original draft preparation, J.A.L., M.S., X.Y., C.N.B. and D.M.L.; writing—Review and editing, J.A.L., X.Y., C.N.B., D.M.L., X.V.Z., B.S.T., P.S., J.J.V. and M.J.B. All authors have read and agreed to the published version of the manuscript.

Funding: This research was funded by USDA NRCS Conservation Innovation Grant Project No. 69-3A75-11-177, USDA Hatch Project TN TEN00442 and agricultural research institutions at Louisiana State University, Mississippi State University, University of Missouri, and University of Tennessee.

Conflicts of Interest: The authors declare no conflict of interest.

References

1. U.S. Department of Agriculture (USDA). Natural Resources Conservation Service (NRCS). Available online: http://www.nrcs.usda.gov/wps/portal/nrcs/detailfull/national/programs/initiatives/?cid=stelprdb1048200 (accessed on 2 October 2020).
2. USDA, National Agricultural Statistics Service (NASS). Quick Stats. Cotton, Planted Area (ha). Available online: http://quickstats.nass.usda.gov (accessed on 2 October 2020).
3. Main, C.L.; Barber, L.T.; Boman, R.K.; Chapman, K.; Dodds, D.M.; Duncan, S.; Edmisten, K.L.; Horn, P.; Jones, M.A.; Morgan, G.D.; et al. Effects of nitrogen and planting seed size on cotton growth, development, and yield. *Agron. J.* **2013**, *105*, 1853–1859. [CrossRef]
4. MacDonald, B.C.T.; Rochester, I.J.; Nadelko, A. High yielding cotton produced without excessive nitrous oxide emissions. *Agron. J.* **2015**, *107*, 1673–1681. [CrossRef]
5. Hake, K.; Cassman, K.; Ebelhar, W. Cotton Physiology Today. 1991. Available online: https://www.cotton.org/tech/physiology/cpt/upload/CPT-Jan91-v2-3.pdf (accessed on 2 October 2020).
6. Lund, E.D.; Wolcott, M.C.; Hanson, G.P. Applying nitrogen site-specifically using soil electrical conductivity maps and precision agriculture technology. In Proceedings of the 2nd International Nitrogen Conference on Science and Policy, Potomac, MD, USA, 14–18 October 2001; pp. 767–776.
7. U.S. Environmental Protection Agency. *National Water Quality Inventory: Report to Congress, 2004 Reporting Cycle (EPA 841-R-08-001)*; Environmental Protection Agency, Office of Water: Washington, DC, USA, 2014.
8. USDA, NRCS. Environmental Quality Incentives Program. Available online: https://www.nrcs.usda.gov/wps/portal/nrcs/main/national/programs/financial/eqip/ (accessed on 2 October 2020).
9. Zhou, X.; English, B.C.; Larson, J.A.; Lambert, D.M.; Roberts, R.K.; Boyer, C.; Velandia, M.; Falconer, L.L.; Martin, S.W. Precision farming adoption trends in the southern U.S. *J. Cotton Sci.* **2017**, *21*, 143–155.
10. Isik, M.; Khanna, M. Stochastic technology, risk preferences, and adoption of site-specific technologies. *Amer. J. Agric. Econ.* **2003**, *85*, 305–317. [CrossRef]
11. Rütting, T.; Aronsson, H.; Delin, S. Efficient use of nitrogen in agriculture. *Nutr. Cycl. Agroecosyst.* **2018**, *110*, 1–5. [CrossRef]
12. Wang, X.; Miao, Y.; Dong, R.; Chen, Z.; Kusnierek, K.; Mi, G.; Mulla, D.J. Economic Optimal Nitrogen Rate Variability of Maize in Response to Soil and Weather Conditions: Implications for Site-Specific Nitrogen Management. *Agronomy* **2020**, *10*, 1237. [CrossRef]
13. Walker, E.R.; Mengistu, A.; Bellaloui, N.; Koger, C.H.; Roberts, R.K.; Larson, J.A. Plant population and row-spacing effects on maturity group III soybean. *Agron. J.* **2010**, *102*, 821–826. [CrossRef]
14. Quine, T.A.; Zhang, Y. An Investigation of Spatial Variation in Soil Erosion, soil Properties, and Crop Production in Agricultural fields in Devon, United Kingdom. *J. Soil Water Cons.* **2002**, *57*, 55–65.
15. Breitenbeck, G.A. Use of soil nitrate tests for nitrogen recommendations: Research perspective. In *Nitrogen Nutrition in Cotton: Practical Issues*; Miley, W.N., Oosterhuis, D.M., Eds.; American Society of Agronomy: Madison, WI, USA, 1990; pp. 77–87. [CrossRef]
16. Duncan, L.; Raper, T. Cotton Nitrogen Management in Tennessee. University of Tennessee Extension, Publication W 783. 2019. Available online: http://www.utcrops.com/cotton/PDF%20files/W783.pdf (accessed on 2 October 2020).
17. Raper, T.B.; Varco, J.J.; Hubbard, K.J. Canopy-based normalized difference vegetation index sensors for monitoring cotton nitrogen status. *Agron. J.* **2013**, *105*, 1345–1354. [CrossRef]
18. Biermacher, J.T.; Brorsen, B.W.; Epplin, F.M.; Solie, J.B.; Raun, W.R. Economic feasibility of site-specific optical sensing for managing nitrogen fertilizer for growing wheat. *Precis. Agric.* **2009**, *10*, 213–230. [CrossRef]
19. Biermacher, J.T.; Brorsen, B.W.; Epplin, F.M.; Solie, J.B.; Raun, W.R. The Economic potential of precision nitrogen application with wheat based on plant sensing. *Agric. Econ.* **2009**, *40*, 397–407. [CrossRef]
20. Boyer, C.N.; Brorsen, B.W.; Solie, J.B.; Raun, W.R. Profitability of variable rate nitrogen application in wheat production. *Precis. Agric.* **2011**, *12*, 473–487. [CrossRef]
21. Butchee, K.S.; May, J.; Arnall, B. Sensor based nitrogen management reduced nitrogen and maintained yield. *Crop Manag.* **2011**, *10*, 1–5. [CrossRef]

22. Ortiz-Monasterio, J.I.; Raun, W.R. Reduced nitrogen and improved farm income for irrigated spring wheat in the Yaqui Valley, Mexico, using sensor based nitrogen management. *J. Agric. Sci.* **2007**, *145*, 1–8. [CrossRef]
23. Stamatiadis, S.; Schepers, J.S.; Evangelou, E.; Glampedakis, A.; Glampedakis, M.; Dercas, N.; Tsadilas, C.; Tserlikakis, N.; Tsadila, E. Variable-rate application of high spatial resolution can improve cotton N-use efficiency and profitability. *Precis. Agric.* **2020**, *21*, 695–712. [CrossRef]
24. Stefanini, M.; Larson, J.A.; Lambert, D.M.; Yin, X.; Boyer, C.; Scharf, P.; Tubaña, B.S.; Varco, J.J.; Dunn, D.; Savoy, H.J.; et al. Effects of optical sensing based variable rate nitrogen management on yields, nitrogen use, and profitability for cotton. *Precis. Agric.* **2019**, *20*, 591–610. [CrossRef]
25. Scharf, P.C.; Shannon, D.K.; Palm, H.L.; Sudduth, K.A.; Drummond, S.T.; Kitchen, N.R.; Mueller, L.J.; Hubbard, V.C.; Oliveira, L.F. Sensor-based nitrogen applications out-performed producer-chosen rates for corn in on-farm demonstrations. *Agron. J.* **2011**, *103*, 1683–1691. [CrossRef]
26. Lowenberg-DeBoer, J. Risk management potential of precision farming technologies. *J. Agric. Appl. Econ.* **1999**, *31*, 275–285. [CrossRef]
27. Karatay, Y.N.; Meyer-Aurich, A. Profitability and downside risk implications of site-specific nitrogen management with respect to wheat grain quality. *Precis. Agric.* **2020**, *21*, 449–472. [CrossRef]
28. Lowenberg-DeBoer, J.; Erickson, B. Setting the record straight on precision agriculture adoption. *Agron. J.* **2019**, *111*, 1552–1569. [CrossRef]
29. U.S. Geology Survey. National Elevation Dataset: Metadata. Available online: http://ned.usgs.gov/ (accessed on 2 October 2020).
30. USDA NRCS. Soil Survey Geographic (SSURGO) Database. Available online: http://www.nrcs.usda.gov/wps/portal/nrcs/detail/soils/survey/?cid=nrcs142p2_053627 (accessed on 2 October 2020).
31. USDA NRCS. Soil Texture Calculator. Available online: http://www.nrcs.usda.gov/wps/portal/nrcs/detail/soils/survey/?cid=nrcs142p2_054167 (accessed on 2 October 2020).
32. USDA Agricultural Research Service. Revised Universal Soil Loss Equation, Version 2. Available online: http://fargo.nserl.purdue.edu/rusle2_dataweb/RUSLE2_Index.htm (accessed on 2 October 2020).
33. PRISM Climate Group. Northwest Alliance for Computational Science and Engineering. Oregon State University. Available online: http://www.prism.oregonstate.edu/recent (accessed on 2 October 2020).
34. Federal Reserve Bank of St. Louis. Gross Domestic Product: Implicit Price Deflator. Available online: http://research.stlouisfed.org/fred2/series/GDPDEF/ (accessed on 24 November 2020).
35. USDA NASS. Quick Stats. Cotton, Price Received, Measured in $ lb^{-1}. Available online: http://quickstats.nass.usda.gov/ (accessed on 2 October 2020).
36. USDA NASS. Mississippi. 2014 Statewide EQIP Practice, Ranking and Rate Information. Available online: https://www.nrcs.usda.gov/wps/portal/nrcs/detail/ms/programs/financial/eqip/?cid=stelprdb1193441 (accessed on 2 October 2020).
37. Coreil, C. *Louisiana EQIP Cost Share Payments NRCS Precision Nutrient Management Practice Code Number 590*; USDA NRCS: Alexandria, LA, USA, 2014.
38. Turman, P. *Tennessee EQIP Cost Share Payments NRCS Precision Nutrient Management Practice Code Number 590*; USDA NRCS: Murfreesboro, TN, USA, 2014.
39. Reisner, J. *Tennessee EQIP Cost Share Payments NRCS Precision Nutrient Management Practice Code Number 590*; USDA NRCS: Columbia, MO, USA, 2014.
40. USDA NRCS. Quick Stats. Nitrogen, Price Paid, Measured in $ ton^{-1}. Available online: http://quickstats.nass.usda.gov (accessed on 2 October 2020).
41. Mooney, D.F.; Roberts, R.K.; English, B.C.; Lambert, D.M.; Larson, J.A.; Velandia, M.; Larkin, S.L.; Marra, M.C.; Matin, S.W.; Mishra, A.; et al. *Precision Farming by Cotton Producers in Twelve Southern States: Results From the 2009 Southern Cotton Precision Farming Survey*; Department of Agricultural and Resource Economics, University of Tennessee: Knoxville, TN, USA, 2010. [CrossRef]
42. Schabenberger, O.; Pierce, F.J. *Contemporary Statistical Models for the Plant and Soil Sciences*; CRC Press: Boca Raton, FL, USA, 2001; pp. 474–479.
43. Littell, R.C.; Milliken, G.A.; Stroup, W.W.; Wolfinger, R.D.; Schabenberger, O. *SAS® for Mixed Models*, 2nd ed.; SAS Institute Inc: Cary, NC, USA, 2006.
44. Chatterjee, S.; Price, P. *Regression Analysis by Example*, 2nd ed.; Wiley-Interscience: New York, NY, USA, 1991.

45. Tiessen, H.; Cuevas, E.; Chacon, P. The role of soil organic matter in sustaining soil fertility. *Nature* **1994**, *371*, 783–785. [CrossRef]
46. Madden, R.A.; Williams, J. The correlation between temperature and precipitation in the United States and Europe. *Mon. Wea. Rev.* **1978**, *106*, 142–147. [CrossRef]

MDPI
St. Alban-Anlage 66
4052 Basel
Switzerland
Tel. +41 61 683 77 34
Fax +41 61 302 89 18
www.mdpi.com

Agronomy Editorial Office
E-mail: agronomy@mdpi.com
www.mdpi.com/journal/agronomy

www.ingramcontent.com/pod-product-compliance
Lightning Source LLC
LaVergne TN
LVHW080504180726
843515LV00016B/1391
* 9 7 8 3 0 3 6 5 1 3 4 4 7 *